CHAS. C. IVES,
Talking Machine
SUPPLIES,
29 WALL STREET,
NORWALK, CT.
GREEN STREET
WORCESTER, MASS.
EDISON RECORD
MUSICAL MERCHANDISE
PHONOGRAPHS, BICYCLES
GUNS, RIFLES,
FISHING TACKLE & SPORTING GOODS,
8724
EDWARD A. NELSON
214 MAIN ST
SPRINGFIELD, MASS.
WENDHEISERS
MUSIC STORE,
ROCKVILLE,
CONN.
GEO. H. SHARP & SON
MUSIC DEALERS
WESTFIELD
When Autumn leaves
SMITH BROS. PIANO CO. N.H.
1116 ELM ST. MANCHESTER,
EDISON & VICTOR RECORDS
PIANOS & SHEET MUSIC
BEAUD

PHONOGRAPHICA

The Early History of Recorded Sound Observed

Sheridan

PHONOGRAPHICA

The Early History of Recorded Sound Observed

TIMOTHY C. FABRIZIO AND GEORGE F. PAUL

WITH PHOTOGRAPHS
BY THE AUTHORS

Schiffer Publishing Ltd®
4880 Lower Valley Road, Atglen, PA 19310 USA

DEDICATION

To our children, whom we entrust with the kaleidoscopic legacy of the talking machine.

Library of Congress Cataloging-in-Publication Data

Fabrizio, Timothy C.
Phonographica : the early history of recorded sound observed / by Timothy C. Fabrizio and George F. Paul.
p. cm.
ISBN 0-7643-1985-X
1. Phonograph--History. 2. Phonograph--Collectors and collecting. I. Paul, George F. II. Title.
TS2301.P3 F2994 2004
621.389'33--dc22
2003021845

Designed by Bonnie M. Hensley
Cover design by Bruce Waters
Cover photos by the authors
Type set in EngravrsRoman Bd BT/Korinna BT

ISBN: 0-7643-1985-X
Printed in China
1 2 3 4

Published by Schiffer Publishing Ltd.
4880 Lower Valley Road
Atglen, PA 19310
Phone: (610) 593-1777; Fax: (610) 593-2002
E-mail: Info@schifferbooks.com
Please visit our web site catalog at **www.schifferbooks.com**
We are always looking for people to write books on new and related subjects. If you have an idea for a book please contact us at the above address.

This book may be purchased from the publisher.
Include $3.95 for shipping.
Please try your bookstore first.
You may write for a free catalog.

In Europe, Schiffer books are distributed by
Bushwood Books
6 Marksbury Ave.
Kew Gardens
Surrey TW9 4JF England
Phone: 44 (0) 20 8392-8585; Fax: 44 (0) 20 8392-9876
E-mail: Bushwd@aol.com
Free postage in the U.K., Europe; air mail at cost.

Contents

Introduction: The Image of the Phonograph ___ 6

Chapter One: 1877-1900 ___ 8

Chapter Two: 1901-1910 ___ 54

Chapter Three: 1911-1920 ___ 103

Chapter Four: 1921-1930 ___ 146

Chapter Five: Real Life Images ___ 175

Value Code Key ___ 219

Bibliography ___ 220

Glossary ___ 222

Index ___ 224

Introduction:

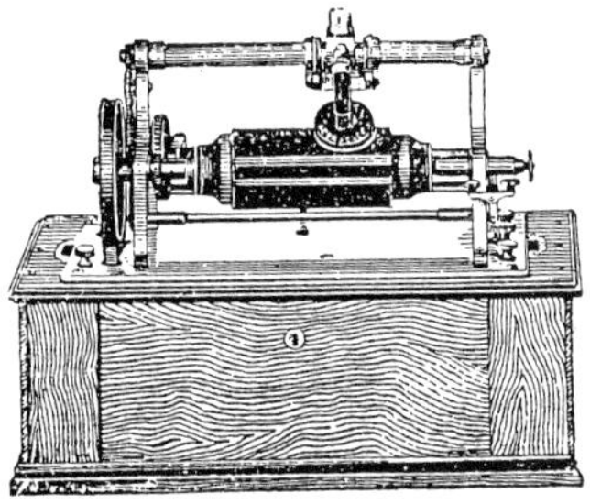

THE IMAGE OF THE PHONOGRAPH

The Development of Public Perception

Objects of enduring beauty continue to command our attention though they are frequently observed. One might stroll past the cathedral of Notre-Dame de Paris every day, and never fail to be struck by its majesty. Assuming one had the good fortune to hang a Monet study of water lilies over one's couch, would it ever be possible to enter the room without admiring the genius of the artist? Utilitarian objects, however, affect us quite differently.

Practical items fascinate us in the immediate. They flatter us by their "up-to-dateness," and make us the temporary envy of others. Yet, an automobile, for example, is "new" for only so long. The average car quickly becomes ordinary, blending into undistinguished others on the road. After decades have passed, a car might become a "classic," and once again inspire notice. It is true of primarily useful objects that they may survive a long period of obscurity to become again the objects of our fancy.

When one of the authors accompanied his family to a rented summer cottage in the Adirondack region of New York State during the 1950s, a neglected Victrola stood in a corner of the living room. Except for the kids, to whom it was a wonderfully strange toy to be abused, no one paid the Victrola even enough attention to carry it to the dump. The decrepit talking machine had been invisible for years.

When does something utterly, engagingly new become comfortably familiar, and, finally, transparent? The transition, though inevitable, does not move at any predictable pace. The other author of this book had a great aunt and uncle whose rural mid-western farm in 1958 still relied on hand-pumped water and an outdoor privy. There was a Victrola in the parlor, too — and the old folks occasionally played it. By studying historical images, advertising and ephemera, we can observe individual phonographs as they retreat from importance to obscurity in people's perceptions.

The Early Impact of the Phonograph

Beginning with the studio poses of Thomas Edison accompanying his newest invention early in 1878, the first images of or references to the Phonograph make obvious its formidable impact on the Victorian world. Printed descriptions of Edison Phonograph demonstrations fairly drip with hyperbole, mentioning such suspicious events as "the instrument clearly audible from three blocks away" (perhaps only by bats or other sharp-eared animals). During the first 30 years of its existence, from 1877 to the end of first decade of the twentieth century, the phonograph, or talking machine, inspired considerable awe and, for those who could afford to own one, pride of possession.

Images

Early images that include talking machines can be loosely arranged into three categories. First, there are scenes where the phonograph is the prized new possession, being ostentatiously celebrated and shown off. Next, are the images in which the phonograph still occupies pride of place, but the other elements of the composition suggest equal value – for instance, a parlor in which the talking machine and the new suite of furniture evidence a parity of importance. Lastly, are the scenes of everyday life in which the phonograph has blended into the background. In some instances, we may see only a part of the instrument, or it may be barely recognizable in the shadows of a room. It is from these images, however, that we may infer the most about the talking machine's place in the lives of its owners.

The literal-minded might prefer images from the first-named group: a grinning gentleman standing next to his easily identifiable Edison "Home" Phonograph, Model "B," with Hawthorne and Sheble flower horn and support. Yet, these pictures only confirm what we already know. It is the third category, where the talking machine has devolved to the ordinary, which requires the discernment of the viewer. From the subtleties of the scene, one must intuit the details of the moment when the shutter snapped. This is the effort of stretching, relaxing, reaching back to experience the everyday essence of the casual past.

Chapter One:

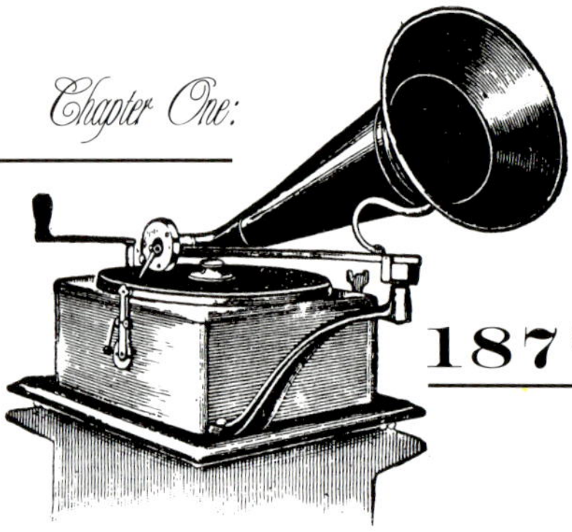

1877-1900

1-1. On April 18, 1878 Thomas Edison demonstrated his recent invention, the Phonograph, in Washington, D.C. before President Rutherford B. Hayes and both houses of Congress. While in the Capital, Edison visited the renowned photographic studio of Mathew Brady. The "Tinfoil Phonograph" shown in this historic image is sometimes referred to as the "Brady Model" due to its association with the event that recorded it for posterity. *Courtesy of the Julien Anton collection. (Value code: VR)*

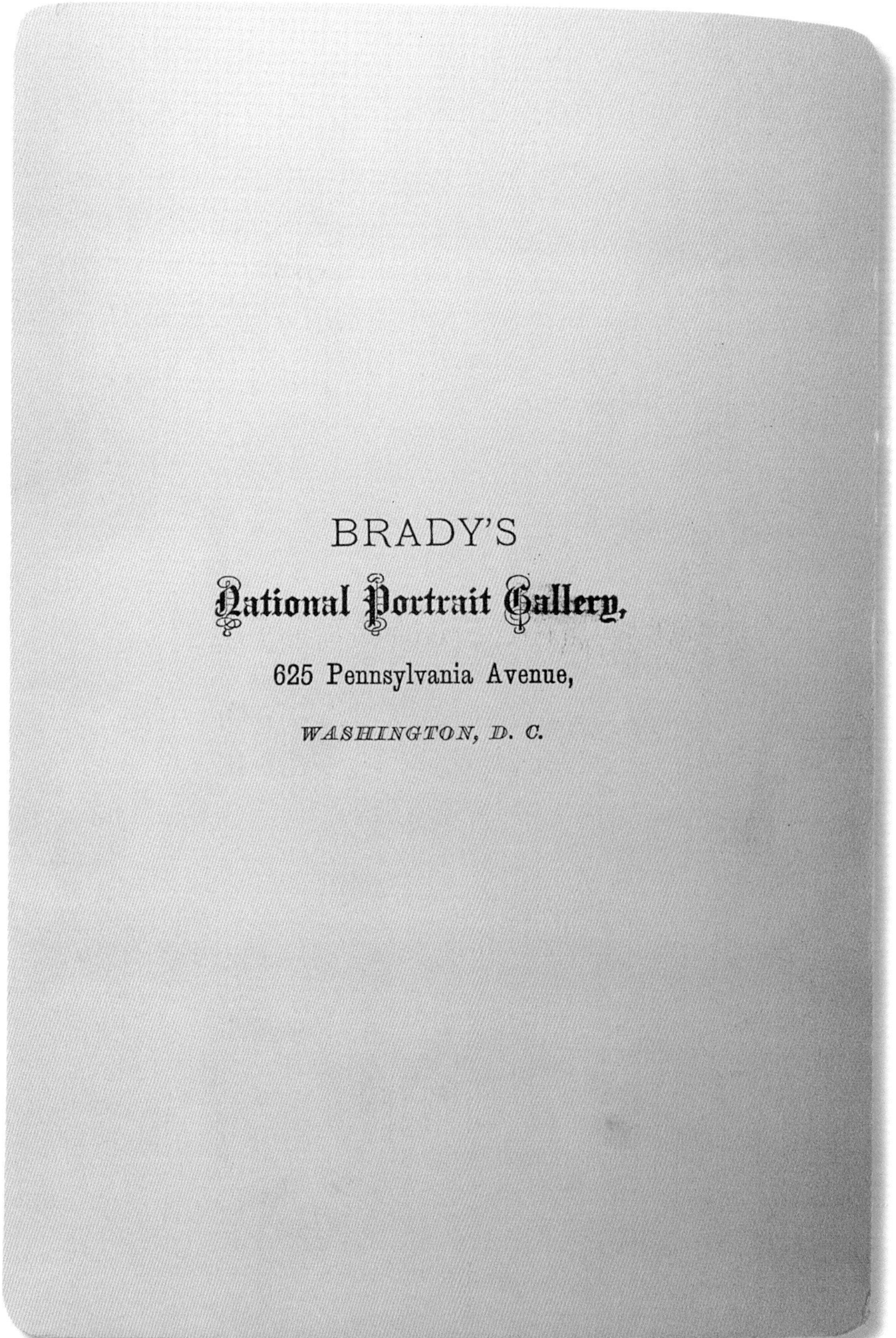

1-2. The impact of Mathew Brady's name was equal to Edison's in 1878. Today, Mathew Brady is well-remembered for the extraordinary American Civil War photographs taken by him and his studio staff. In 1878, despite the publicity that had been heaped upon him, Brady was in serious financial straits. He had spent a fortune documenting the Civil War in the hope that the United States government would purchase the photographs. When the government refused, Brady was forced to declare bankruptcy. The general public also failed to support his documentarian efforts; people wanted to forget the recent, traumatic conflict. Brady, however, was able to keep his Washington D.C. studio in operation. Although Congress awarded him $25,000, three years before Edison's visit, Brady remained deeply in debt. In 1896, he would die in a hospital charity ward. *Courtesy of the Julien Anton collection.*

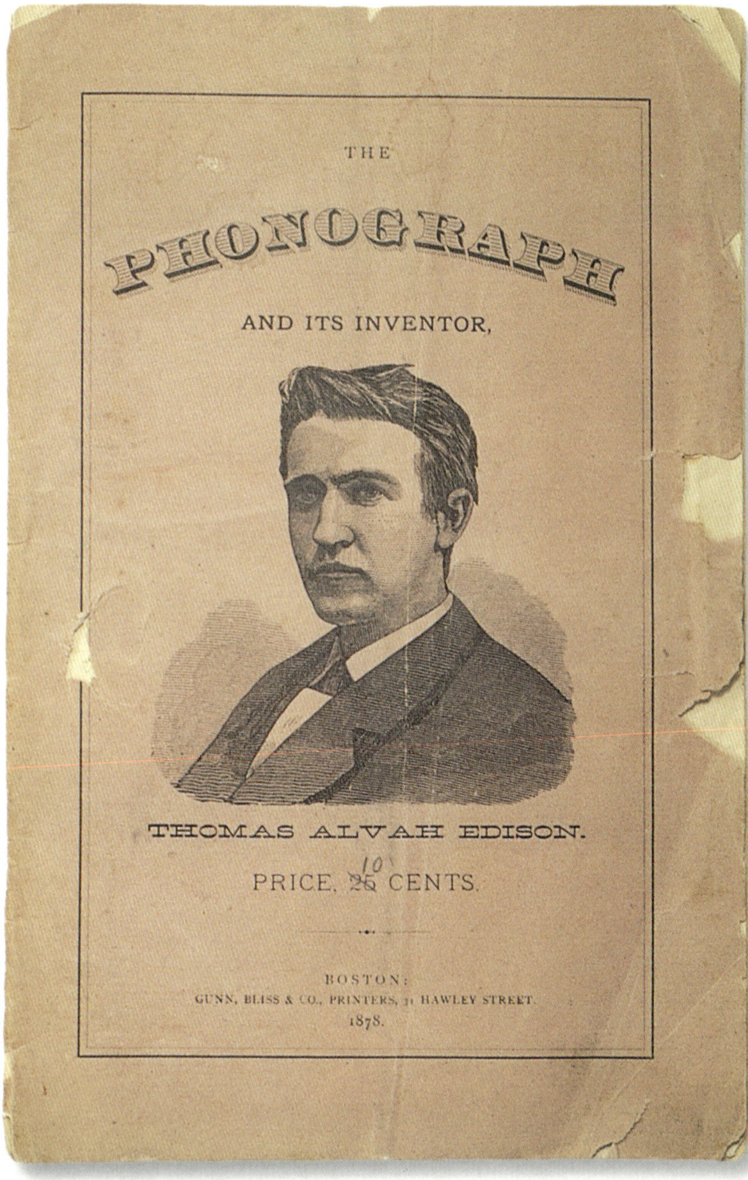
THE

PHONOGRAPH

AND ITS INVENTOR,

THOMAS ALVAH EDISON.

PRICE, ~~25~~ 10 CENTS.

BOSTON:
GUNN, BLISS & CO., PRINTERS, 31 HAWLEY STREET.
1878.

1-3. When the Edison Phonograph became well publicized in 1878, the public wanted to know more about the man who would soon be known as the "Wizard of Menlo Park." This 6" x 9" pamphlet was one of the first treatises on the Phonograph, as well as an early biographical sketch of Thomas A. Edison. *Courtesy of Tom and Sandi McCarthy. (Value code: VR)*

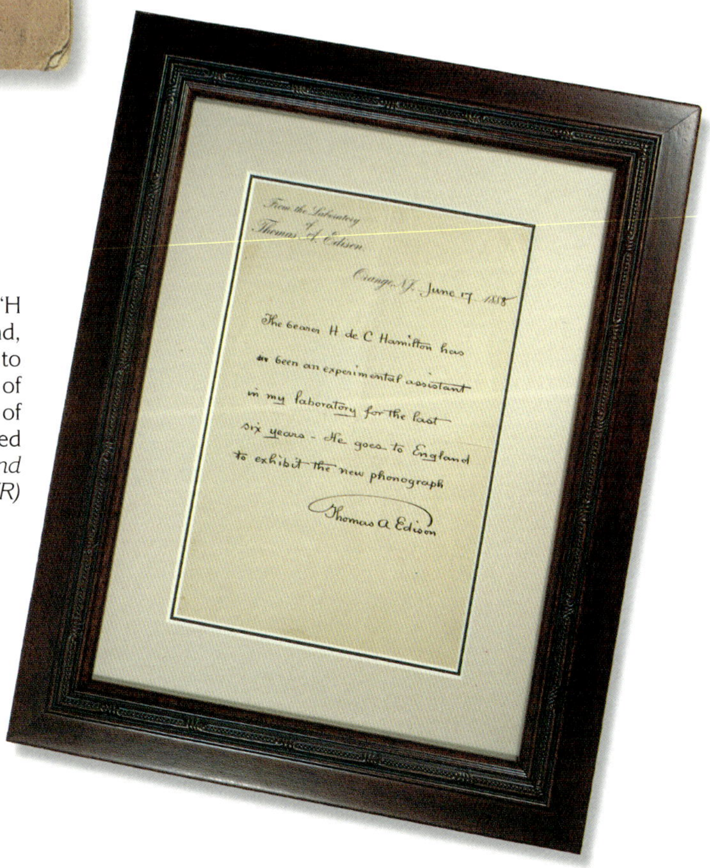
From the Laboratory of Thomas A. Edison.

Orange N.J. June 17 1888

The bearer H de C Hamilton has been an experimental assistant in my laboratory for the last six years - He goes to England to exhibit the new phonograph

Thomas A. Edison

1-4. This letter was given by Thomas Edison to "H de C Hamilton" upon his departure to England, carrying the first of the Phonographs destined to become the Class "M." The famous photograph of Edison and his assistants after three days of nonstop experimenting shows them gathered around that first example. *Courtesy of Nina and Dave Heitz. (Value code: VR)*

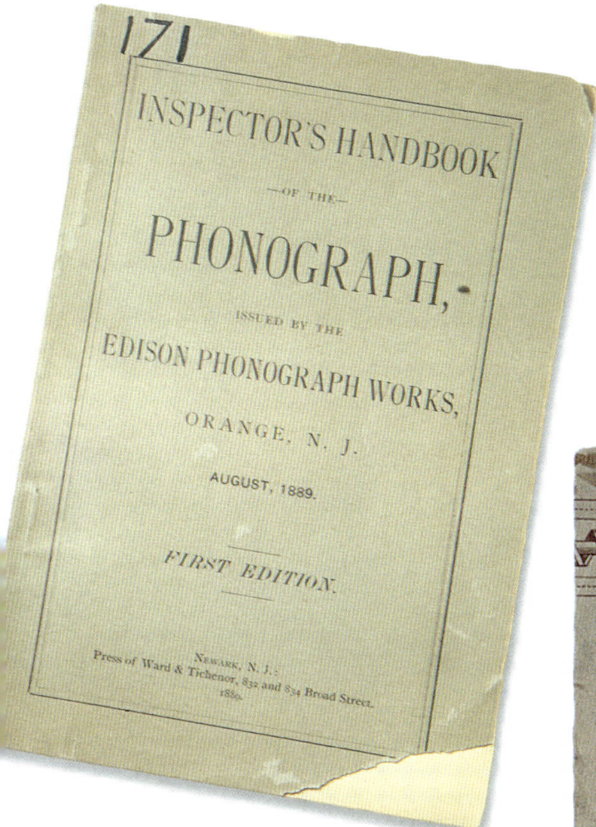

171

INSPECTOR'S HANDBOOK

—OF THE—

PHONOGRAPH,

ISSUED BY THE

EDISON PHONOGRAPH WORKS,

ORANGE, N. J.

AUGUST, 1889.

FIRST EDITION.

NEWARK, N. J.:
Press of Ward & Tichenor, 832 and 834 Broad Street.
1889.

1-5. In mid-1888, the Edison Phonograph Works began production of the Class "M" Phonograph. It was a heavy affair with an electric motor driven by a 2 1/2 volt battery. Although the Class "M" was the first practical talking machine available to the public, it still required a certain amount of skill to adjust and operate it. The entire output of Class "M" Phonographs at this time was being purchased by the North American Phonograph Company, which leased them to users for $40.00 per year. It was important that North American's agents knew how to install, maintain, and operate the machines so that lessees could be properly instructed. In response to requests, the Edison Phonograph Works provided this *Inspector's Handbook of the Phonograph* in August 1889. *Courtesy of Ray Phillips. (Value code: VR)*

INSTRUCTIONS

FOR THE

Phonograph-Graphophone.

THESE instructions embody the essential features necessary to use the Phonograph-Graphophone intelligently. They are based on the assumption that the user has some knowledge of the instrument, although it is believed that they are sufficiently explicit to enable any one of ordinary intelligence to operate the machine.

1-6. The Phonograph-Graphophone (as the first Graphophones were called) was an impractical device whose failings were not (as implied by these 1889 instructions) due to the substandard intelligence of the user! *Courtesy of the collection of Howard Hazelcorn. (Value code: VR)*

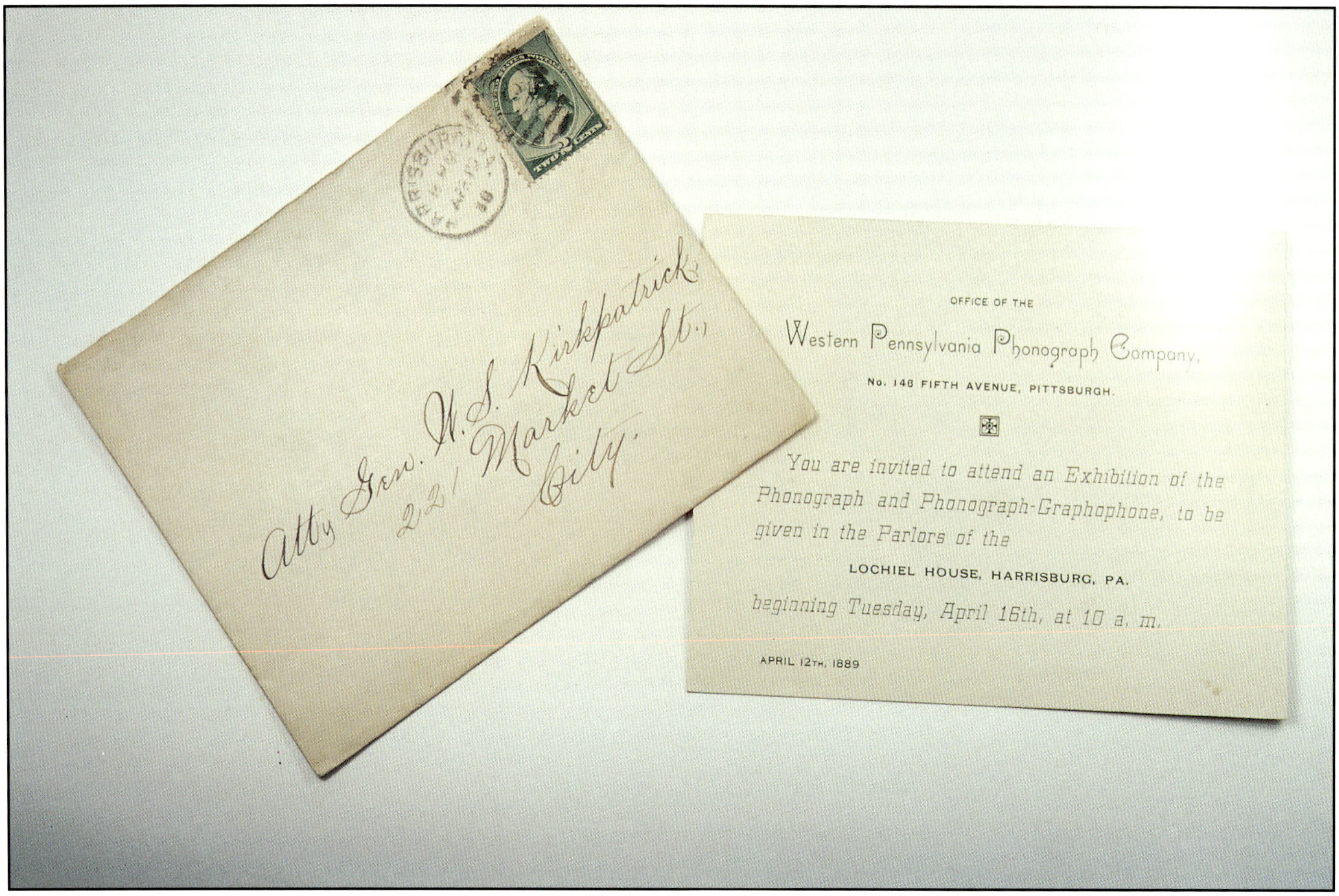

1-7. Pennsylvania Attorney General W.S. Kirkpatrick received an invitation to an exhibit of the Edison Phonograph and the Phonograph-Graphophone beginning April 16, 1889. The Western Pennsylvania Phonograph Company was one of the local subsidiaries of the North American Phonograph Company, which offered these machines (the Edison Class "M" and the Type "B" treadle-driven Graphophone) for lease to businessmen and government officials. *(Value code: VR)*

Oppsosite page: 1-8. Edison's recognition as an inventive genius, churning out scientific marvels, was of incalculable aid to his varied business enterprises. Here, from the time of Edison's introduction of the "Improved Phonograph" at the Paris Exposition of 1889, sheet music for an eponymous waltz, depicting a rather inebriated-looking inventor. *Courtesy of the Scott and Denise Corbett collection. (Value code: E)*

Dédiée à Mr Thomas Elva EDISON
RÉPERTOIRE DES BALS DE PARIS
EDISON
Grande Valse Electrique
Brillante et Facile
Pour PIANO
PAR
ALBERT DE LA GRAVELIÈRE
PRIX: 7f50
Paris, Léopold CERF, 25, Rue Notre-Dame de Nazareth
Propriété pour tous pays

1-9. This 11" x 7" stock certificate is from the International Graphophone Company of the early 1890s. *Courtesy of the collection of Howard Hazelcorn. (Value code: F)*

1-10. The North American Phonograph Company was in desperate financial straits by 1892, when this bond was issued. Needless to say, it was not a "sound" investment. Measures 14" x 17". *Courtesy of the collection of Howard Hazelcorn. (Value code: E)*

1-11. By 1893, the North American Phonograph Company had abandoned the idea of leasing machines, and Thomas Edison, the firm's chief creditor, had assumed control. This dealer certificate (measuring 11" x 14"), dated May 18, 1893, affirmed that John F. Barber was legally entitled to sell Edison Class "M" Phonographs. By this time, the treadle Graphophones had been given up, their playing mechanisms converted by the American Graphophone Company to run with small, battery-powered motors, and play Edison-style cylinders. *Courtesy of Nina and Dave Heitz. (Value code: VR)*

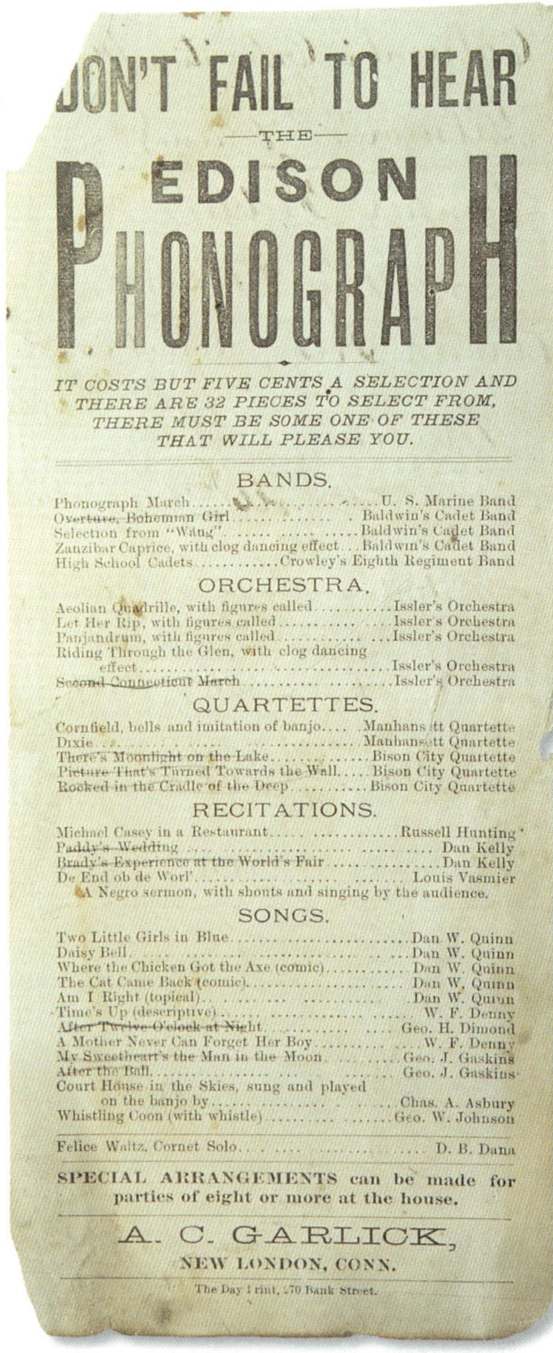

DON'T FAIL TO HEAR

THE

EDISON PHONOGRAPH

IT COSTS BUT FIVE CENTS A SELECTION AND THERE ARE 32 PIECES TO SELECT FROM, THERE MUST BE SOME ONE OF THESE THAT WILL PLEASE YOU.

BANDS.

Phonograph March.......... U. S. Marine Band
Overture, Bohemian Girl.......... Baldwin's Cadet Band
Selection from "Wang".......... Baldwin's Cadet Band
Zanzibar Caprice, with clog dancing effect... Baldwin's Cadet Band
High School Cadets.......... Crowley's Eighth Regiment Band

ORCHESTRA.

Aeolian Quadrille, with figures called.......... Issler's Orchestra
Let Her Rip, with figures called.......... Issler's Orchestra
Panjandrum, with figures called.......... Issler's Orchestra
Riding Through the Glen, with clog dancing effect.......... Issler's Orchestra
Second Connecticut March.......... Issler's Orchestra

QUARTETTES.

Cornfield, bells and imitation of banjo.... Manhans tt Quartette
Dixie.......... Manhansett Quartette
There's Moonlight on the Lake.......... Bison City Quartette
Picture That's Turned Towards the Wall.... Bison City Quartette
Rocked in the Cradle of the Deep.......... Bison City Quartette

RECITATIONS.

Michael Casey in a Restaurant.......... Russell Hunting
Paddy's Wedding.......... Dan Kelly
Brady's Experience at the World's Fair.......... Dan Kelly
De End ob de Worl'.......... Louis Vasmier
A Negro sermon, with shouts and singing by the audience.

SONGS.

Two Little Girls in Blue.......... Dan W. Quinn
Daisy Bell.......... Dan W. Quinn
Where the Chicken Got the Axe (comic).......... Dan W. Quinn
The Cat Came Back (comic).......... Dan W. Quinn
Am I Right (topical).......... Dan W. Quinn
Time's Up (descriptive).......... W. F. Denny
After Twelve O'clock at Night.......... Geo. H. Dimond
A Mother Never Can Forget Her Boy.......... W. F. Denny
My Sweetheart's the Man in the Moon.......... Geo. J. Gaskins
After the Ball.......... Geo. J. Gaskins
Court House in the Skies, sung and played on the banjo by.......... Chas. A. Asbury
Whistling Coon (with whistle).......... Geo. W. Johnson

Felice Waltz, Cornet Solo.......... D. B. Dana

SPECIAL ARRANGEMENTS can be made for parties of eight or more at the house.

A. C. GARLICK,

NEW LONDON, CONN.

The Day Print, 270 Bank Street.

1-12. A broadside, 4" x 9 1/2". A.C. Garlick was an enterprising showman with a good stock of cylinder records from which his patrons could choose. The selections and their performers tell us a great deal about the time period. Dan W. Quinn recalled beginning his long recording career in 1892, and one of his listed songs, "Daisy Bell" (commonly and incorrectly known as "A Bicycle Built For Two") was a considerable success that year. An even bigger hit the same year was "After the Ball" from the theatrical production "A Trip to Chinatown." Here we see it listed as performed by George J. Gaskin (misspelled with an extra "s" on the broadside). George W. Johnson, the first African-American to record commercially, was also a popular artist of the early 1890s. So, we may surmise that Mr. Garlick's nickel-a-pop business was being conducted in 1892 or slightly thereafter. At that time, the only feasible instrument for reproducing records was an Edison Class "M" battery-powered Phonograph. Please note that "special arrangements" could be made for larger parties. Chances are this involved the use of an amplifying horn, rather than the normally-used ear-tubes, which limited the performance to each nickel-paying customer. *Courtesy of Tom Sideleau. (Value code: VR)*

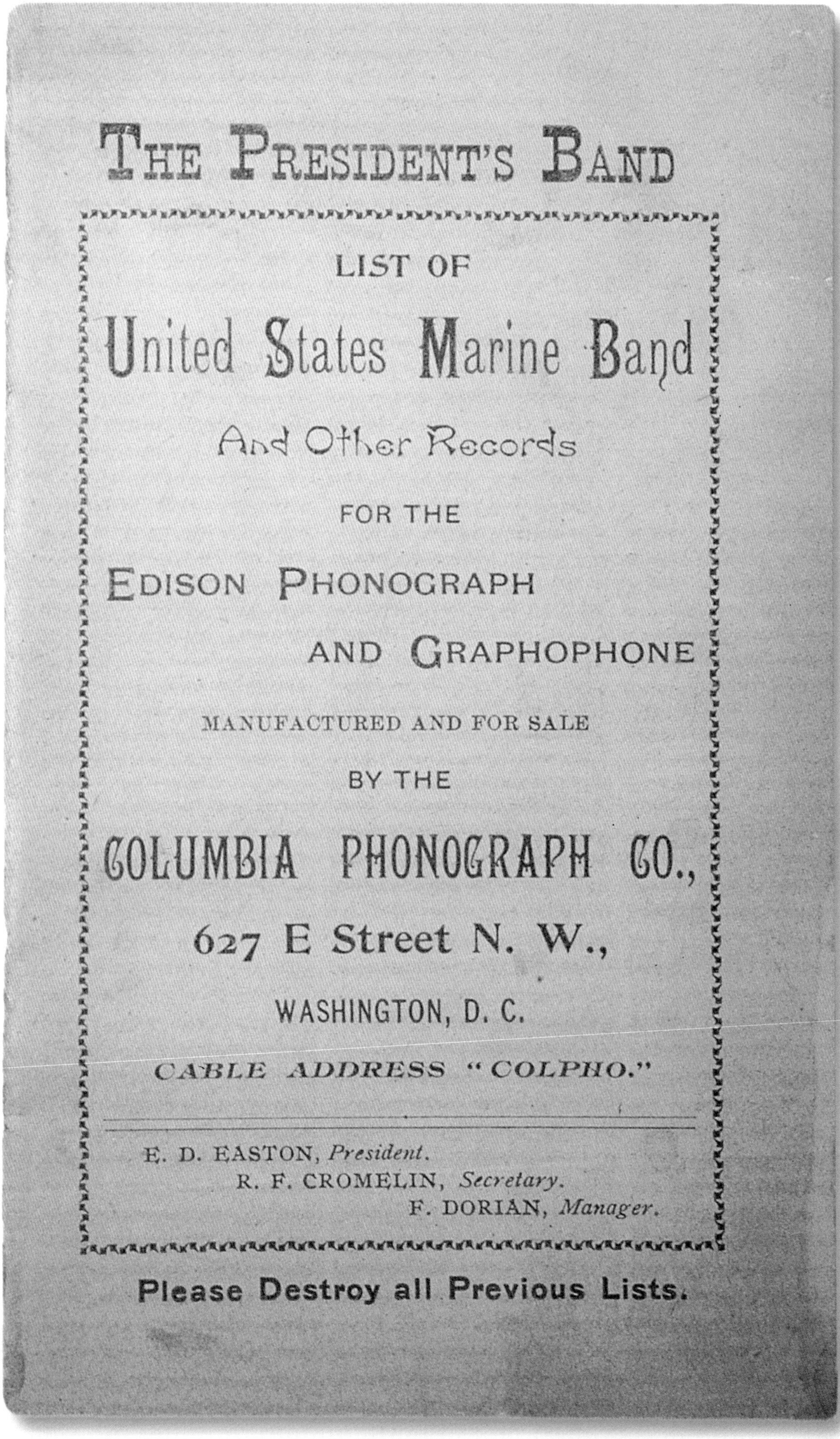

THE PRESIDENT'S BAND

LIST OF

United States Marine Band

And Other Records

FOR THE

EDISON PHONOGRAPH

AND GRAPHOPHONE

MANUFACTURED AND FOR SALE

BY THE

COLUMBIA PHONOGRAPH CO.,

627 E Street N. W.,

WASHINGTON, D. C.

CABLE ADDRESS "COLPHO."

E. D. EASTON, *President.*
R. F. CROMELIN, *Secretary.*
F. DORIAN, *Manager.*

Please Destroy all Previous Lists.

1-13. The happenstance of its location in Washington, D.C., which gave it access to the services of the United States Marine Band, and the resourcefulness of its chief, Edward D. Easton, gave the Columbia Phonograph Company the edge it needed to survive the dark days of the talking machine industry 1890-1893. During this period, few Phonographs and even fewer Graphophones were in the hands of the public, however there was a consistent need for cylinder records. Stepping to the fore, Columbia grasped the record business for dear life and held on with both hands, as its market for cylinders slowly expanded. By September 1893, when this catalog was issued, the popularity of and respect for the Marine Band had been a huge factor in making Columbia a success in an industry riddled with failure. 3 1/2" x 6". *Courtesy of Ray Phillips. (Value code: VR)*

1-14. The youthful face of Thomas Edison, as he appeared about the time of the Phonograph's inception. Edison's visage was used to promote a wide variety of items from cigars (as seen here) to meat extract. These little acclamations of his genius drew attention to his own business enterprises, and the objects that bore his famous name, such as Phonographs. 4 1/2" x 6 1/2". *Courtesy of the Charles Hummel collections. (Value code: F)*

1-15. These French children's blocks assemble into four scenes, one of which includes the little phonograph drama shown. Note the girl rushing away, holding her ears at the lower left. We can think of several records that could elicit the same reaction from most anyone. *Courtesy of Philippe Le Ray. (Value code: VR)*

Opposite page: 1-16. From its inception in 1893 until its sale to the Columbia Phonograph Company on August 1, 1897, the Chicago Talking Machine Company made its presence known in the trade. A method of cylinder recording duplication was patented by one of its founders, Leon F. Douglass, and subsequently sold to Columbia. In 1894, when this cover letter was written, the firm was obviously very much focused on the business applications of talking machines. *Courtesy of Anthony Wellman.*

MAIN EXPRESS 128, LONG DISTANCE TELEPHONE.

LEON F. DOUGLASS, VICE-PRESIDENT.

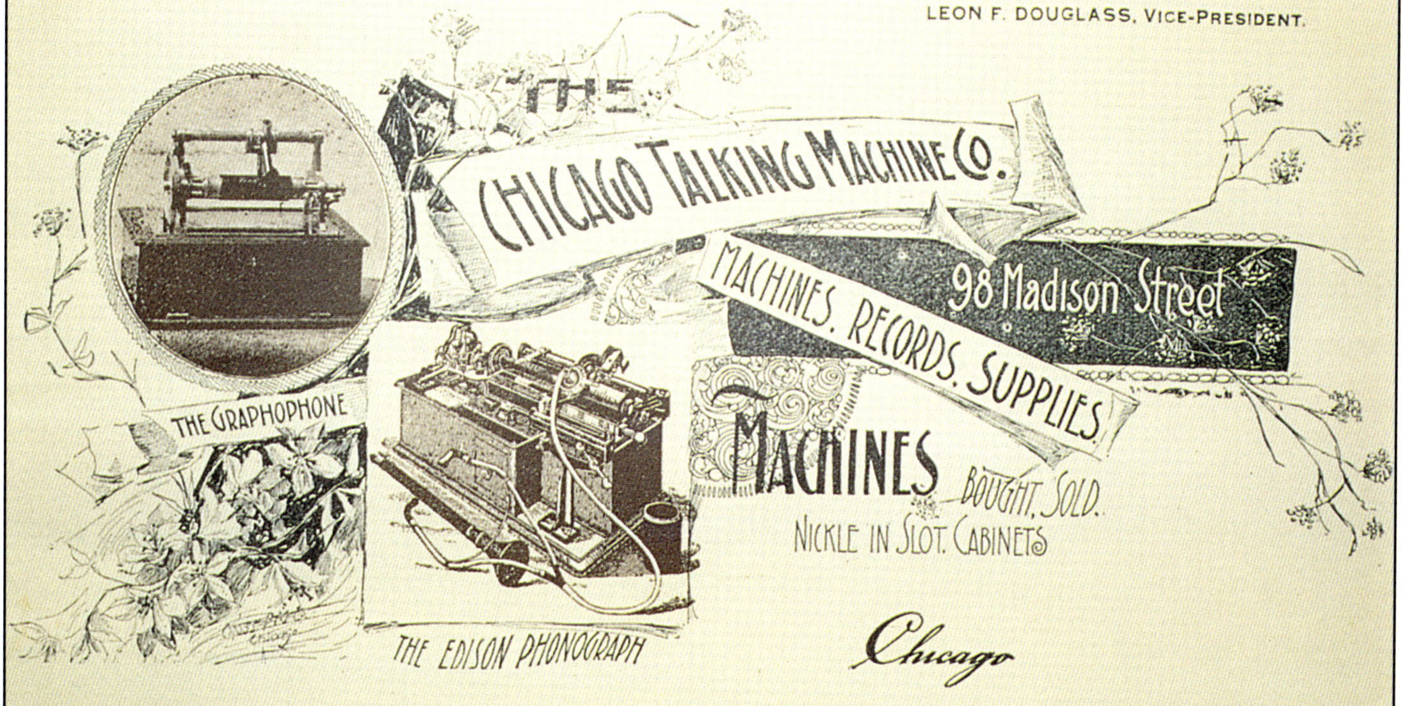

To whom it may concern:

We call your attention to the list of names on the card enclosed which embraces a few of those using our ''Talking Machines''.

We make the assertion and can demonstrate its truth, that under our system of conducting correspondence, two Talking Machines and one GOOD transcriber on the typewriter, can do the work of any two average stenographers, thus saving in your office the salary of one or more persons, in proportion to the number employed, besides increasing the efficiency of your dictators.

Our Talking Machines are simple in mechanism and the ''GRAPHOPHONE, '95 MODEL'' is perfect in every respect.

Our represenitive will call and explain upon request.

Respectfully,

CHICAGO TALKING MACHINE CO.,

98 Madison Street.

Chicago users of
COMMERCIAL TALKING MACHINES.

Marshal Field & Co.,	~~13~~ 16	Machines.
J. V. Farwell Co.,	8	"
Brown Bros., Unity Building,	5	"
Chicago Record,	4	"
Carson, Pire, Scott & Co.,	4	"
Chicago Tribune,	4	"
Lord & Thomas, 45 Randolph Street,	5	"
Casey & Day, Rookery Building,	3	"
Oxford Mfg.. Co., 342 Wabash Avenue,	3	"
Washburn Moen Mfg., Co., 107 Lake St.,	3	"
Lyon & Healy, Adams & Wabash Ave.,	2	"
Franklin Engr. and Electro Co.,	2	"
Rams Horn, The Temple,	2	"
Street Ry. Review, 269 Dearborn Street	2	"
Yale & Towne Mfg., Co., 152 Wabash Ave.,	2	"
North Western Lumberman,	2	"
Chicago Daily News,	2	"
Geo. P. Bent, 333 S. Canal Street,	2	"

The above firms and many others are constantly using "Talking Machines" in their offices and will testify personally or by telephone. Any one wishing to investigate further, address,

The Chicago Talking Machine Co.,
98 MADISON STREET,

Main Express, 128, Long Distance Telephone. OVER.

1-17. The "...card enclosed..." alluded to in the previous letter, contained a list of users of commercial talking machines – the topmost number updated by hand to reflect recent sales. *Courtesy of Anthony Wellman.*

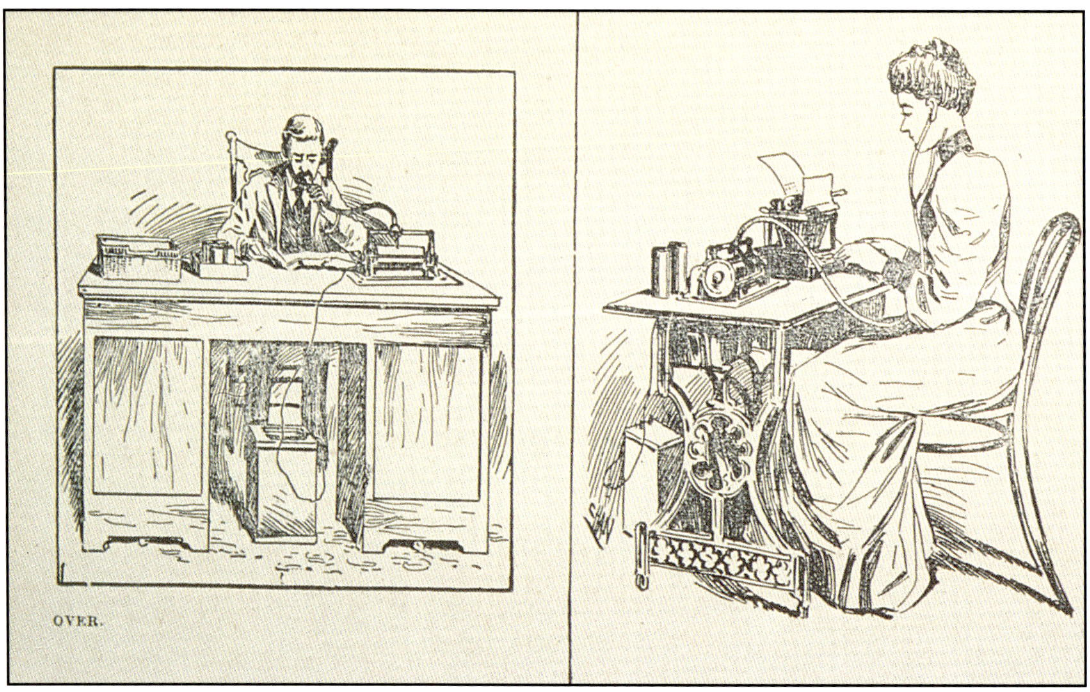

1-19. The reverse of the card featured cuts of a businessman dictating, and a stenographer transcribing from a Graphophone. *Courtesy of Anthony Wellman.*

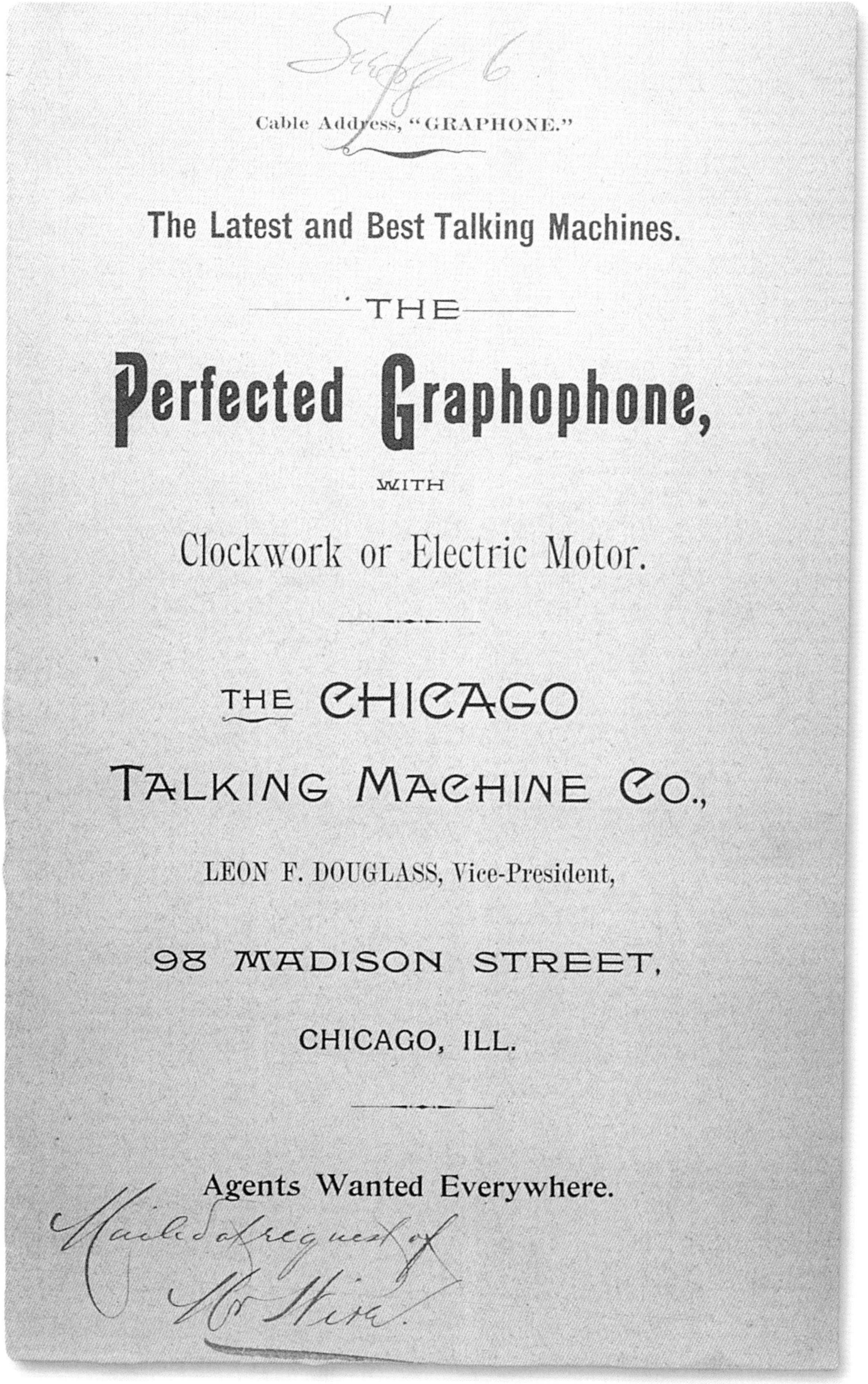
Cable Address, "GRAPHONE."

The Latest and Best Talking Machines.

THE

Perfected Graphophone,

WITH

Clockwork or Electric Motor.

THE CHICAGO

TALKING MACHINE CO.,

LEON F. DOUGLASS, Vice-President,

98 MADISON STREET,

CHICAGO, ILL.

Agents Wanted Everywhere.

1-19. This (late 1894) catalog of "Perfected Graphophones" was printed by the American Graphophone Company and distributed by the Chicago Talking Machine Company between its own covers. This very early document chronicled the Type "C," "E," "F," "G," "H," "I," "K," "R," "S," "T," and "U" Graphophones. Note that none of these machines yet bore a decal on its cabinet, nor was the Columbia Phonograph Company (American Graphophone's distributor) mentioned at all. Not until 1895 was American Graphophone to relinquish *all* sales activity to the Columbia Phonograph Company. *Courtesy of Anthony Wellman.*

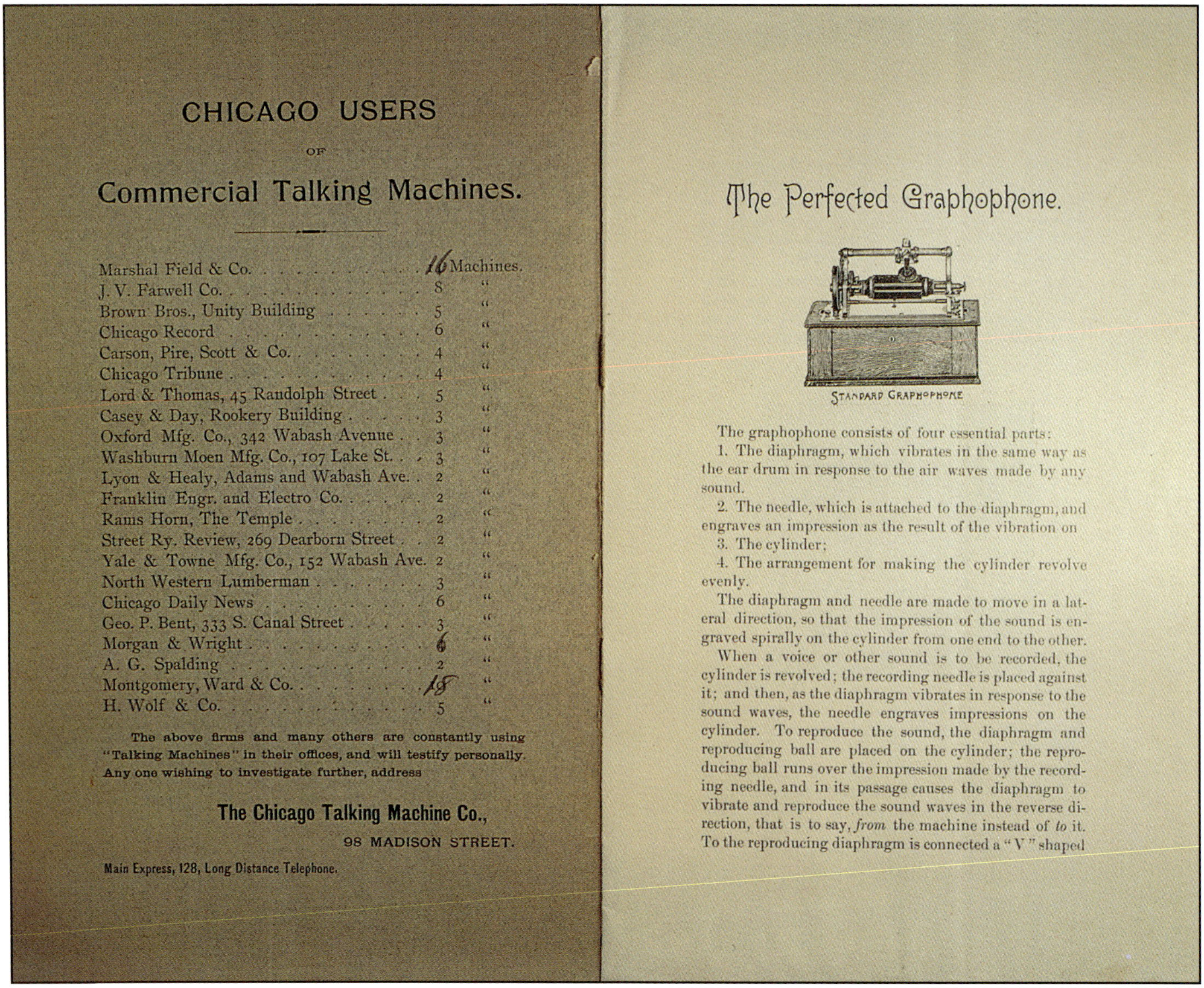

CHICAGO USERS

OF

Commercial Talking Machines.

Marshal Field & Co.	16	Machines.
J. V. Farwell Co.	8	"
Brown Bros., Unity Building	5	"
Chicago Record	6	"
Carson, Pire, Scott & Co.	4	"
Chicago Tribune	4	"
Lord & Thomas, 45 Randolph Street	5	"
Casey & Day, Rookery Building	3	"
Oxford Mfg. Co., 342 Wabash Avenue	3	"
Washburn Moen Mfg. Co., 107 Lake St.	3	"
Lyon & Healy, Adams and Wabash Ave.	2	"
Franklin Engr. and Electro Co.	2	"
Rams Horn, The Temple	2	"
Street Ry. Review, 269 Dearborn Street	2	"
Yale & Towne Mfg. Co., 152 Wabash Ave.	2	"
North Western Lumberman	3	"
Chicago Daily News	6	"
Geo. P. Bent, 333 S. Canal Street	3	"
Morgan & Wright	6	"
A. G. Spalding	2	"
Montgomery, Ward & Co.	18	"
H. Wolf & Co.	5	"

The above firms and many others are constantly using "Talking Machines" in their offices, and will testify personally. Any one wishing to investigate further, address

The Chicago Talking Machine Co.,
98 MADISON STREET.

Main Express, 128, Long Distance Telephone.

The Perfected Graphophone.

STANDARD GRAPHOPHONE

The graphophone consists of four essential parts:

1. The diaphragm, which vibrates in the same way as the ear drum in response to the air waves made by any sound.

2. The needle, which is attached to the diaphragm, and engraves an impression as the result of the vibration on

3. The cylinder;

4. The arrangement for making the cylinder revolve evenly.

The diaphragm and needle are made to move in a lateral direction, so that the impression of the sound is engraved spirally on the cylinder from one end to the other.

When a voice or other sound is to be recorded, the cylinder is revolved; the recording needle is placed against it; and then, as the diaphragm vibrates in response to the sound waves, the needle engraves impressions on the cylinder. To reproduce the sound, the diaphragm and reproducing ball are placed on the cylinder; the reproducing ball runs over the impression made by the recording needle, and in its passage causes the diaphragm to vibrate and reproduce the sound waves in the reverse direction, that is to say, *from* the machine instead of *to* it. To the reproducing diaphragm is connected a "V" shaped

1-20. The first catalog spread made a pitch for commercial machines, and included an elementary description of the basic Graphophone mechanism for the neophyte. *Courtesy of Anthony Wellman.*

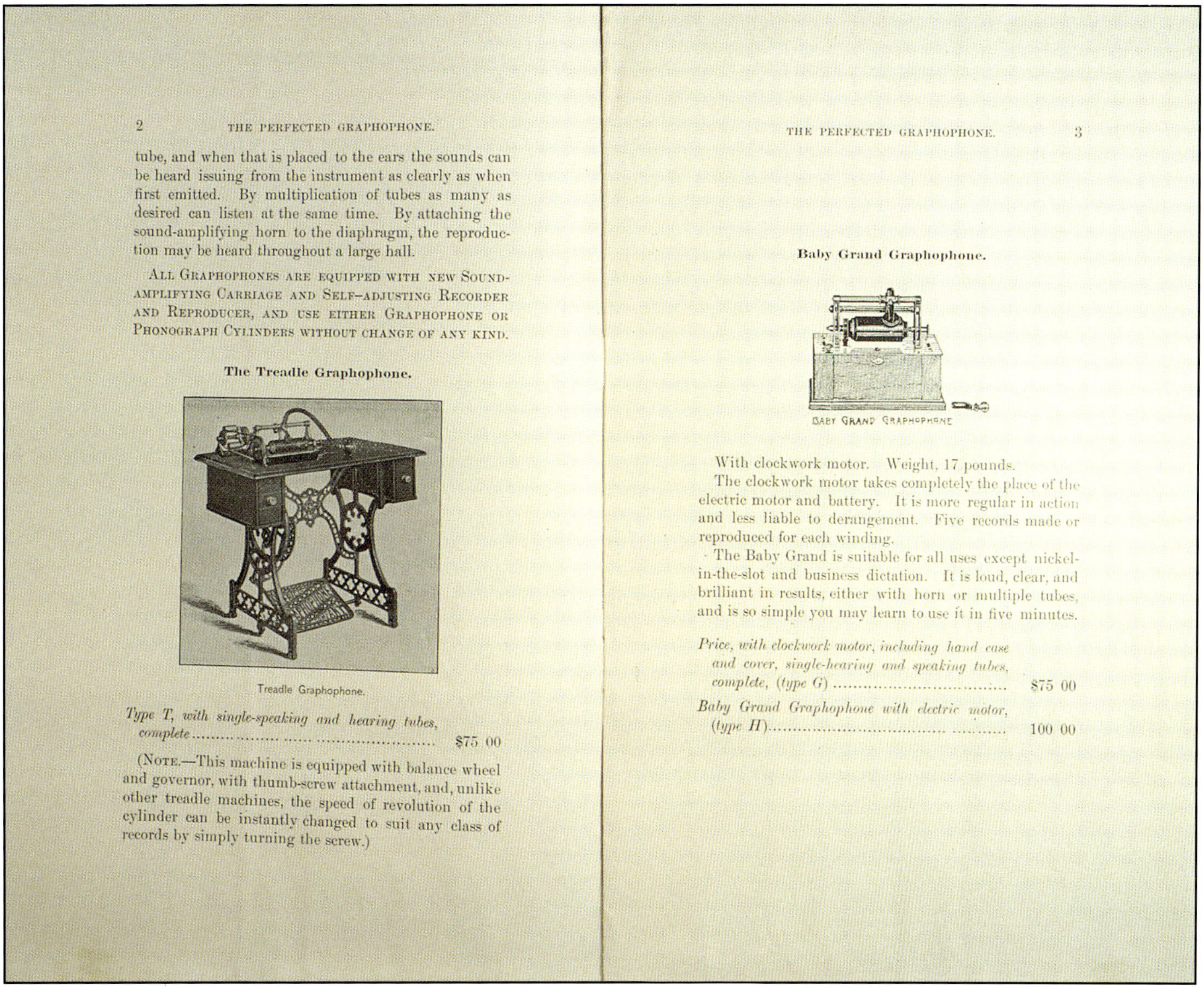

2 THE PERFECTED GRAPHOPHONE.

tube, and when that is placed to the ears the sounds can be heard issuing from the instrument as clearly as when first emitted. By multiplication of tubes as many as desired can listen at the same time. By attaching the sound-amplifying horn to the diaphragm, the reproduction may be heard throughout a large hall.

ALL GRAPHOPHONES ARE EQUIPPED WITH NEW SOUND-AMPLIFYING CARRIAGE AND SELF-ADJUSTING RECORDER AND REPRODUCER, AND USE EITHER GRAPHOPHONE OR PHONOGRAPH CYLINDERS WITHOUT CHANGE OF ANY KIND.

The Treadle Graphophone.

Treadle Graphophone.

Type T, with single-speaking and hearing tubes, complete $75 00

(NOTE.—This machine is equipped with balance wheel and governor, with thumb-screw attachment, and, unlike other treadle machines, the speed of revolution of the cylinder can be instantly changed to suit any class of records by simply turning the screw.)

THE PERFECTED GRAPHOPHONE. 3

Baby Grand Graphophone.

BABY GRAND GRAPHOPHONE

With clockwork motor. Weight, 17 pounds.

The clockwork motor takes completely the place of the electric motor and battery. It is more regular in action and less liable to derangement. Five records made or reproduced for each winding.

The Baby Grand is suitable for all uses except nickel-in-the-slot and business dictation. It is loud, clear, and brilliant in results, either with horn or multiple tubes, and is so simple you may learn to use it in five minutes.

Price, with clockwork motor, including hand case and cover, single-hearing and speaking tubes, complete, (type G) $75 00

Baby Grand Graphophone with electric motor, (type H) 100 00

1-21. The Type "T" ("Treadle") Graphophone and the Types "G" and "H" ("Baby Grand") Graphophones were illustrated here. The "Baby Grand" was the first talking machine designed solely for entertainment purposes. At $75.00, it was still too expensive for most consumers. *Courtesy of Anthony Wellman.*

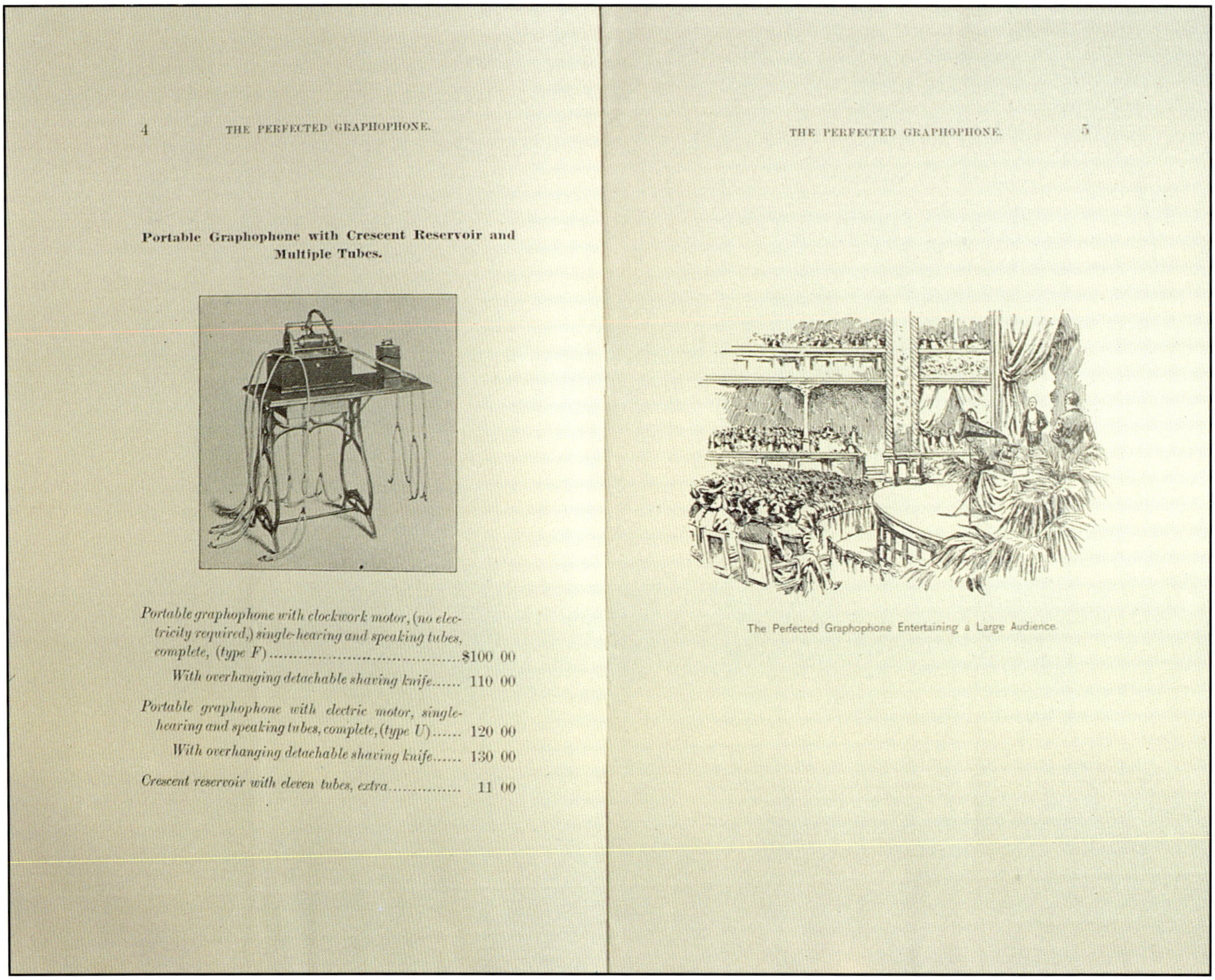

4 THE PERFECTED GRAPHOPHONE.

Portable Graphophone with Crescent Reservoir and Multiple Tubes.

Portable graphophone with clockwork motor, (no electricity required,) single-hearing and speaking tubes, complete, (type F) $100 00

With overhanging detachable shaving knife...... 110 00

Portable graphophone with electric motor, single-hearing and speaking tubes, complete, (type U)...... 120 00

With overhanging detachable shaving knife...... 130 00

Crescent reservoir with eleven tubes, extra.............. 11 00

THE PERFECTED GRAPHOPHONE. 5

The Perfected Graphophone Entertaining a Large Audience.

1-22. The basic Graphophone apparatus could be equipped with a spring motor (Type "F") or a 2-volt battery-driven electric motor (Type "U"), both apparently recommended for exhibition use. *Courtesy of Anthony Wellman.*

6 THE PERFECTED GRAPHOPHONE.

Business Graphophone.

Dictating to the Commercial Graphophone, the Modern Substitute for the Stenographer.

Commercial Graphophone, in hand case or table, with fixed knife for shaving cylinders, electric motor to run from storage battery, speaking and hearing tubes, complete, (type K)........................$150 00

Combination Treadle and Motor Graphophone, on table, to be operated either by electric motor or by treadle, fixed shaving knife, single-hearing tube and speaking tube and battery cord, (type C)............ 160 00

To run from 110-*volt* CONTINUOUS *electric current, (type I)*........................ 160 00

Type I is so arranged that it is only necessary to screw into your electric lamp socket the plug attached to the graphophone, and turn the button. This machine is especially recommended where the electric light current (continuous, *not* alternating) can conveniently be obtained.

THE PERFECTED GRAPHOPHONE. 7

Continuously Running Graphophone.

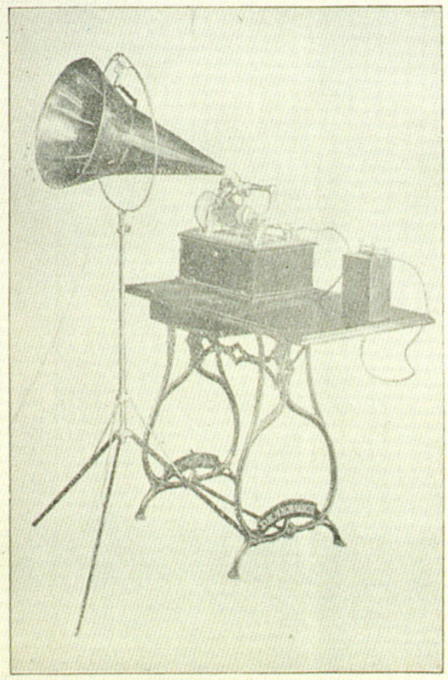

This machine will travel and play continuously without an attendant, until stopped.

Price, with single-hearing and speaking tubes, complete, (type R)........................$150 00

1-23. The Type "K" ("Commercial") Graphophone included a fixed shaver, and was equipped with a 2-volt battery motor. The Type "C" ("Combination") was mounted in a foot-treadle table and supplied also with a 2-volt electric motor. Either power source could be engaged at the discretion of the user. The Type "I" ("Incandescent") was equipped with a 110-volt DC mains current motor. *Courtesy of Anthony Wellman.*

8 THE PERFECTED GRAPHOPHONE.

Automatic Nickel-in-the-Slot Graphophone.

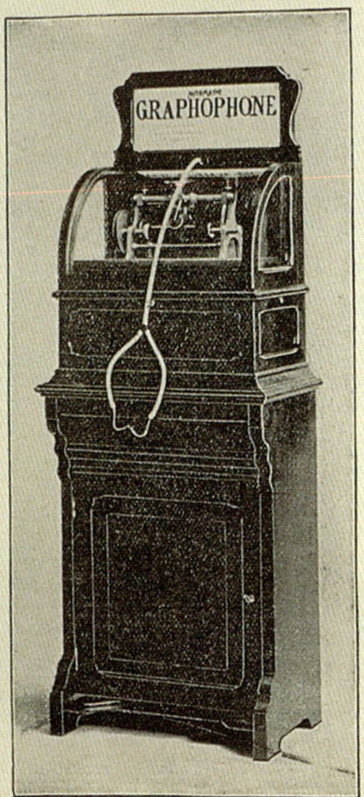

Automatic graphophone, with QUICK RETURN *nickel-in-the-slot mechanism, handsome cabinet, complete, as illustrated, (type S)*................................$160 00

THE PERFECTED GRAPHOPHONE. 9

Interchangeable slot and exhibition graphophone with nickel-in-the-slot mechanism, etc., same as type S, also hand cabinet, recorder, single-speaking and hearing tubes, and 11-way hearing tubes, complete, (type E)................................$175 00

This outfit is so arranged that it may either be used as a nickel-in-the-slot machine, earning money and taking care of itself, while the owner is otherwise engaged, or may be converted into a portable exhibition machine to carry from place to place in the hand.

Cylinders.

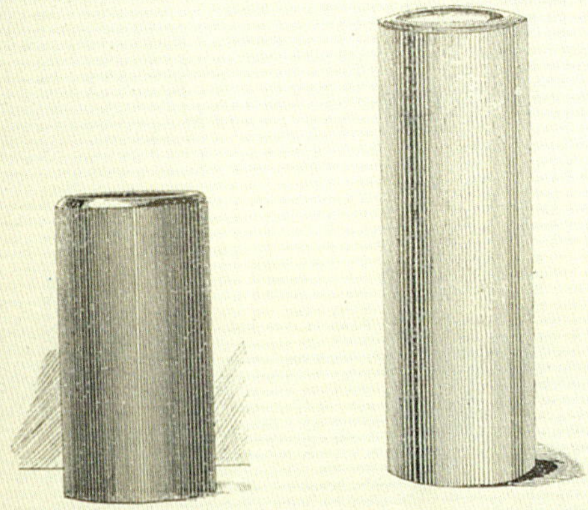

Cylinder P. Cylinder E.

The above illustrations show the types of cylinders that may be used on the graphophone.

Cylinder P is slipped on the mandrel furnished with every machine.

1-24. The Type "S" ("Slot") was ensconced in a handsome floor cabinet with curved glass. The Type "E" ("Exhibitor") Graphophone was the same machine with an additional portable "hand cabinet," a recorder, speaking tube, and hearing tubes. The Columbia "P" cylinder would be manufactured in various materials for 16 more years. The "E" cylinder (described on the following catalog page), was discontinued two years later. *Courtesy of Anthony Wellman.*

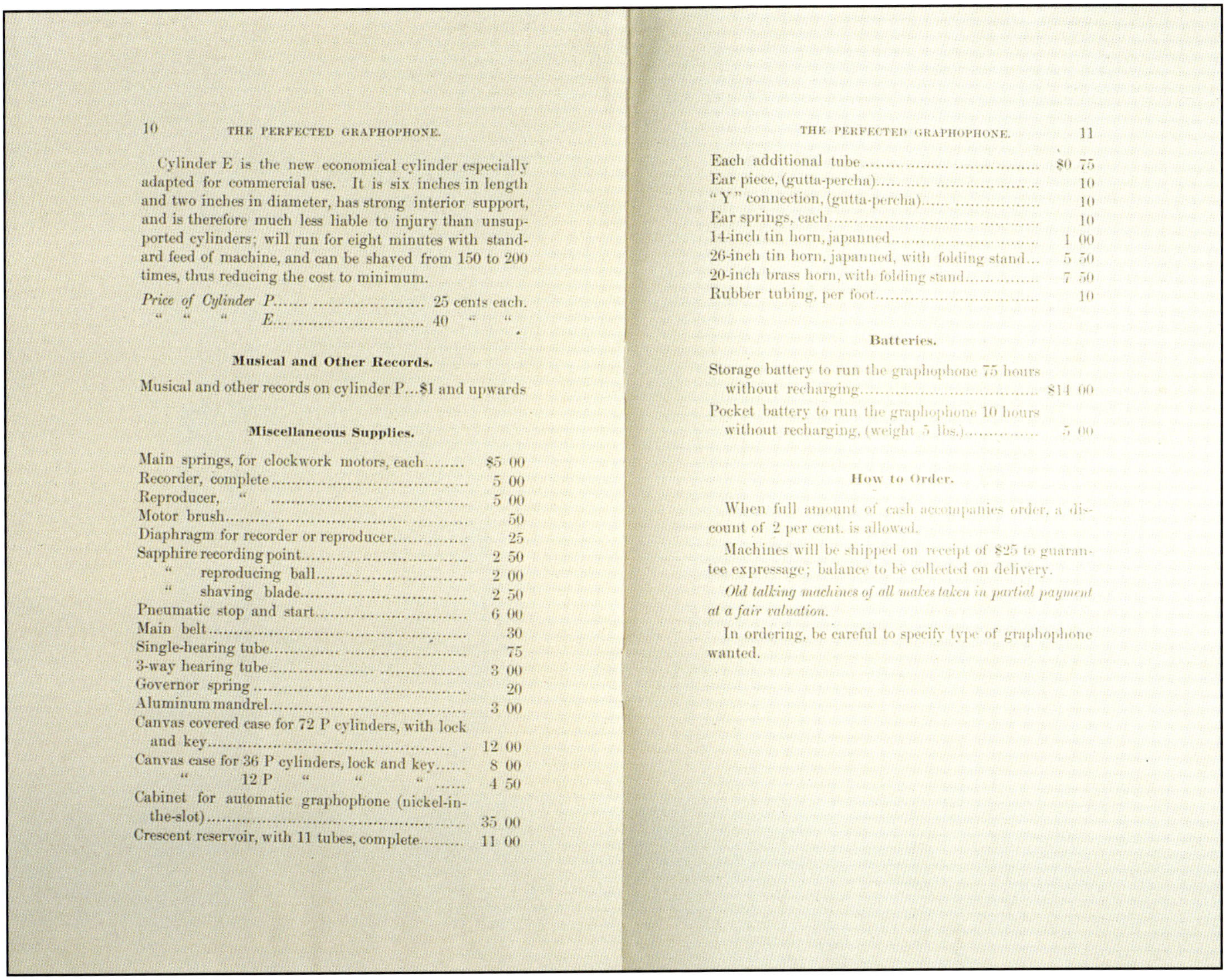

10 THE PERFECTED GRAPHOPHONE.

Cylinder E is the new economical cylinder especially adapted for commercial use. It is six inches in length and two inches in diameter, has strong interior support, and is therefore much less liable to injury than unsupported cylinders; will run for eight minutes with standard feed of machine, and can be shaved from 150 to 200 times, thus reducing the cost to minimum.

Price of Cylinder P....... 25 cents each.
" " " *E*....... 40 " "

Musical and Other Records.

Musical and other records on cylinder P...$1 and upwards

Miscellaneous Supplies.

Main springs, for clockwork motors, each	$5 00
Recorder, complete	5 00
Reproducer, "	5 00
Motor brush	50
Diaphragm for recorder or reproducer	25
Sapphire recording point	2 50
" reproducing ball	2 00
" shaving blade	2 50
Pneumatic stop and start	6 00
Main belt	30
Single-hearing tube	75
3-way hearing tube	3 00
Governor spring	20
Aluminum mandrel	3 00
Canvas covered case for 72 P cylinders, with lock and key	12 00
Canvas case for 36 P cylinders, lock and key	8 00
" 12 P " " "	4 50
Cabinet for automatic graphophone (nickel-in-the-slot)	35 00
Crescent reservoir, with 11 tubes, complete	11 00

THE PERFECTED GRAPHOPHONE. 11

Each additional tube	$0 75
Ear piece, (gutta-percha)	10
"Y" connection, (gutta-percha)	10
Ear springs, each	10
14-inch tin horn, japanned	1 00
26-inch tin horn, japanned, with folding stand	5 50
20-inch brass horn, with folding stand	7 50
Rubber tubing, per foot	10

Batteries.

Storage battery to run the graphophone 75 hours without recharging	$14 00
Pocket battery to run the graphophone 10 hours without recharging, (weight 5 lbs.)	5 00

How to Order.

When full amount of cash accompanies order, a discount of 2 per cent. is allowed.

Machines will be shipped on receipt of $25 to guarantee expressage; balance to be collected on delivery.

Old talking machines of all makes taken in partial payment at a fair valuation.

In ordering, be careful to specify type of graphophone wanted.

1-25. The short-lived "E" cylinder was described here, along with the claim of eight minutes of playing time. The huge entertainment potential of this product was either not recognized or ignored. Any of the Graphophones in this catalog except the "Baby Grand" could accommodate this 6" long cylinder. No Edison Phonographs could play it, thus American Graphophone could have wielded a powerful advantage over its rival – had the company directors possessed that vision. *Courtesy of Anthony Wellman.*

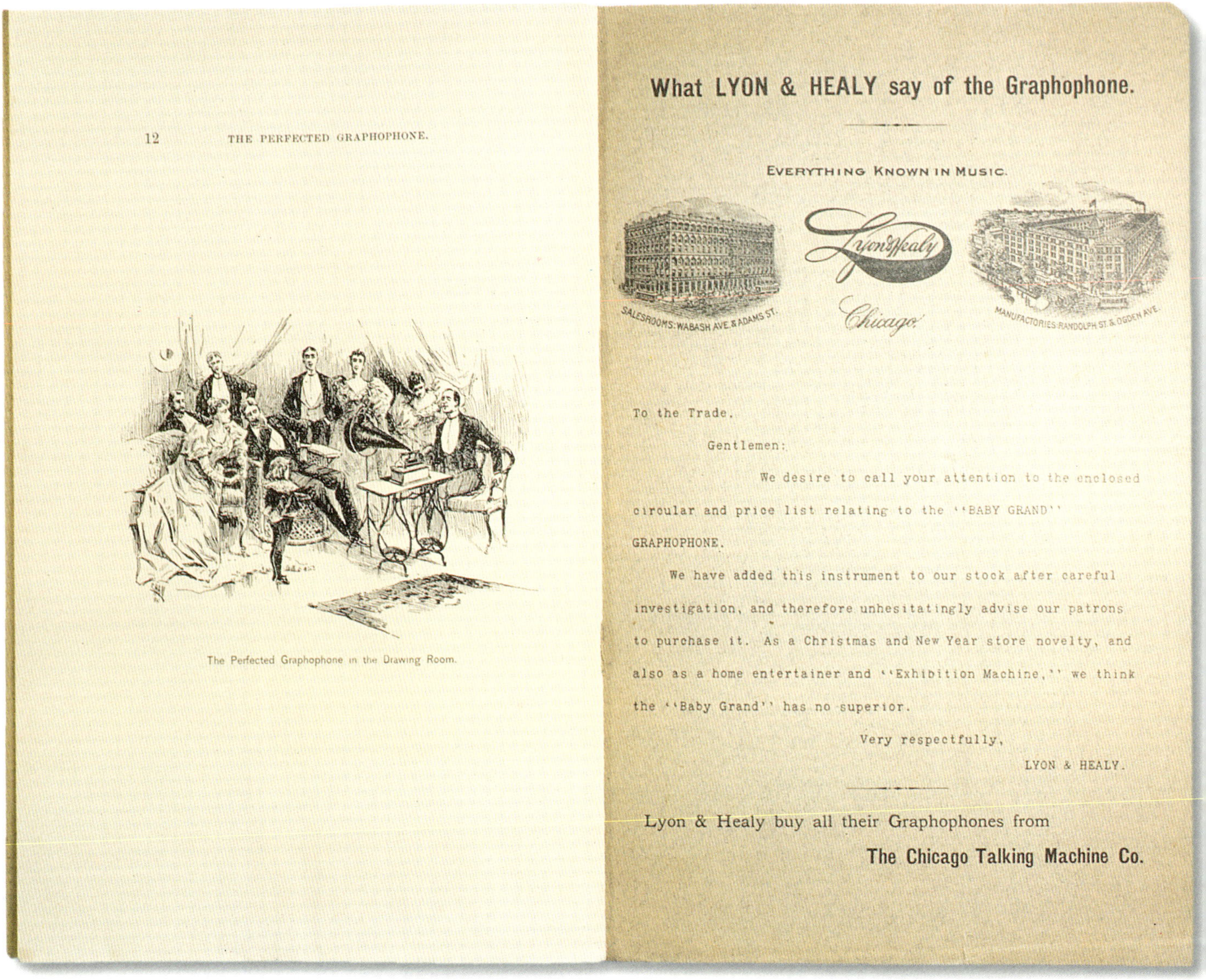
12 THE PERFECTED GRAPHOPHONE.

The Perfected Graphophone in the Drawing Room.

What LYON & HEALY say of the Graphophone.

EVERYTHING KNOWN IN MUSIC.

SALESROOMS: WABASH AVE. & ADAMS ST.

Lyon & Healy

Chicago

MANUFACTORIES: RANDOLPH ST. & OGDEN AVE.

To the Trade,

Gentlemen:

We desire to call your attention to the enclosed circular and price list relating to the ''BABY GRAND'' GRAPHOPHONE.

We have added this instrument to our stock after careful investigation, and therefore unhesitatingly advise our patrons to purchase it. As a Christmas and New Year store novelty, and also as a home entertainer and ''Exhibition Machine,'' we think the ''Baby Grand'' has no superior.

Very respectfully,

LYON & HEALY.

Lyon & Healy buy all their Graphophones from

The Chicago Talking Machine Co.

1-26. The final catalog spread featured a testimonial from the prestigious firm of Lyon & Healy. The mention of a "...Christmas and New Year store novelty..." suggests a late 1894 date for this remarkable catalog. *Courtesy of Anthony Wellman.*

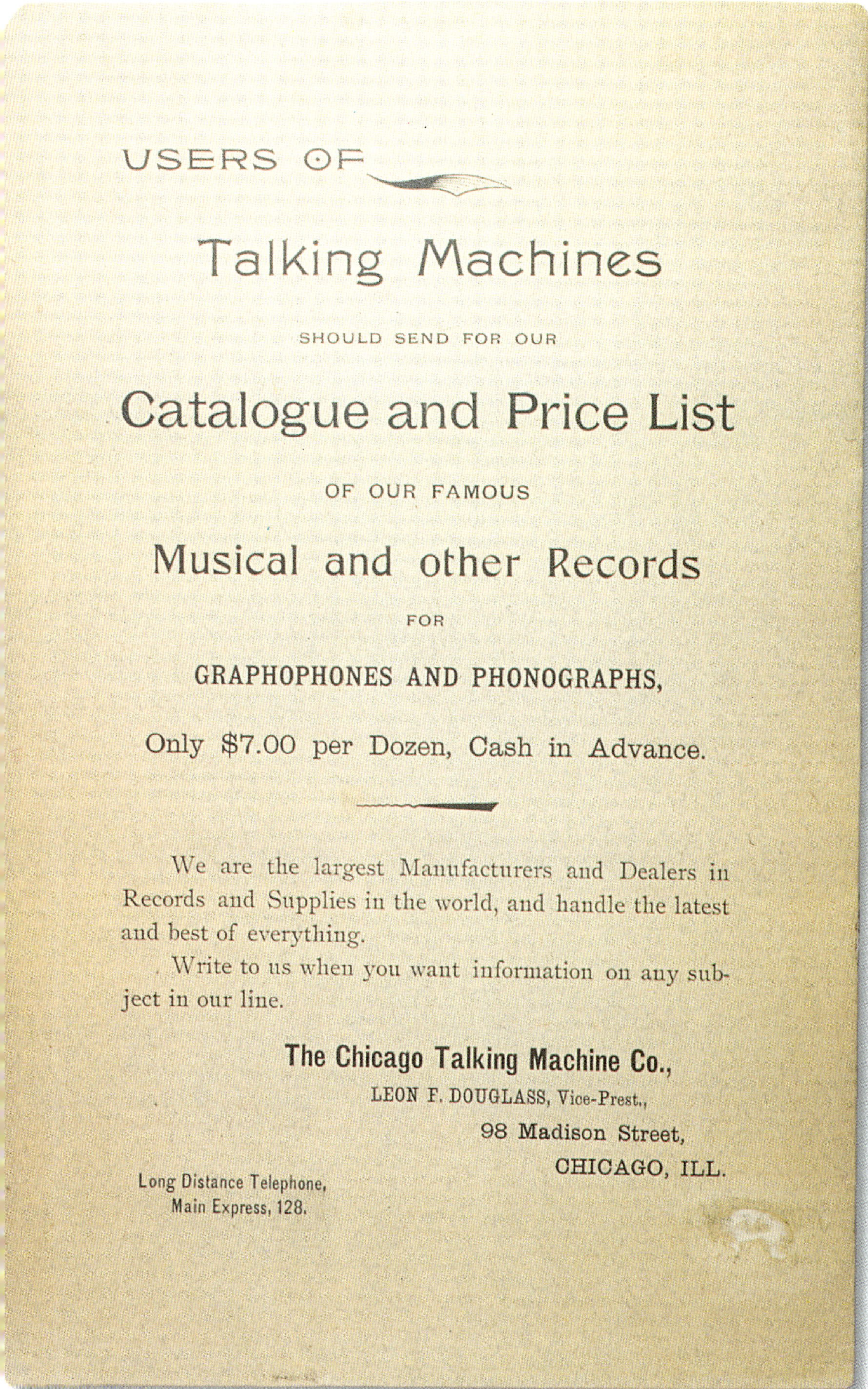

USERS OF

Talking Machines

SHOULD SEND FOR OUR

Catalogue and Price List

OF OUR FAMOUS

Musical and other Records

FOR

GRAPHOPHONES AND PHONOGRAPHS,

Only $7.00 per Dozen, Cash in Advance.

We are the largest Manufacturers and Dealers in Records and Supplies in the world, and handle the latest and best of everything.

Write to us when you want information on any subject in our line.

The Chicago Talking Machine Co.,
LEON F. DOUGLASS, Vice-Prest.,
98 Madison Street,
CHICAGO, ILL.

Long Distance Telephone,
Main Express, 128.

1-27. The back cover of the 1894 Chicago Talking Machine Company catalog. Note the "Cash in Advance" admonition regarding records. A large segment of the cylinder market at this time would have included arcade owners and itinerant showmen, cash businessmen. *Courtesy of Anthony Wellman.*

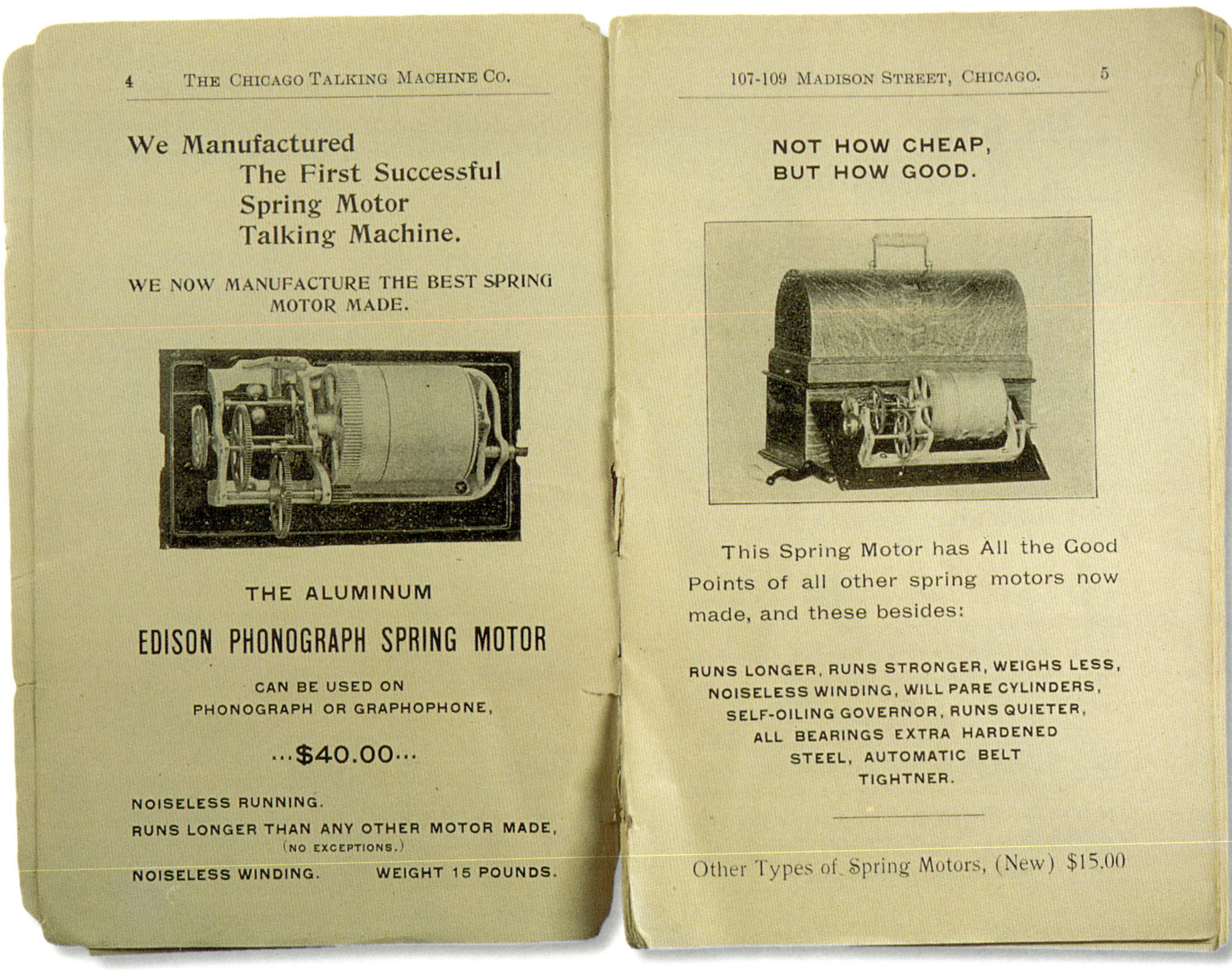
4 THE CHICAGO TALKING MACHINE CO.

We Manufactured
The First Successful
Spring Motor
Talking Machine.

WE NOW MANUFACTURE THE BEST SPRING MOTOR MADE.

THE ALUMINUM

EDISON PHONOGRAPH SPRING MOTOR

CAN BE USED ON
PHONOGRAPH OR GRAPHOPHONE,

...$40.00...

NOISELESS RUNNING.
RUNS LONGER THAN ANY OTHER MOTOR MADE,
(NO EXCEPTIONS.)
NOISELESS WINDING. WEIGHT 15 POUNDS.

107-109 MADISON STREET, CHICAGO. 5

NOT HOW CHEAP,
BUT HOW GOOD.

This Spring Motor has All the Good Points of all other spring motors now made, and these besides:

RUNS LONGER, RUNS STRONGER, WEIGHS LESS, NOISELESS WINDING, WILL PARE CYLINDERS, SELF-OILING GOVERNOR, RUNS QUIETER, ALL BEARINGS EXTRA HARDENED STEEL, AUTOMATIC BELT TIGHTNER.

Other Types of Spring Motors, (New) $15.00

1-28. The Chicago Talking Machine Company was a "live wire outfit." It offered, apparently on an exclusive basis, the spring motors developed by Edward H. Amet – the first practical spring motors in the industry. By late 1896, Amet's star was falling (at the hands of Columbia's lawyers), and the Chicago Talking Machine Company offered its own powerful spring motor designed to compete with the three-spring Capps motor that powered Edison "Spring Motor" Phonographs. This page from a late 1896 to early 1897 Chicago Talking Machine Company catalog illustrates the short-lived three-spring aluminum motor whose designer remains unknown. Catalog measures 4" x 6" closed. *Courtesy of Ernest Carl Allen. (Value code: VR)*

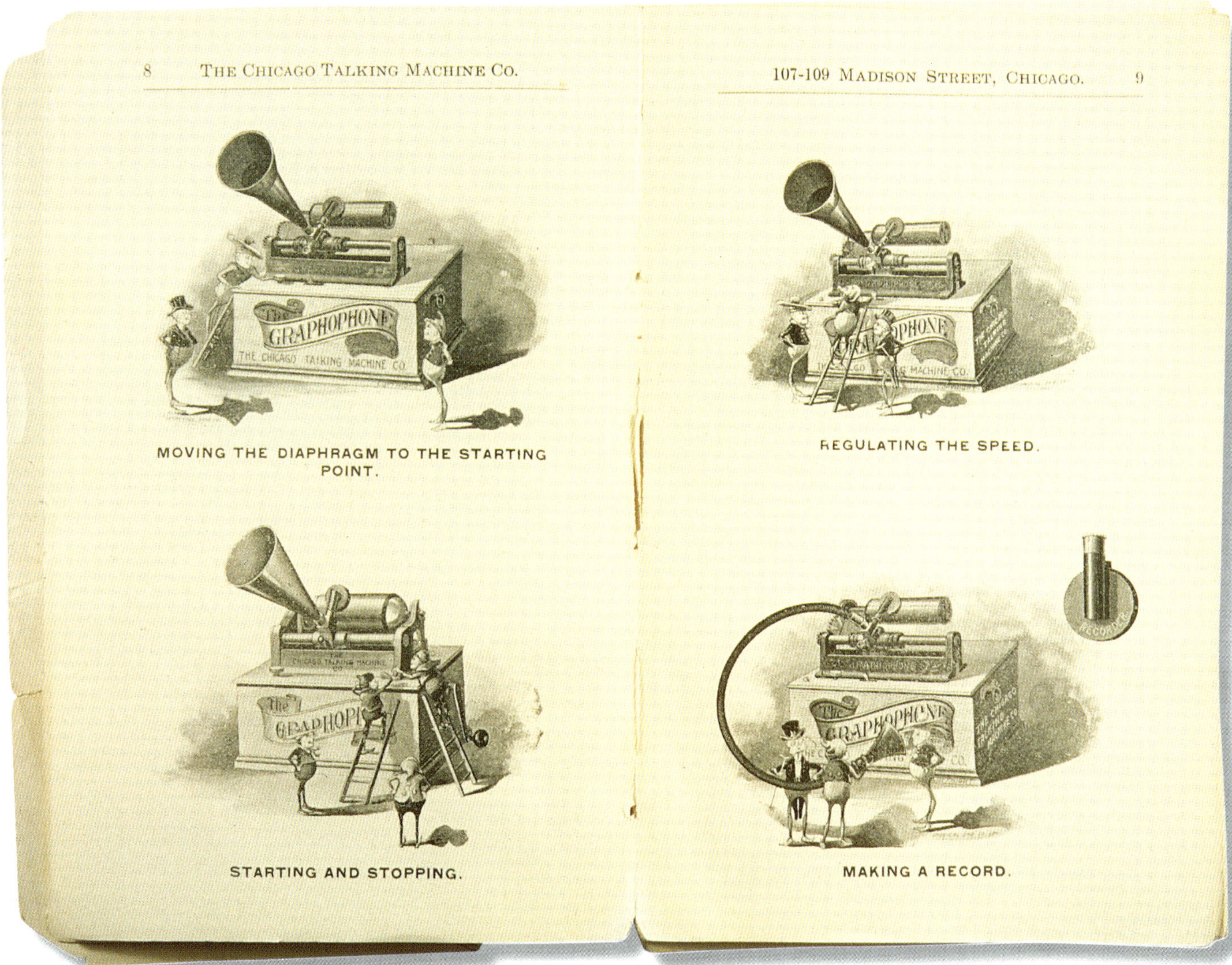

1-29. On this spread, the currently-popular Palmer Cox "Brownies" demonstrated the ease with which a Type "A" Graphophone ($25.00) could be operated. The "Chicago Talking Machine Co." lettering has not been seen on cabinets, but the firm appended its own small decals to the lower right-hand portion of the "Graphophone" banner, as seen in these illustrations. *Courtesy of Ernest Carl Allen.*

18 THE CHICAGO TALKING MACHINE CO.

STANDARD GRAPHOPHONE.

THE BEST TALKING MACHINE FOR EXHIBITION, HOME ENTERTAINMENT AND COMMERCIAL DICTATION.

Makes records, reproduces them and pares the cylinder, making it ready for another record. One cylinder will hold one thousand words or more and can be shaved about two hundred times. Run by spring motor, electric light or storage battery. Furnished with No. 1 Outfit with the Spring Motor.

107-109 MADISON STREET, CHICAGO. 19

TYPE K. F.

WITH HORN ATTACHED. SEE OUTFIT NO. 1.

1-30. The last of the old Bell-Tainter treadle playing mechanisms was used in the Type "K" (or "Standard") Graphophones. The Chicago Talking Machine Company applied its own logical nomenclature to this spring motor (invented by Thomas Macdonald) version, called the "KF," although elsewhere in the catalog it is identified more traditionally as the "F." The 110-volt DC motor was supplied for the Type "I," and the 2-volt battery motor was found in the Type "K." *Courtesy of Ernest Carl Allen.*

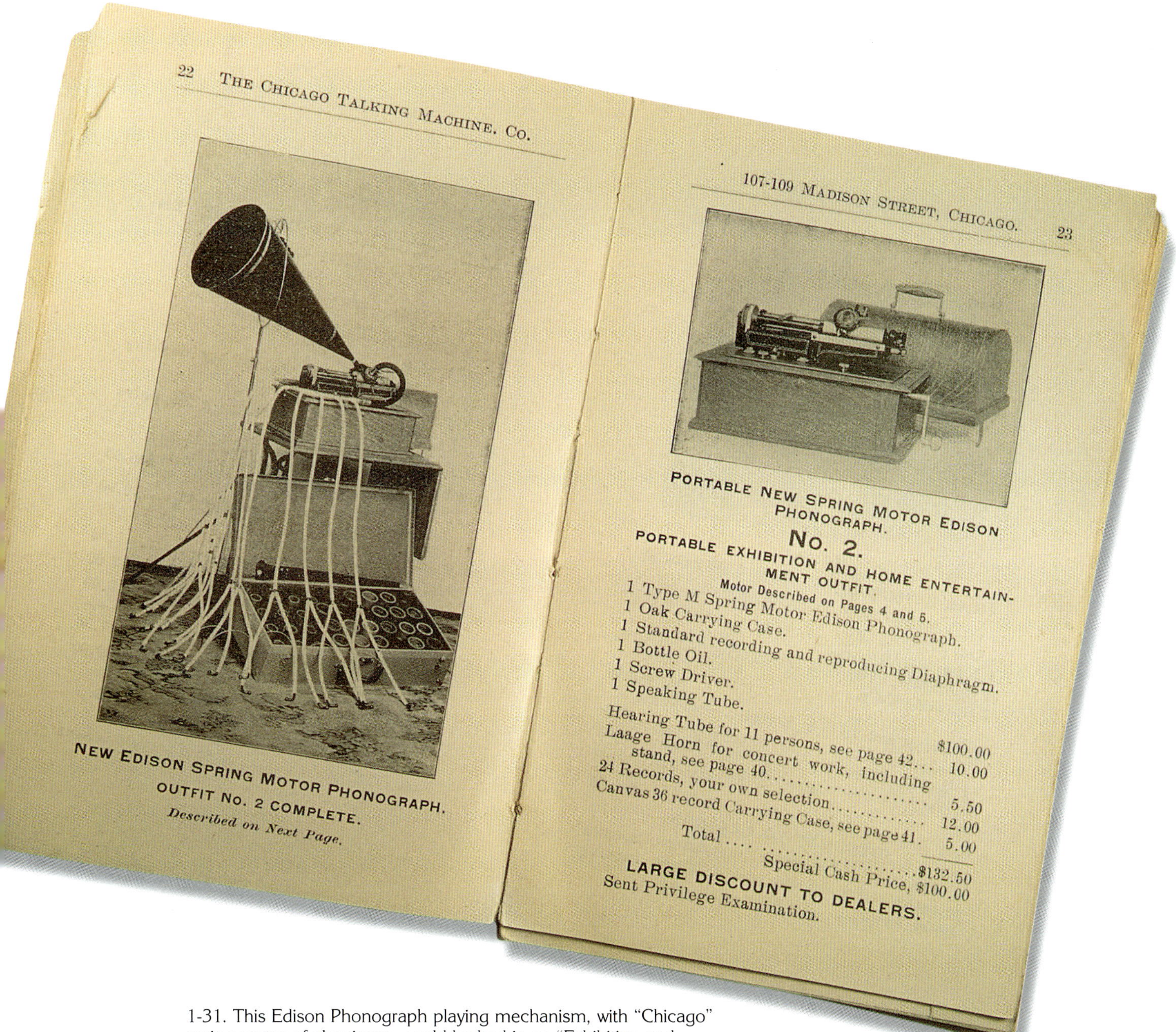

22 THE CHICAGO TALKING MACHINE. CO.

NEW EDISON SPRING MOTOR PHONOGRAPH.
OUTFIT No. 2 COMPLETE.
Described on Next Page.

107-109 MADISON STREET, CHICAGO. 23

PORTABLE NEW SPRING MOTOR EDISON PHONOGRAPH.

No. 2.

PORTABLE EXHIBITION AND HOME ENTERTAINMENT OUTFIT.

Motor Described on Pages 4 and 5.

1 Type M Spring Motor Edison Phonograph.
1 Oak Carrying Case.
1 Standard recording and reproducing Diaphragm.
1 Bottle Oil.
1 Screw Driver.
1 Speaking Tube.

Hearing Tube for 11 persons, see page 42...	$100.00
Laage Horn for concert work, including stand, see page 40..................	10.00
24 Records, your own selection..........	5.50
Canvas 36 record Carrying Case, see page 41.	12.00
Total	5.00
	$132.50
Special Cash Price,	$100.00

LARGE DISCOUNT TO DEALERS.
Sent Privilege Examination.

1-31. This Edison Phonograph playing mechanism, with "Chicago" spring motor of aluminum, could be had in an "Exhibition and Home Entertainment Outfit." *Courtesy of Ernest Carl Allen.*

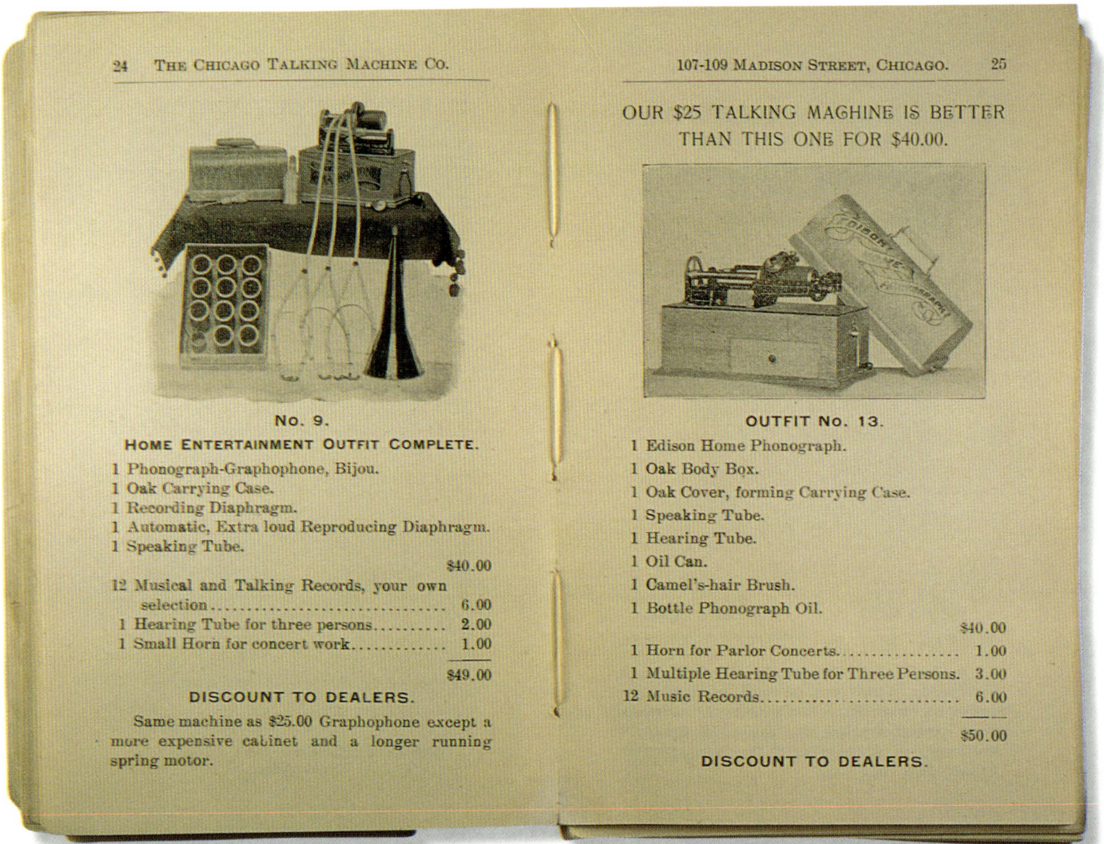

24 THE CHICAGO TALKING MACHINE CO.

No. 9.

HOME ENTERTAINMENT OUTFIT COMPLETE.

1 Phonograph-Graphophone, Bijou.
1 Oak Carrying Case.
1 Recording Diaphragm.
1 Automatic, Extra loud Reproducing Diaphragm.
1 Speaking Tube.

	$40.00
12 Musical and Talking Records, your own selection	6.00
1 Hearing Tube for three persons	2.00
1 Small Horn for concert work	1.00
	$49.00

DISCOUNT TO DEALERS.

Same machine as $25.00 Graphophone except a more expensive cabinet and a longer running spring motor.

107-109 MADISON STREET, CHICAGO. 25

OUR $25 TALKING MACHINE IS BETTER THAN THIS ONE FOR $40.00.

OUTFIT No. 13.

1 Edison Home Phonograph.
1 Oak Body Box.
1 Oak Cover, forming Carrying Case.
1 Speaking Tube.
1 Hearing Tube.
1 Oil Can.
1 Camel's-hair Brush.
1 Bottle Phonograph Oil.

	$40.00
1 Horn for Parlor Concerts	1.00
1 Multiple Hearing Tube for Three Persons.	3.00
12 Music Records	6.00
	$50.00

DISCOUNT TO DEALERS.

1-32. The Type "AN" Graphophone was featured on page 24, while the early Edison "Home" Phonograph was reluctantly offered on page 25. The Chicago Talking Machine Company's stock of Edison merchandise was primarily drawn from its purchase of the defunct North American Phonograph Company's inventory. The inclusion of the recently released Edison "Home" Phonograph was probably meant only to demonstrate the superiority of the competing Graphophones. *Courtesy of Ernest Carl Allen.*

1-33. This 1896 booklet promoted winter travel to New York City on the New York Central Railroad. Its 64 pages contain timetables and advertisements, one of which we illustrate next. *Courtesy of John Nagy. (Value code: VR)*

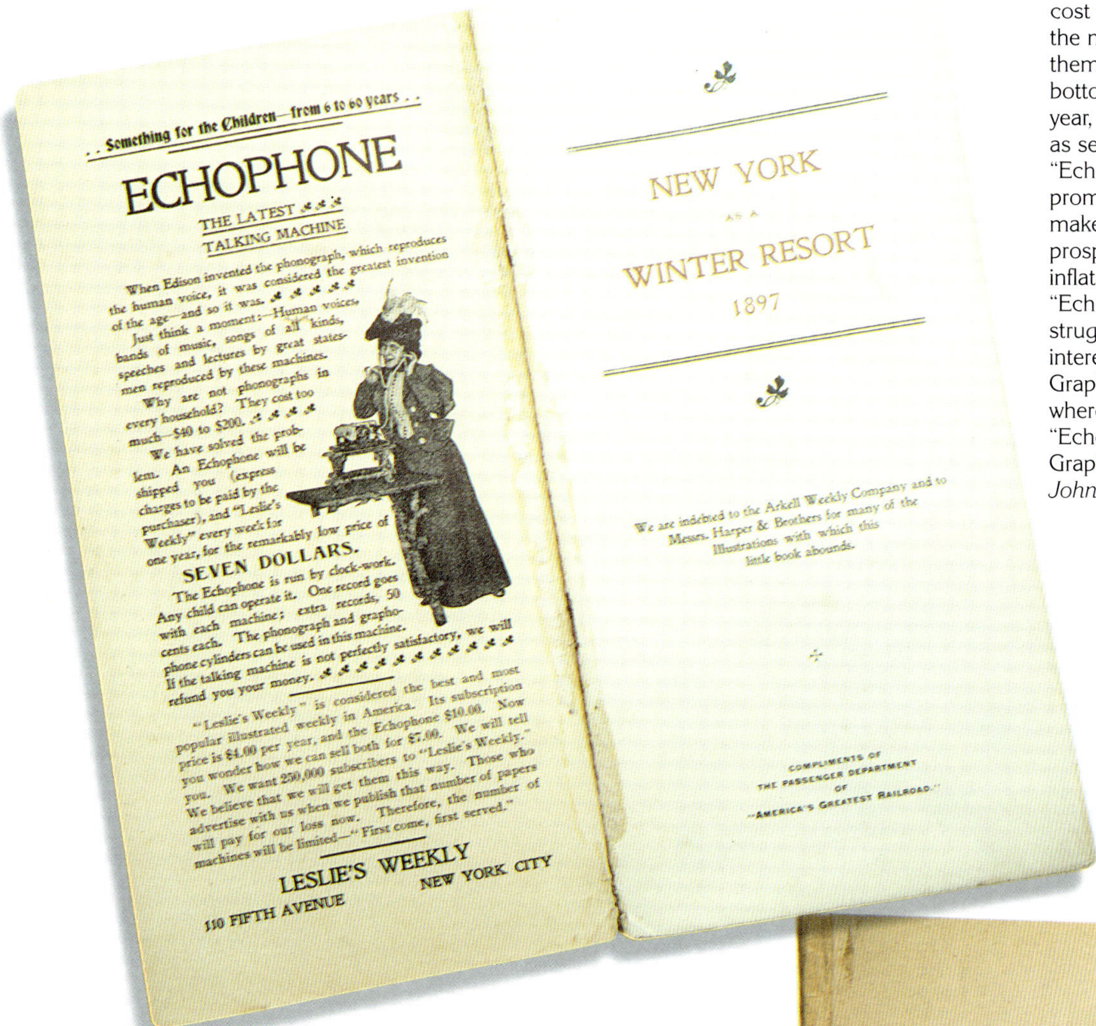

1-34. The "Echophone," invented by Edward H. Amet, was the first low-cost talking machine to be put on the market. Amet began distributing them in late 1895 at the rock-bottom price of $5.00. The following year, the periodical *Leslie's Weekly*, as seen here, was using the "Echophone" as an incentive to promote subscriptions. In order to make the deal seem attractive to the prospective customer, *Leslie's* inflated the "value" of the "Echophone" to $10.00! Amet struggled against the better-funded interests of the American Graphophone Company in court, where it was decided that the "Echophone" infringed Graphophone patents. *Courtesy of John Nagy.*

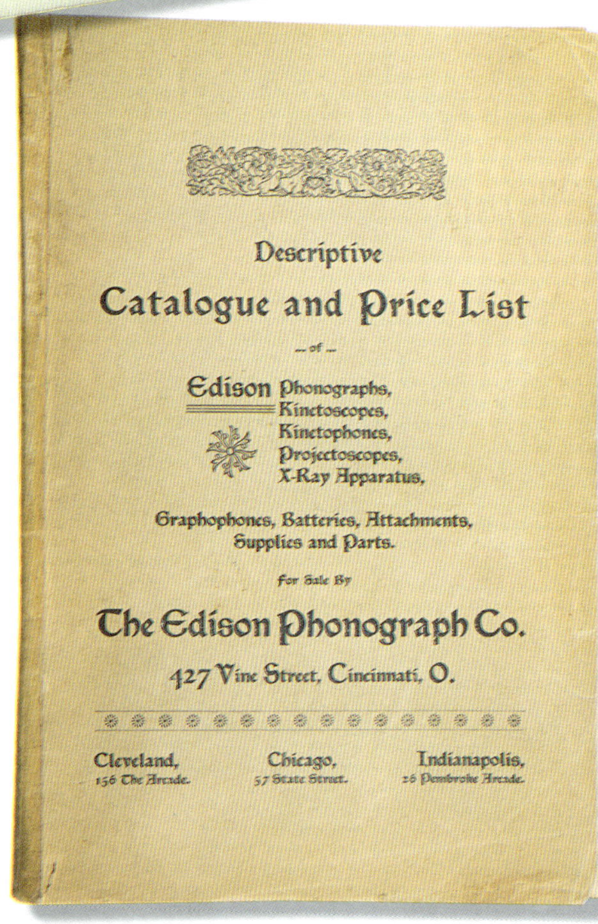

1-35. An example of an Edison catalog from the spring or summer of 1897. No "Standard," no "Gem," no "Concert." The $100.00 "Spring Motor" and the $40.00 "Home" were the only domestic offerings of a Phonograph line badly in need of expansion. Measures 5 3/8" x 7 7/8". *Courtesy of Ernest Carl Allen. (Value code: VR)*

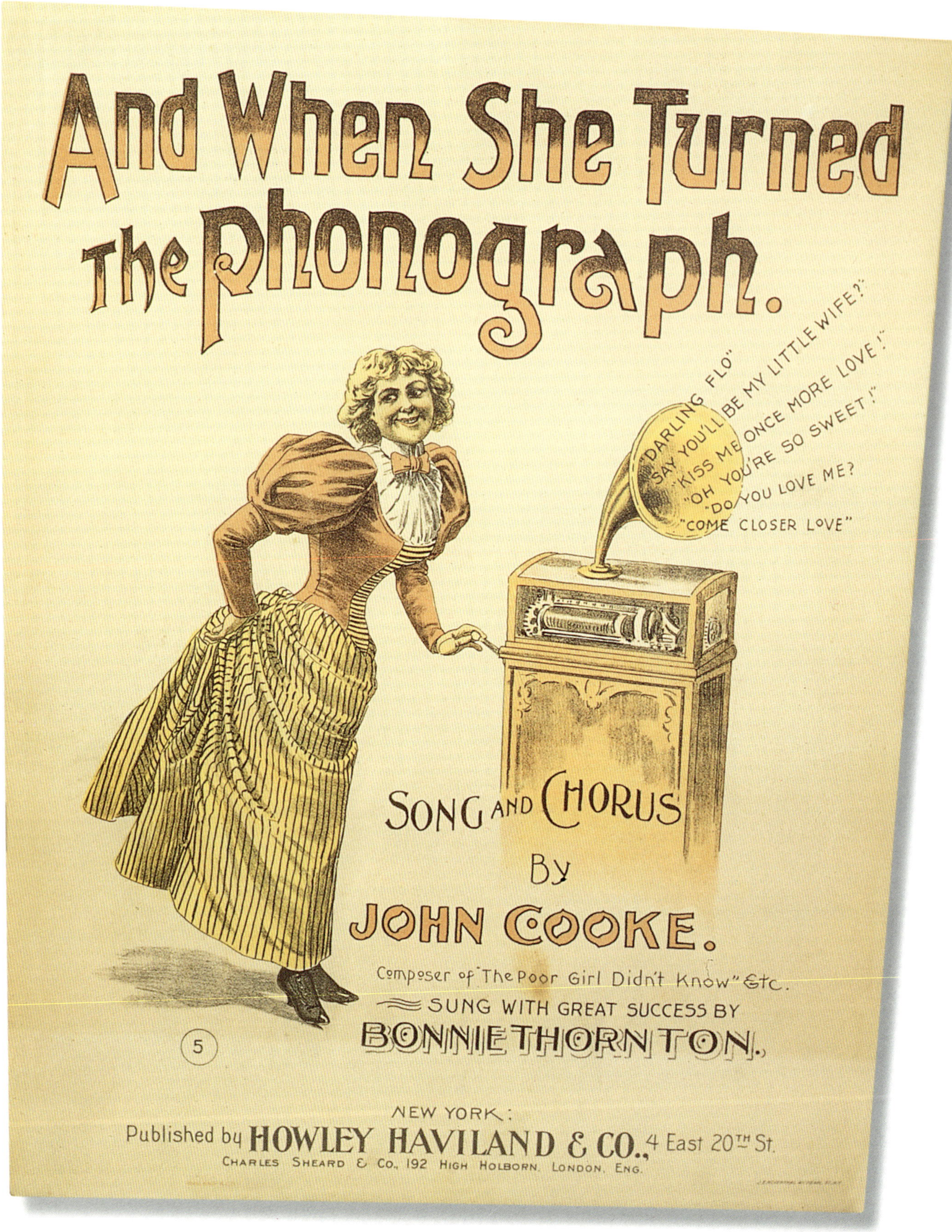

1-36. This sheet music, copyrighted in 1897, showed a mischievous lass playing ("turning") what appears to be a coin-operated phonograph – and no doubt scandalizing the crowd with its racy recording! *Courtesy of Tom and Sandi McCarthy. (Value code: F)*

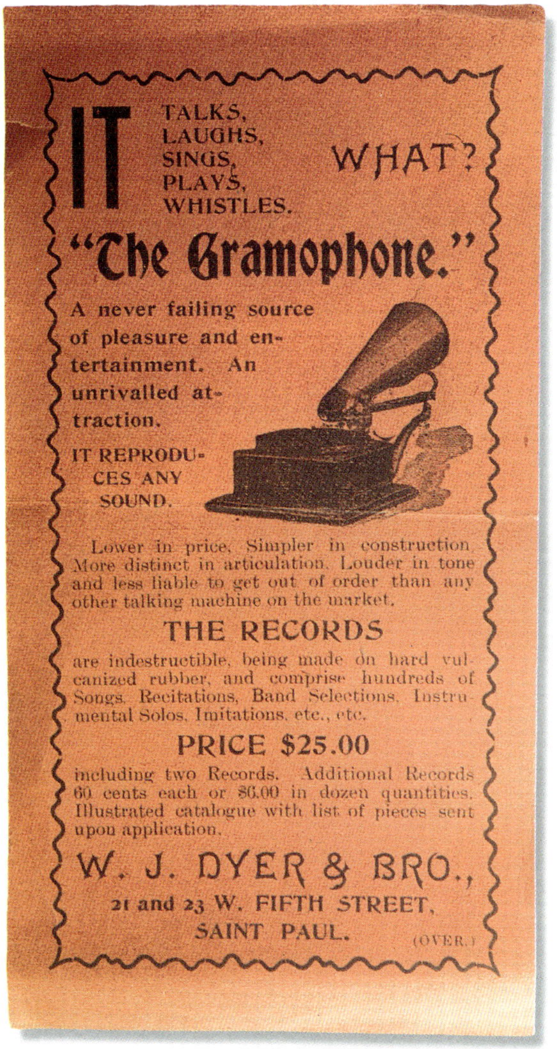

1-37. The first spring-driven Gramophone employed an unsophisticated motor wound by a lever at the side of the cabinet. This 3 1/4" x 6 1/4" sheet from an enterprising Minnesota merchant illustrated this machine, which was available from December 1896 until September 1897. *Courtesy of Ernest Carl Allen. (Value code: F)*

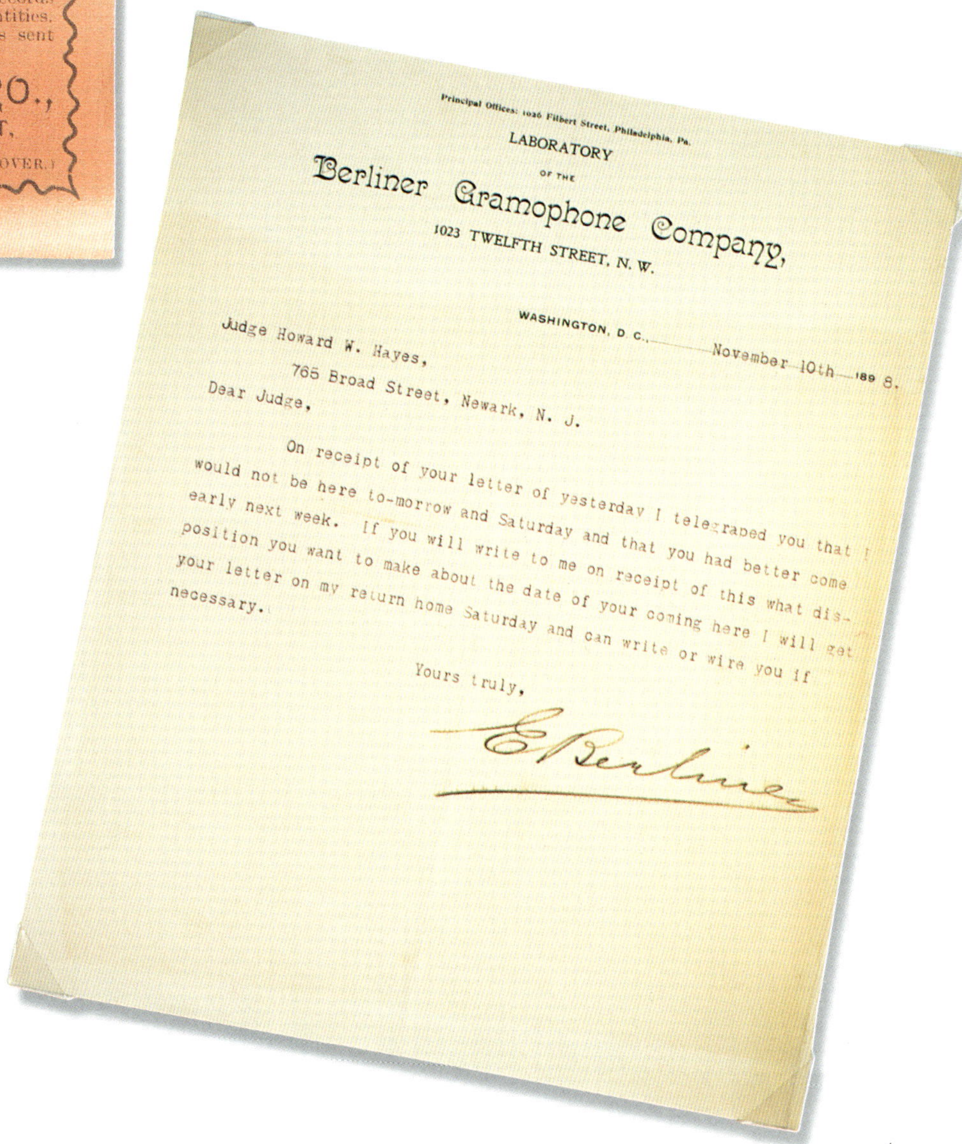

Principal Offices: 1026 Filbert Street, Philadelphia, Pa.

LABORATORY
OF THE
Berliner Gramophone Company,
1023 TWELFTH STREET, N. W.

WASHINGTON, D. C., November 10th 1898.

Judge Howard W. Hayes,
765 Broad Street, Newark, N. J.

Dear Judge,

On receipt of your letter of yesterday I telegraped you that I would not be here to-morrow and Saturday and that you had better come early next week. If you will write to me on receipt of this what disposition you want to make about the date of your coming here I will get your letter on my return home Saturday and can write or wire you if necessary.

Yours truly,

E Berliner

1-38. The signature of Emile Berliner, inventor of the Gramophone, on an unexceptional business letter. *(Value code: C)*

1-39. Three brochures from the late 1890s. In the center, the "Directions for Operating" the "Columbia" (Type "A") and "Bijou" (Type "AN") Graphophones dates from 1897. The other two catalog pages illustrate products of the Hawthorne & Sheble company of Philadelphia. The "Penn Horn Crane" is simply the H&S product sold under a retailer's name. *Courtesy of Martin F. Bryan. (Value code: G)*

1-40. Sears, Roebuck & Company began carrying Graphophones in 1897, and by fall 1898, when this catalog was issued, the firm offered a full line of machines and sundries. Measures 4 7/8" x 6 5/8" closed. *Courtesy of Ernest Carl Allen. (Value code: E, complete catalog)*

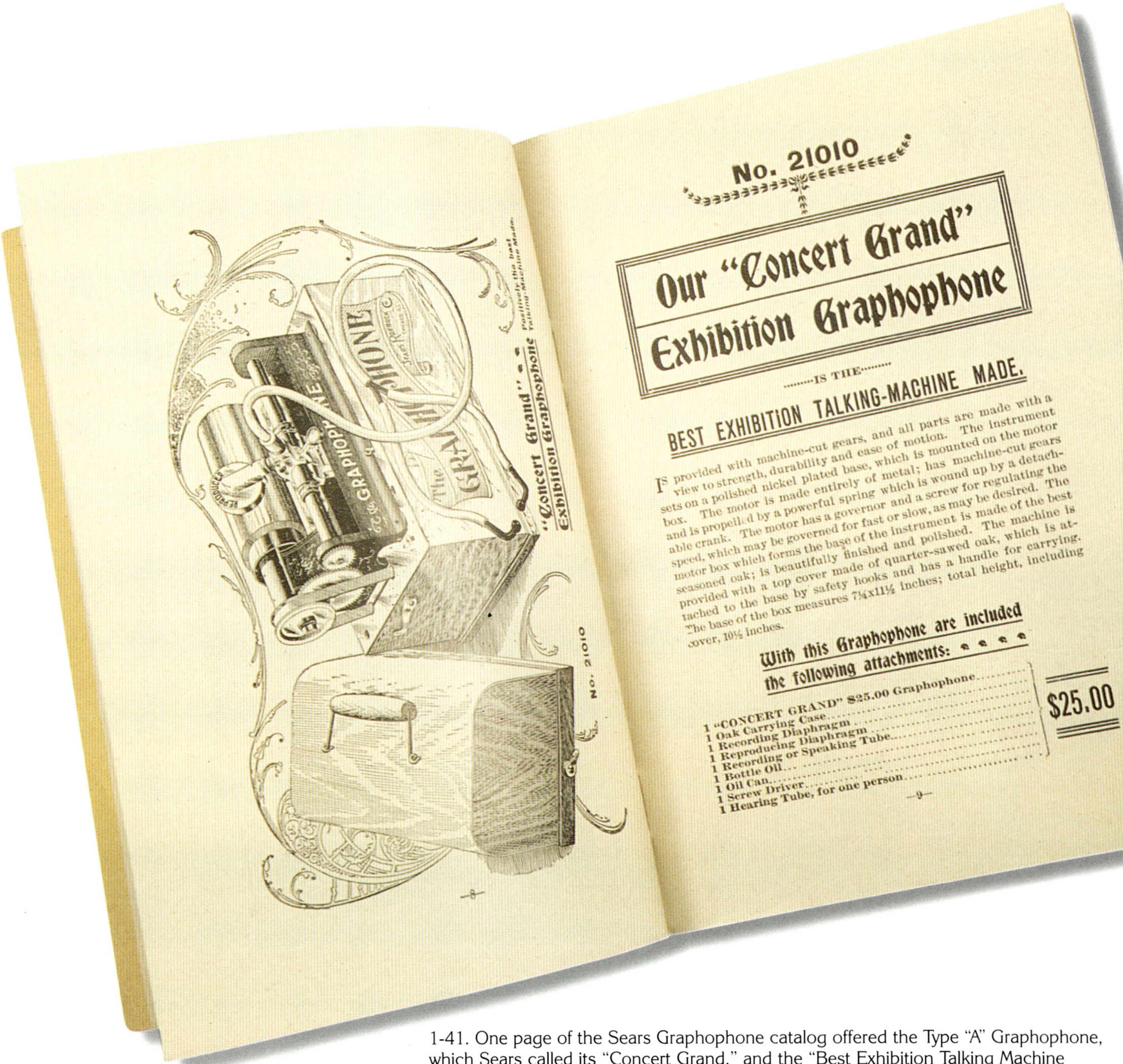

No. 21010

Our "Concert Grand" Exhibition Graphophone

IS THE

BEST EXHIBITION TALKING-MACHINE MADE.

Is provided with machine-cut gears, and all parts are made with a view to strength, durability and ease of motion. The instrument sets on a polished nickel plated base, which is mounted on the motor box. The motor is made entirely of metal; has machine-cut gears and is propelled by a powerful spring which is wound up by a detachable crank. The motor has a governor and a screw for regulating the speed, which may be governed for fast or slow, as may be desired. The motor box which forms the base of the instrument is made of the best seasoned oak; is beautifully finished and polished. The machine is provided with a top cover made of quarter-sawed oak, which is attached to the base by safety hooks and has a handle for carrying. The base of the box measures 7¼x11½ inches; total height, including cover, 10½ inches.

With this Graphophone are included the following attachments:

1 "CONCERT GRAND" $25.00 Graphophone
1 Oak Carrying Case
1 Recording Diaphragm
1 Reproducing Diaphragm
1 Recording or Speaking Tube
1 Bottle Oil
1 Oil Can
1 Screw Driver
1 Hearing Tube, for one person

$25.00

1-41. One page of the Sears Graphophone catalog offered the Type "A" Graphophone, which Sears called its "Concert Grand," and the "Best Exhibition Talking Machine Made." The catalog description failed to explain how a single set of hearing-tubes could stand up to that claim! (Exhibition outfits usually included multibranched listening-tubes, see illustration 5-3.) Note the Sears, Roebuck identification on the Graphophone decal. Some machines of this type were indeed marked this way – a practice that Sears ended by 1902. *Courtesy of Ernest Carl Allen.*

1-42. Peter Bacigalupi was surely the most successful talking machine dealer in San Francisco at the turn of the twentieth century. The pasteboard cylinder record carrying box shown here, circa 1899, is literally crammed with advertising for both Edison Phonographs and Columbia Graphophones. *Courtesy of Gilbert "Bert" Pasley. (Value code: VR)*

1-43. Menier chocolate was a popular brand of confection in *Belle Epoque* France when this miniature "phonograph kiosk" was produced. Outside, quaint lithographed Menier advertising of the period; inside, a simple Lioret phonograph mechanism (of the type used in Lioret's Phonographe "Le Merveilleux" and the "Bébé Jumeau" talking doll). *Courtesy of the Julien Anton collection. (Value code: VR)*

1-44. Despite the penciled inscription, this catalog dates from 1899. The "Edison Talking Machine Company" nomenclature suggests that this firm predated the Talking Machine Company of Chicago, which is the better-known distributor of the Polyphone equipment mentioned at the bottom. Polyphone attachments consisted of a carrier arm incorporating two reproducers whose styli tracked within the same rotation of the record groove. This supposedly made the reproduction "Twice as Loud, Twice as Clear, Twice as Sweet" when heard through the two separate Polyphone horns. *Courtesy of Ernest Carl Allen. (Value code: E)*

1-45. Gianni Bettini designed and sold special carrier arms, recorders, and reproducers ("Micro-Diaphragms") for improving the reproduction of stock Edison Phonographs and Columbia Graphophones. Through his social connections, Bettini also persuaded many first-rate operatic performers to visit his studio on Fifth Avenue, New York City, and record their performances on cylinders. These products were sold in the United States from the late 1890s until 1901, after which Bettini moved his retail operation to Paris. This 11" x 15" embossed hanger was made in 1899, and illustrated one of Bettini's reproducers (sans carrier arm) inverted to show its details. The card was manufactured by Robert Hartmann, 289 Canal Street, New York City, and was illustrated in one of the firm's catalogs of 1900. Bettini even wrote a testimonial, stating, "I have received your last shipment of signs, and am glad to say that I am very much pleased with their appearance. They have been entirely satisfactory in every respect." *Courtesy of the Johnson Victrola Museum/Delaware State Museums. (Value code: VR)*

1-46. Cigars were as popular at the end of the nineteenth century as they became at the end of the twentieth. The Kelley Cigar Company kept a sideline of Graphophones and Columbia cylinder records. The customer collected coupons toward the "value" of a Graphophone or music box, whichever he preferred. *(Value code: H)*

1-47. The retailer who affixed his label to this late 1890s cylinder record box included a helpful tip on "The Way To Hold A Record." *Courtesy of Gilbert (Bert) Pasley. (Value code: H)*

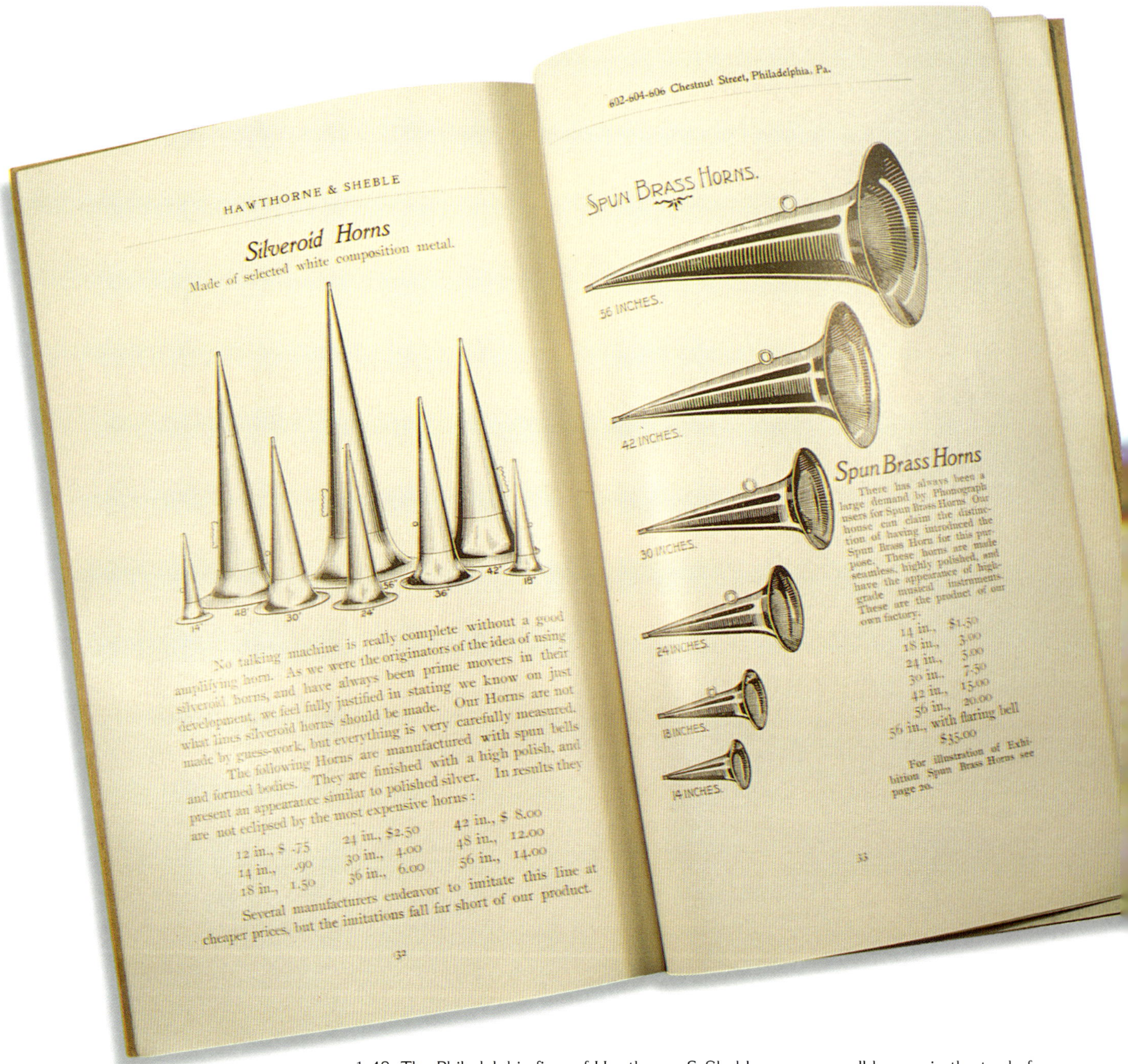

1-48. The Philadelphia firm of Hawthorne & Sheble was very well known in the trade for its quality "aftermarket" items. Edison, Columbia, Victor, and smaller talking machine companies stocked H&S merchandise in their respective catalogs, sometimes claiming them as "our own" or "our special design." H&S conducted a retail business as well, and this 1899 catalog chronicles its wide range of products at the close of the nineteenth century. This spread shows one of its best-known lines: talking machine horns. Catalog measures 6" x 9" closed. *Courtesy of Ernest Carl Allen. (Value code: E)*

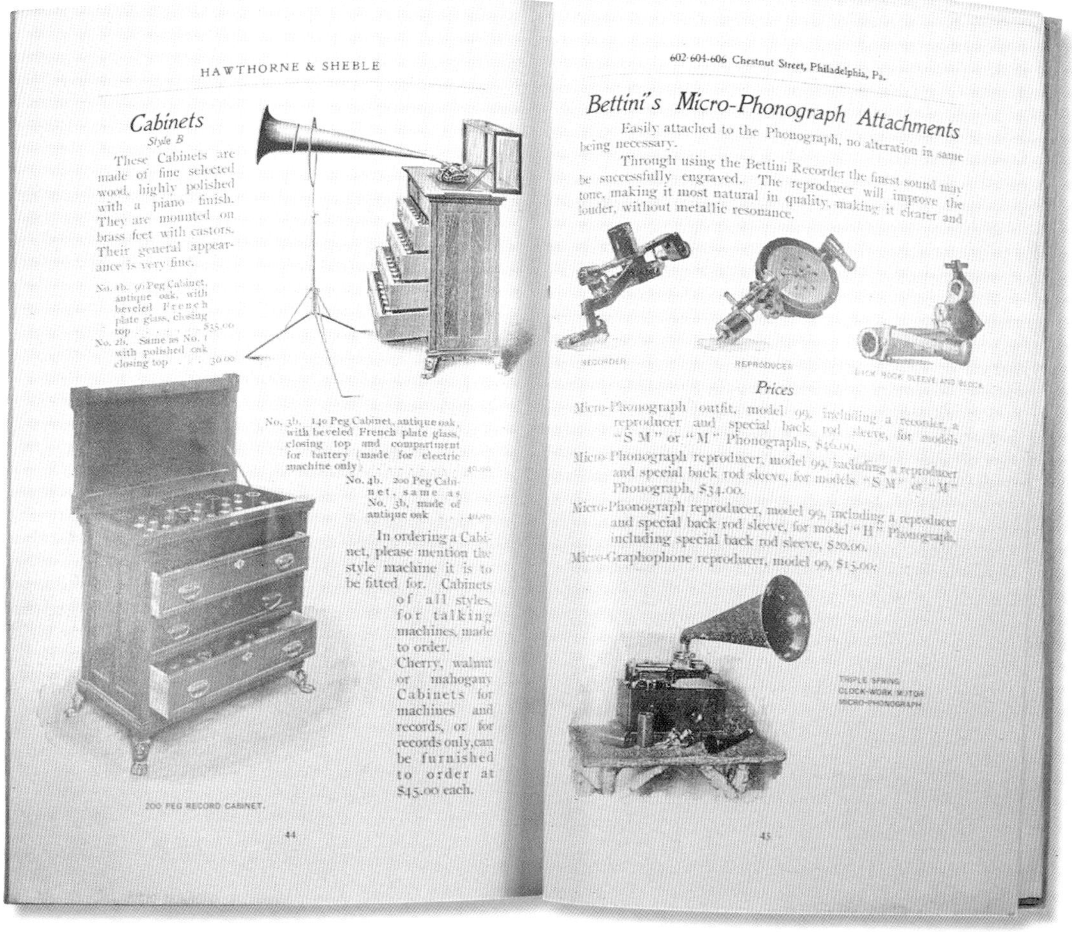

HAWTHORNE & SHEBLE

602-604-606 Chestnut Street, Philadelphia, Pa.

Cabinets

Style B

These Cabinets are made of fine selected wood, highly polished with a piano finish. They are mounted on brass feet with castors. Their general appearance is very fine.

No. 1b. 90 Peg Cabinet, antique oak, with beveled French plate glass, closing top . . . $35.00
No. 2b. Same as No. 1 with polished oak closing top . . . 30.00

No. 3b. 140 Peg Cabinet, antique oak, with beveled French plate glass, closing top and compartment for battery (made for electric machine only) . . . 40.00
No. 4b. 200 Peg Cabinet, same as No. 3b, made of antique oak . . . 40.00

In ordering a Cabinet, please mention the style machine it is to be fitted for. Cabinets of all styles, for talking machines, made to order. Cherry, walnut or mahogany Cabinets for machines and records, or for records only, can be furnished to order at $45.00 each.

200 PEG RECORD CABINET.

Bettini's Micro-Phonograph Attachments

Easily attached to the Phonograph, no alteration in same being necessary.

Through using the Bettini Recorder the finest sound may be successfully engraved. The reproducer will improve the tone, making it most natural in quality, making it clearer and louder, without metallic resonance.

RECORDER REPRODUCER BACK HOOK SLEEVE AND BLOCK

Prices

Micro-Phonograph outfit, model 99, including a recorder, a reproducer and special back rod sleeve, for models "S M" or "M" Phonographs, $46.00.

Micro-Phonograph reproducer, model 99, including a reproducer and special back rod sleeve, for models "S M" or "M" Phonograph, $34.00.

Micro-Phonograph reproducer, model 99, including a reproducer and special back rod sleeve, for model "H" Phonograph, including special back rod sleeve, $20.00.

Micro-Graphophone reproducer, model 99, $15.00.

TRIPLE SPRING CLOCK-WORK MOTOR MICRO-PHONOGRAPH

44 45

1-49. To the left are two H&S cabinets, neither of which would be catalogued beyond 1900. On the right are products from Gianni Bettini of New York City – expensive and sold on a very small scale. *Courtesy of Ernest Carl Allen.*

HAWTHORNE & SHEBLE

602-604-606 Chestnut Street, Philadelphia, Pa.

Bettini's Micro-Phonograph Attachments—Continued

ELECTRIC MOTOR TYPE M MICRO-PHONOGRAPH

BETTINI'S MICRO-GRAPHOPHONE REPRODUCER ON THE COLUMBIA GRAPHOPHONE

Horns for Bettini Attachments

The ordinary horns will not fit the Bettini attachment. We can alter your present horn to fit, or we manufacture the following special horns:

For Micro-Phonograph Reproducer

No. 1 Reproducing Horn, in corrugated tin and aluminum, (no stand required) . . . $3.50
No. 2 Reproducing Horn, all aluminum, (no stand required) 6.00
No. 3 Reproducing Horn, in brass, 30 in. long . . . 8.00
No. 4 Reproducing Horn, in brass, 48 in. long . . . 15.00
No. 8 Reproducing Horn, Aluminum, 36 in., bell 19½ in., Special Exhibition Horn . . . 15.00

For the Home Phonograph and Graphophone Micro-Reproducer

No. 5 Reproducing Horn, in corrugated tin and aluminum, (no stand required) . . . $2.50
No. 6 Reproducing Horn, all aluminum, (no stand required) 5.00
No. 7 Reproducing Horn, in brass, 24 in. long . . . 6.00
Recording Horn with connection, (stand required) . . . 4.50
No. 9 Reproducing all aluminum Horn, special for Exhibition, 36 x 19½ in. . . . 15.00

Bettini Phonograph Speed Indicator

The average speed for making records is about one hundred and twenty revolutions a minute.

In order to obtain the best and most natural reproductions of sound, it is absolutely necessary to have the recorded cylinder revolving during its reproduction at the same speed at which the record was taken; for this purpose we offer the Bettini Speed Indicator, $6.00

To place Speed Indicator on Phonograph, clamp it on upright post holding the left-hand centre of the mandrel.

Speed Indicator in position

H. & S. Electric Lubricant

For Spring Motor Phonographs and Graphophones

10 cents per tube

This lubricant is manufactured expressly for us from carefully selected graphite, oil, etc., and cannot be recommended too highly to those using machines operated by spring motors. It not only reduces the amount of friction and wear to a minimum, but at the same time eradicates any grinding or rumbling noise connected with the running of the clockwork.

Special Hearing Tubes

These are made up entirely without the use of any nickel connections, hard rubber pieces being substituted instead. They will be found very serviceable and easily handled.

1-way hearing tube . . . $0.50
2-way hearing tube . . . 1.00
3-way hearing tube . . . 1.50
4-way hearing tube . . . 2.00
5-way hearing tube . . . 2.50
6-way hearing tube . . . 3.00

Made for all models of Phonographs and Graphophones

46 47

1-50. The following page illustrates more of the Bettini line, including its distinctive horns. *Courtesy of Ernest Carl Allen.*

HAWTHORNE & SHEBLE

Storage Batteries

Numerous cheap and non-efficient batteries have been placed on the market and the public have suffered in consequence. There is nothing regarding which the general public is so ignorant about as a Storage Battery. It resolves itself down to purely a matter of confidence in the seller. We have experimented with all styles until we have secured one which best answers the necessary requirements.

It was formerly a serious question to get a Secondary Battery properly recharged, but to-day any Electric Light Plant is equipped to do the work, and the expensive Primary Cell has been almost entirely done away with.

Price, 100 Amp. hour Battery, running Phonograph about 40 hours and Graphophone about 50 hours, $14.50.

Price, 150 Amp. hour Battery, running Phonograph about 65 hours and Graphophone about 80 hours, $18.00.

Price, 300 Amp. hour Battery, running Phonograph about 120 hours and Graphophone about 140 hours, $25.00.

Recharging—150 amp. hour battery, $1.00.
Recharging—300 amp. hour battery, $1.50.

The above cut will give an idea of the general appearance of the Chloride Batteries.

22

602-604-606 Chestnut Street, Philadelphia, Pa.

Edison Phonograph Battery

For Stationary Work

Price, complete, $15.00

Size, 25½ in. x 8 in. x 15 in.

This Battery consists of four Edison-Lalande cells, Type "S." Porcelain Jars (capacity of 300 ampere-hours), fitted in metal-lined polished oak box, and with Battery cord.

It will run the Phonograph for 100 hours before needing to be recharged, and as there is no waste except when the Battery is doing actual work, it will be found the most efficient and economical Battery of this type for this class of work, in cases where the Phonograph is used in private houses, business offices, etc. Where the Phonograph is required for portable exhibition work, we can furnish special Batteries for this purpose.

Cost of Renewal Charge, for Complete Battery of four Cells, Type "S," $4.76

The Edison Exhibition Phonograph Battery

Consisting of four Liquid Tight Enameled Steel Cells, Type "V" (capacity 150 ampere-hours), in polished oak case, with strap for carrying same.

Price, complete, $12.00

This Battery will run the Phonograph for 50 hours with single charge, and as it is absolutely liquid tight, it is specially recommended for traveling exhibition work.

Size over all, 5¾ in. x 8 in.

Price of Renewal Parts for 4 "V" cells, $3.00

Hydrometer and Glass for Testing Storage Battery, $2.00
Battery Cords, each - - - - - - - - 1.00

23

1-51. From the sublime to the utilitarian – H&S carried everything for talking machines, including batteries for electric-motor machines (notably the Edison Class "M"). *Courtesy of Ernest Carl Allen.*

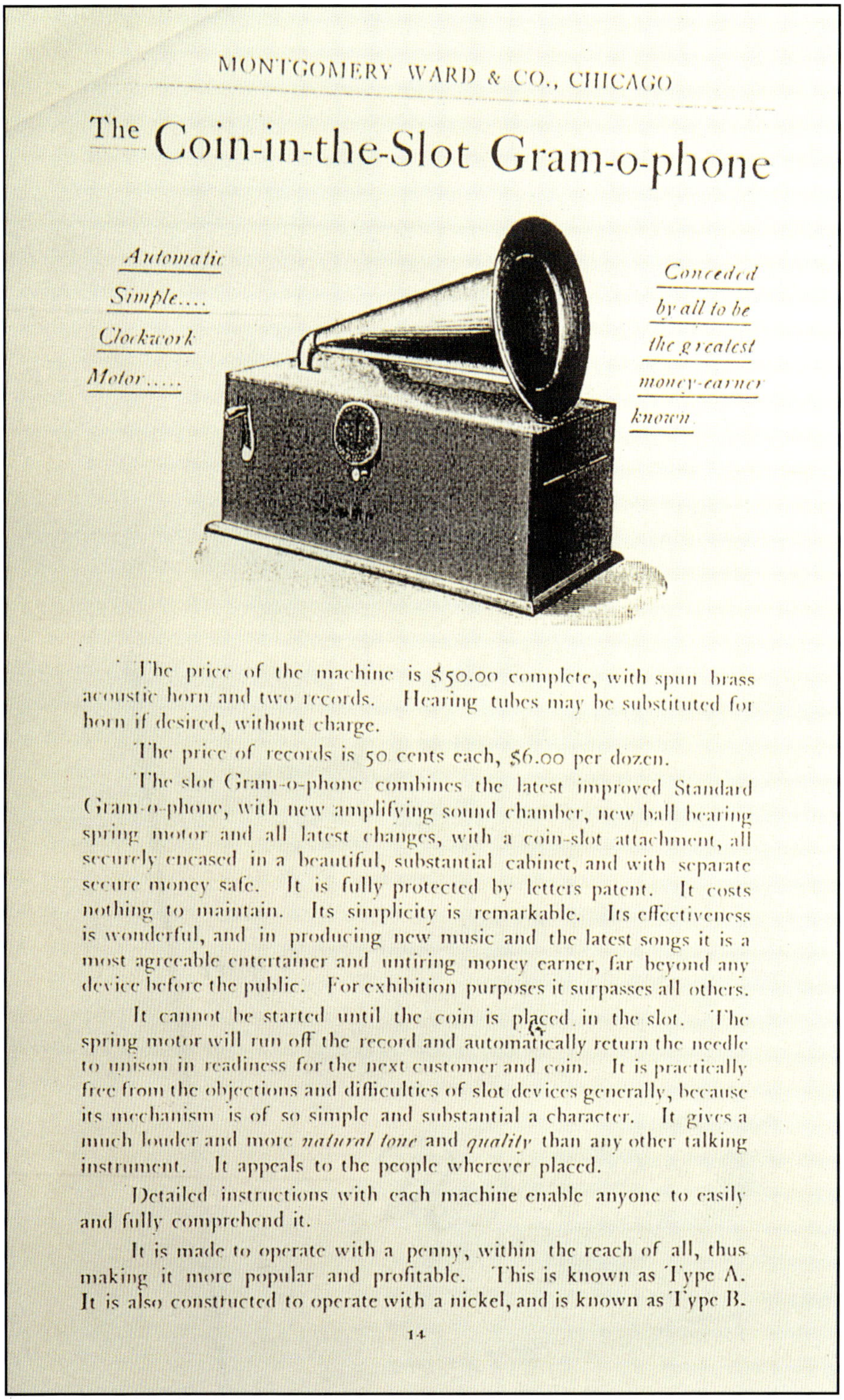
MONTGOMERY WARD & CO., CHICAGO

The Coin-in-the-Slot Gram-o-phone

Automatic
Simple....
Clockwork
Motor.....

Conceded
by all to be
the greatest
money-earner
known.

The price of the machine is $50.00 complete, with spun brass acoustic horn and two records. Hearing tubes may be substituted for horn if desired, without charge.

The price of records is 50 cents each, $6.00 per dozen.

The slot Gram-o-phone combines the latest improved Standard Gram-o-phone, with new amplifying sound chamber, new ball bearing spring motor and all latest changes, with a coin-slot attachment, all securely encased in a beautiful, substantial cabinet, and with separate secure money safe. It is fully protected by letters patent. It costs nothing to maintain. Its simplicity is remarkable. Its effectiveness is wonderful, and in producing new music and the latest songs it is a most agreeable entertainer and untiring money earner, far beyond any device before the public. For exhibition purposes it surpasses all others.

It cannot be started until the coin is placed in the slot. The spring motor will run off the record and automatically return the needle to unison in readiness for the next customer and coin. It is practically free from the objections and difficulties of slot devices generally, because its mechanism is of so simple and substantial a character. It gives a much louder and more *natural tone* and *quality* than any other talking instrument. It appeals to the people wherever placed.

Detailed instructions with each machine enable anyone to easily and fully comprehend it.

It is made to operate with a penny, within the reach of all, thus making it more popular and profitable. This is known as Type A. It is also constructed to operate with a nickel, and is known as Type B.

14

1-52. The extreme rarity of this type of coin-slot Berliner Gramophone (versions sold in Europe by the Gramophone Company, Limited and its sister firms are far more frequently seen) can be attributed in large part to the lofty price: $50.00 in 1898. Contained within this discreet oak box was an intact Berliner "Trade Mark" Gramophone, with a few external mechanical modifications to allow coin operation. Rather ironically, the Universal Talking Machine Company was created in February 1898, ancillary to Emile Berliner's distributor, in order to convert Gramophones to coin use. Ultimately, this secondary firm would become the vehicle of Berliner's ruin in the United States, but not until it stopped adapting Gramophones and began manufacturing a competing product: the Zonophone. The automatic Gramophones, seen here offered by Montgomery Ward, proved a failure. Berliner company officials noted that "only a very few have been sold." Curiously, the artist has *reversed* the normal position of the horn. On the actual machine, the horn would extend *away* from the viewer. *Courtesy of Ernest Allen. (Value code: E, complete catalog)*

1-53. Coin-operated phonographs suffered the depredations of public use, and often lost parts and pieces that could easily be broken off, such as the sign boards which promoted the machine or announced the name of the recorded selection. Even more temporal than the signboards were the signs themselves. Here is an extremely rare "blank" to fit a Columbia coin-op such as a "BS" or "AS." In this instance, the novelty of the mechanism is emphasized. Printed area measures 2 5/8" x 12 1/2", circa 1900. *Courtesy of the Charles Hummel collections. (Value code: VR)*

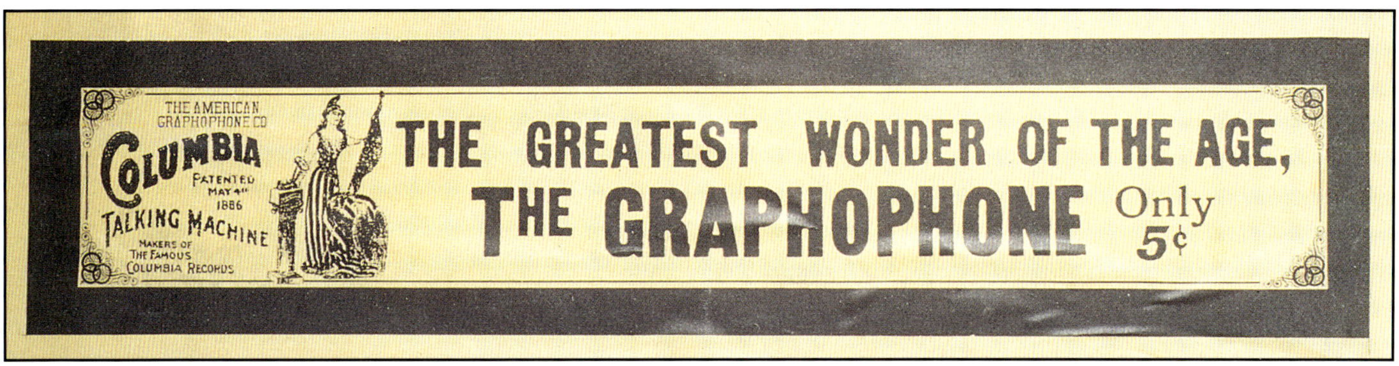

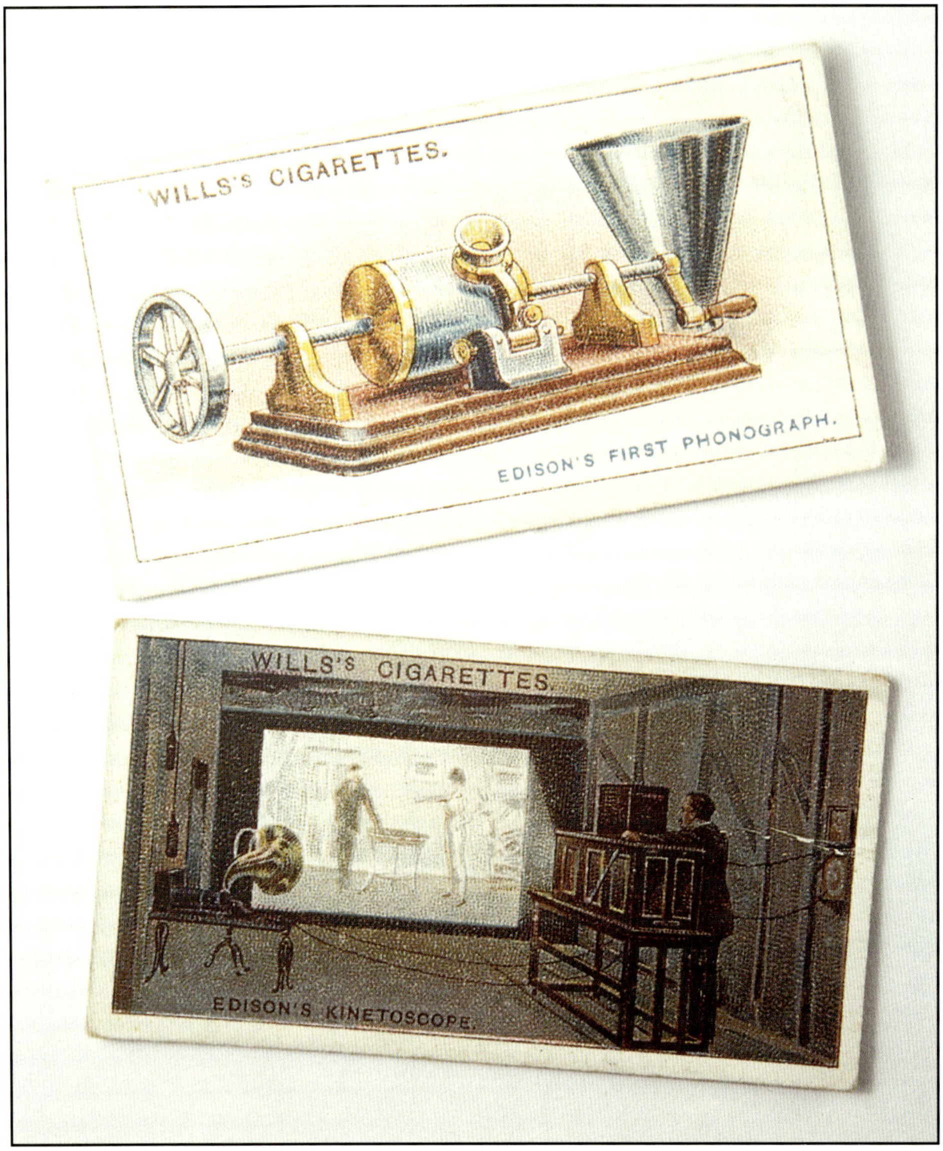

1-54. Two cigarette premium cards from Wills's (British) brand. Above, an imaginative depiction of a "Tin Foil" Phonograph, complete with reproducing funnel. Below, an artist's impression of talking movies, Edison-style. Edison did, in fact, introduce a "Phono-Kinetoscope" in 1895, that combined background music (as opposed to synchronized sound) for "peep show" (individually viewed, as opposed to projected) films. Successful projected cinema with sound accompaniment was a more difficult proposition, and many inventors tried their hands from the 1890s through the end of the 1920s. *Courtesy of Yves Rouchaleau. (Value code: F)*

1-55. A scientific instrument dealer in Lisbon, Portugal pasted his label inside the lid of a Type "B" Graphophone, circa 1900. It's interesting to note that the talking machine doesn't even warrant a mention. *Courtesy of George Glastris.*

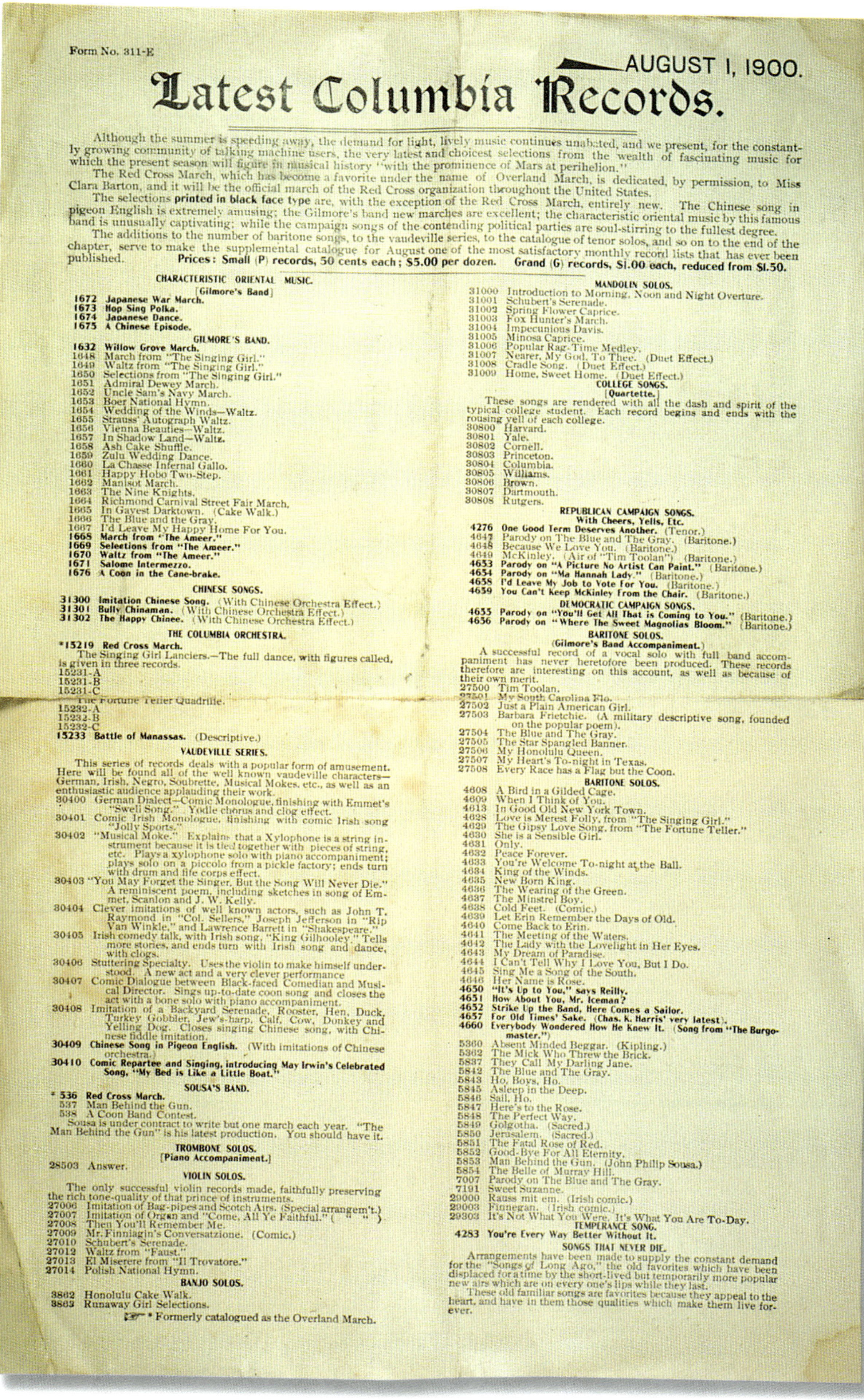

Form No. 311-E

AUGUST 1, 1900.

Latest Columbia Records.

Although the summer is speeding away, the demand for light, lively music continues unabated, and we present, for the constantly growing community of talking machine users, the very latest and choicest selections from the wealth of fascinating music for which the present season will figure in musical history "with the prominence of Mars at perihelion."

The Red Cross March, which has become a favorite under the name of Overland March, is dedicated, by permission, to Miss Clara Barton, and it will be the official march of the Red Cross organization throughout the United States.

The selections **printed in black face type** are, with the exception of the Red Cross March, entirely new. The Chinese song in pigeon English is extremely amusing; the Gilmore's band new marches are excellent; the characteristic oriental music by this famous band is unusually captivating; while the campaign songs of the contending political parties are soul-stirring to the fullest degree.

The additions to the number of baritone songs, to the vaudeville series, to the catalogue of tenor solos, and so on to the end of the chapter, serve to make the supplemental catalogue for August one of the most satisfactory monthly record lists that has ever been published. **Prices: Small (P) records, 50 cents each; $5.00 per dozen. Grand (G) records, $1.00 each, reduced from $1.50.**

CHARACTERISTIC ORIENTAL MUSIC.
[Gilmore's Band]

1672 Japanese War March.
1673 Hop Sing Polka.
1674 Japanese Dance.
1675 A Chinese Episode.

GILMORE'S BAND.

1632 Willow Grove March.
1648 March from "The Singing Girl."
1649 Waltz from "The Singing Girl."
1650 Selections from "The Singing Girl."
1651 Admiral Dewey March.
1652 Uncle Sam's Navy March.
1653 Boer National Hymn.
1654 Wedding of the Winds—Waltz.
1655 Strauss' Autograph Waltz.
1656 Vienna Beauties—Waltz.
1657 In Shadow Land—Waltz.
1658 Ash Cake Shuffle.
1659 Zulu Wedding Dance.
1660 La Chasse Infernal Gallo.
1661 Happy Hobo Two-Step.
1662 Manisot March.
1663 The Nine Knights.
1664 Richmond Carnival Street Fair March.
1665 In Gayest Darktown. (Cake Walk.)
1666 The Blue and the Gray.
1667 I'd Leave My Happy Home For You.
1668 March from "The Ameer."
1669 Selections from "The Ameer."
1670 Waltz from "The Ameer."
1671 Salome Intermezzo.
1676 A Coon in the Cane-brake.

CHINESE SONGS.

31300 Imitation Chinese Song. (With Chinese Orchestra Effect.)
31301 Bully Chinaman. (With Chinese Orchestra Effect.)
31302 The Happy Chinee. (With Chinese Orchestra Effect.)

THE COLUMBIA ORCHESTRA.

***15219 Red Cross March.**

The Singing Girl Lanciers.—The full dance, with figures called, is given in three records.

15231-A
15231-B
15231-C

The Fortune Teller Quadrille.

15232-A
15232-B
15232-C
15233 Battle of Manassas. (Descriptive.)

VAUDEVILLE SERIES.

This series of records deals with a popular form of amusement. Here will be found all of the well known vaudeville characters—German, Irish, Negro, Soubrette, Musical Mokes, etc., as well as an enthusiastic audience applauding their work.

30400 German Dialect—Comic Monologue, finishing with Emmet's "Sweli Song." Yodle chorus and clog effect.
30401 Comic Irish Monologue, finishing with comic Irish song "Jolly Sports."
30402 "Musical Moke." Explains that a Xylophone is a string instrument because it is tied together with pieces of string, etc. Plays a xylophone solo with piano accompaniment; plays solo on a piccolo from a pickle factory; ends turn with drum and fife corps effect.
30403 "You May Forget the Singer, But the Song Will Never Die." A reminiscent poem, including sketches in song of Emmet, Scanlon and J. W. Kelly.
30404 Clever imitations of well known actors, such as John T. Raymond in "Col. Sellers," Joseph Jefferson in "Rip Van Winkle," and Lawrence Barrett in "Shakespeare."
30405 Irish comedy talk, with Irish song, "King Gilhooley." Tells more stories, and ends turn with Irish song and dance, with clogs.
30406 Stuttering Specialty. Uses the violin to make himself understood. A new act and a very clever performance
30407 Comic Dialogue between Black-faced Comedian and Musical Director. Sings up-to-date coon song and closes the act with a bone solo with piano accompaniment.
30408 Imitation of a Backyard Serenade, Rooster, Hen, Duck, Turkey Gobbler, Jew's-harp, Calf, Cow, Donkey and Yelling Dog. Closes singing Chinese song, with Chinese fiddle imitation.
30409 Chinese Song in Pigeon English. (With imitations of Chinese orchestra.)
30410 Comic Repartee and Singing, introducing May Irwin's Celebrated Song, "My Bed is Like a Little Boat."

SOUSA'S BAND.

*** 536 Red Cross March.**
537 Man Behind the Gun.
538 A Coon Band Contest.

Sousa is under contract to write but one march each year. "The Man Behind the Gun" is his latest production. You should have it.

TROMBONE SOLOS.
[Piano Accompaniment.]

28503 Answer.

VIOLIN SOLOS.

The only successful violin records made, faithfully preserving the rich tone-quality of that prince of instruments.

27006 Imitation of Bag-pipes and Scotch Airs. (Special arrangem't.)
27007 Imitation of Organ and "Come, All Ye Faithful." (" ")
27008 Then You'll Remember Me.
27009 Mr. Finniagin's Conversatzione. (Comic.)
27010 Schubert's Serenade.
27012 Waltz from "Faust."
27013 El Miserere from "Il Trovatore."
27014 Polish National Hymn.

BANJO SOLOS.

3862 Honolulu Cake Walk.
3863 Runaway Girl Selections.

☞ * Formerly catalogued as the Overland March.

MANDOLIN SOLOS.

31000 Introduction to Morning, Noon and Night Overture.
31001 Schubert's Serenade.
31002 Spring Flower Caprice.
31003 Fox Hunter's March.
31004 Impecunious Davis.
31005 Minosa Caprice.
31006 Popular Rag-Time Medley.
31007 Nearer, My God, To Thee. (Duet Effect.)
31008 Cradle Song. (Duet Effect.)
31009 Home, Sweet Home. (Duet Effect.)

COLLEGE SONGS.
[Quartette.]

These songs are rendered with all the dash and spirit of the typical college student. Each record begins and ends with the rousing yell of each college.

30800 Harvard.
30801 Yale.
30802 Cornell.
30803 Princeton.
30804 Columbia.
30805 Williams.
30806 Brown.
30807 Dartmouth.
30808 Rutgers.

REPUBLICAN CAMPAIGN SONGS.
With Cheers, Yells, Etc.

4276 One Good Term Deserves Another. (Tenor.)
4647 Parody on The Blue and The Gray. (Baritone.)
4648 Because We Love You. (Baritone.)
4649 McKinley. (Air of "Tim Toolan") (Baritone.)
4653 Parody on "A Picture No Artist Can Paint." (Baritone.)
4654 Parody on "Ma Hannah Lady." (Baritone.)
4658 I'd Leave My Job to Vote For You. (Baritone.)
4659 You Can't Keep McKinley From the Chair. (Baritone.)

DEMOCRATIC CAMPAIGN SONGS.

4655 Parody on "You'll Get All That is Coming to You." (Baritone.)
4656 Parody on "Where The Sweet Magnolias Bloom." (Baritone.)

BARITONE SOLOS.
(Gilmore's Band Accompaniment.)

A successful record of a vocal solo with full band accompaniment has never heretofore been produced. These records therefore are interesting on this account, as well as because of their own merit.

27500 Tim Toolan.
27501 My South Carolina Flo.
27502 Just a Plain American Girl.
27503 Barbara Frietchie. (A military descriptive song, founded on the popular poem).
27504 The Blue and The Gray.
27505 The Star Spangled Banner.
27506 My Honolulu Queen.
27507 My Heart's To-night in Texas.
27508 Every Race has a Flag but the Coon.

BARITONE SOLOS.

4608 A Bird in a Gilded Cage.
4609 When I Think of You.
4613 In Good Old New York Town.
4628 Love is Merest Folly, from "The Singing Girl."
4629 The Gipsy Love Song, from "The Fortune Teller."
4630 She is a Sensible Girl.
4631 Only.
4632 Peace Forever.
4633 You're Welcome To-night at the Ball.
4634 King of the Winds.
4635 New Born King.
4636 The Wearing of the Green.
4637 The Minstrel Boy.
4638 Cold Feet. (Comic.)
4639 Let Erin Remember the Days of Old.
4640 Come Back to Erin.
4641 The Meeting of the Waters.
4642 The Lady with the Lovelight in Her Eyes.
4643 My Dream of Paradise.
4644 I Can't Tell Why I Love You, But I Do.
4645 Sing Me a Song of the South.
4646 Her Name is Rose.
4650 "It's Up to You," says Reilly.
4651 How About You, Mr. Iceman?
4652 Strike Up the Band, Here Comes a Sailor.
4657 For Old Times' Sake. (Chas. K. Harris' very latest).
4660 Everybody Wondered How He Knew It. (Song from "The Burgomaster.")
5360 Absent Minded Beggar. (Kipling.)
5362 The Mick Who Threw the Brick.
5837 They Call My Darling Jane.
5842 The Blue and The Gray.
5843 Ho, Boys, Ho.
5845 Asleep in the Deep.
5846 Sail, Ho.
5847 Here's to the Rose.
5848 The Perfect Way.
5849 Golgotha. (Sacred.)
5850 Jerusalem. (Sacred.)
5851 The Fatal Rose of Red.
5852 Good-Bye For All Eternity.
5853 Man Behind the Gun. (John Philip Sousa.)
5854 The Belle of Murray Hill.
7007 Parody on The Blue and The Gray.
7191 Sweet Suzanne.
29000 Rauss mit em. (Irish comic.)
29003 Finnegan. (Irish comic.)
29303 It's Not What You Were, It's What You Are To-Day.

TEMPERANCE SONG.

4283 You're Every Way Better Without It.

SONGS THAT NEVER DIE.

Arrangements have been made to supply the constant demand for the "Songs of Long Ago," the old favorites which have been displaced for a time by the short-lived but temporarily more popular new airs which are on every one's lips while they last.

These old familiar songs are favorites because they appeal to the heart, and have in them those qualities which make them live forever.

1-56. This August 1900 8 1/4" x 12 5/8" circular for the Latest Columbia Records included eight Republican Campaign Songs, but only two Democratic records. The Republican candidate, William McKinley, eventually won the election, only to be assassinated a year later at the Buffalo Pan-American Exposition. Circular printed on both sides. *(Value code: F)*

The Multiplex Graphophone Grand

THE LATEST DEVELOPMENT OF THE GRAPHOPHONE.

This new instrument was built for exhibition AT THE PARIS EXPOSITION.

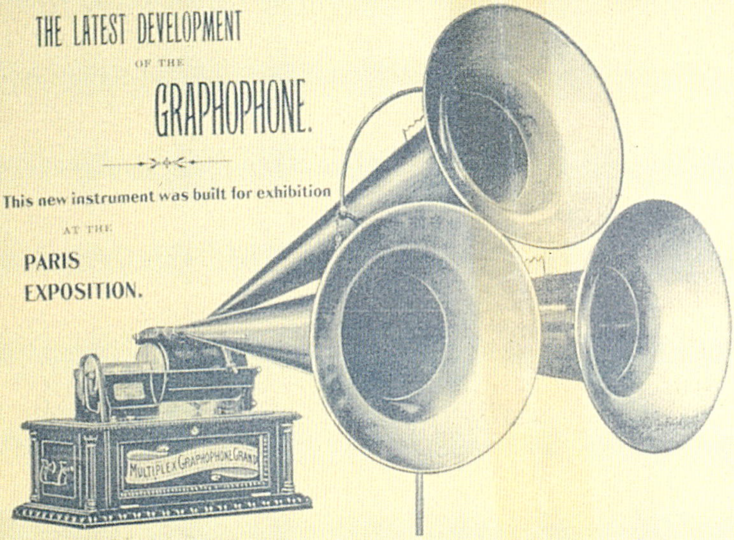

A Talking Machine having the

Volume of Several Grand Instruments

Reproducing in unison.

The most wonderful sound-reproducing mechanism ever constructed. In volume, the voice of the Multiplex overwhelms the tones of the earlier talking machines as the roar of Niagara's cataract drowns the brook's gurgle.

IT CONTAINS NEW FEATURES IN ADDITION TO THOSE THAT CREATED SUCH A PROFOUND SENSATION WHEN EMBODIED IN THE GRAPHOPHONE GRAND.

IT USES

Three Separate Reproducers

ACTING IN ABSOLUTE UNISON WITH THREE SEPARATE AND DISTINCT RECORDS,

Each one of which gives the same loud, pure tone as that of the Graphophone Grand. The combination of all three in unison gives

AN INTENSITY OF VOLUME

AND A

SWEETNESS AND RICHNESS OF TONE

Which seem almost beyond belief.

THE tones of the MULTIPLEX are far more faithful to the original rendition of voice or instrument than those of any other talking machine. This fact is due to greater discrimination in the process of recording, rendered possible only by the use of separate recording horns and styluses.

The three reproducers are entirely independent one from another, yet so arranged as to reproduce in unison. Results are obtained that it is difficult to realize are within the possibilities of sound reproducing mechanism.

The Multiplex can be converted in an instant into

A GRAPHOPHONE GRAND OF TRIPLE SIZE

by using only one reproducer to cover the entire length of the cylinder. Special records may be had which will thus give a reproduction ten minutes long.

Designed especially for exhibition purposes at the Paris Exposition, the Multiplex Graphophone Grand is not listed in our catalogue nor carried in our regular stock of machines. But the results obtained from it are so superb, and its reproductions so marvelously faithful, both in volume and tone, to the real voice or instrument, that we have decided to make machines and records of the Multiplex type to order, confidently assuring those who purchase them that they represent the very highest development in the art of sound reproduction.

MULTIPLEX GRAPHOPHONE GRAND INCLUDING

3 Recorders,
3 Reproducers,
1 Special Triple Horn Stand,
3 56-Inch Brass Horns,
12 Multiplex Grand Records,
6 Multiplex Grand Blanks.

$1,000

COLUMBIA PHONOGRAPH COMPANY,

NEW YORK, 143, 145 Broadway.
RETAIL BRANCH, 1155, 1157, 1159 Broadway.
CHICAGO, 88 Wabash Ave.
ST. LOUIS, 720-722 Olive St.
PHILADELPHIA, 1032 Chestnut St.
BALTIMORE, 110 E. Baltimore St.
CINCINNATI, 121 E. Fourth St.
WASHINGTON, 919 Pennsylvania Ave.
BUFFALO, 313 Main St.
SAN FRANCISCO, 125 Geary St.

LONDON, 122 Oxford Street, W. PARIS, 34 Boulevard des Italiens. BERLIN, 65-A Friedrichstrasse.

Opposite page: 1-57. For some modern-day collectors, the granddaddy of all talking machines, and a "Holy Grail" of sorts, is the Columbia "Multiplex Grand." First shown at the Paris Exhibition of 1900, this immense machine (with its immense price) unintentionally presaged the future, while exemplifying the demerits of contemporary talking machines. Were "wax" cylinder records fragile and inefficient to store? The regular 5" diameter "Grand" cylinders that Columbia sold were delicate enough, but the "Multiplex Grand" cylinders approached absurdity at 5" in diameter and approximately 9" in length! Were horns, supported by inconvenient floor stands, unsightly and always in the way? This machine compounded the nuisance, albeit with a single, larger floor stand. And yet, this "white elephant" embodied capabilities that were a half-century ahead of its time. The three separate tracks on a single "Multiplex Grand" cylinder were able to reproduce what would be later called "stereophonic sound," although existing advertisements make little or no allusion to this possibility. And, if a single track were recorded the entire length of the cylinder, the resultant length of the recording would have been at least five minutes (this broadside claims a length of ten minutes!). The two-minute limitation of ordinary cylinders until late 1908 was one of the medium's weaknesses. (From 1905-1909 Columbia marketed a line of "standard" diameter cylinders that were 6" long, allowing more space for the recording, but these were comparatively pricey, required special Graphophones to play them, and the expense yielded only one additional minute of playing time.) Had the stereophonic and playing-time potential of the "Multiplex Grand" been adapted to an affordable line of Graphophones, who can say how recording history might have been altered? As for the "Multiplex Grand" itself, two examples are known to have been sold at the time (as yet undiscovered), and an additional reproducer carriage was found in the Columbia archives in the 1980s. No other tangible evidence exists relating to this mammoth device – except for rare advertisements such as this 9 3/4" x 14" circular from 1900-1901. *Courtesy of Ernest Carl Allen. (Value code: VR)*

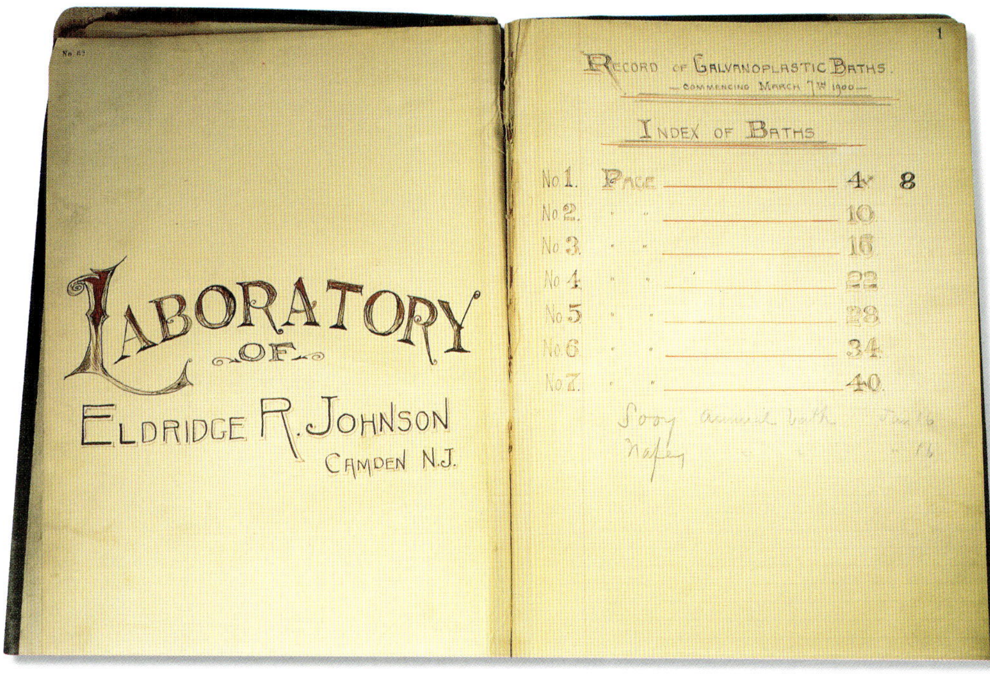

1-58. This extraordinary logbook recorded Eldridge Johnson's experiments with "Galvanoplastic Baths," in other words, creating a metal copy of a disc recording from a wax master. Beginning March 7, 1900, this document reflects Johnson's urgent need to develop recording and pressing facilities of his own, to augment his exclusive contract to manufacture Gramophones for Emile Berliner. Over the next four months, Berliner would be assailed in the courts, culminating in the shutdown of his United States operations on June 25, 1900. This would leave Johnson the weather the storm with a factory full of Gramophones until October 1901, when his new firm, the Victor Talking Machine Company, would emerge. *(Value code: VR)*

1-59. Miss New Year, with a Columbia Graphophone Type "AT" and *cor de chasse* (hunting horn), promises the recipient some happy days.

Chapter Two:

1901-1910

2-1. A delightful Edison trade card. (Value code: G)

2-2. This pole advertiser features an illustration of the early "Home Grand" (or slightly later "Columbia Grand") Graphophone. The piece can be dated to no earlier than 1901 due to the mention of the "NEW STYLE" Graphophone playing "Both Large and Small Records." Such capability was offered only by the "AB," "AD," or "AF," all of which appeared in 1901. Modern collectors may wince at the thought of a "Blind Canvasser" handling fragile 5" diameter cylinders! Measures 5 7/8" x 10 1/2". *Courtesy of Ernest Carl Allen.* (Value code: F)

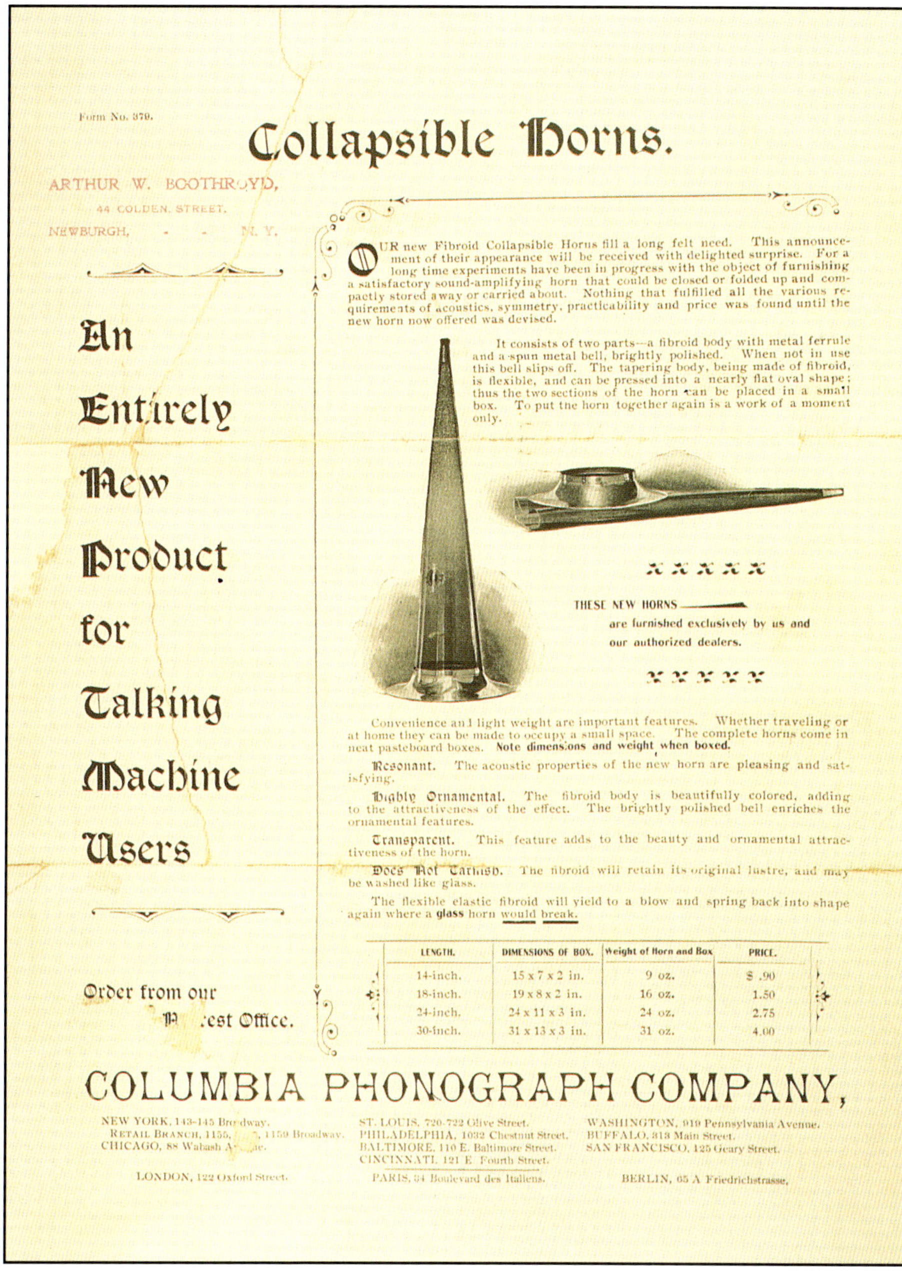

Form No. 879.

Collapsible Horns.

ARTHUR W. BOOTHROYD,
44 COLDEN STREET,
NEWBURGH, N. Y.

An Entirely New Product for Talking Machine Users

OUR new Fibroid Collapsible Horns fill a long felt need. This announcement of their appearance will be received with delighted surprise. For a long time experiments have been in progress with the object of furnishing a satisfactory sound-amplifying horn that could be closed or folded up and compactly stored away or carried about. Nothing that fulfilled all the various requirements of acoustics, symmetry, practicability and price was found until the new horn now offered was devised.

It consists of two parts—a fibroid body with metal ferrule and a spun metal bell, brightly polished. When not in use this bell slips off. The tapering body, being made of fibroid, is flexible, and can be pressed into a nearly flat oval shape; thus the two sections of the horn can be placed in a small box. To put the horn together again is a work of a moment only.

THESE NEW HORNS are furnished exclusively by us and our authorized dealers.

Convenience and light weight are important features. Whether traveling or at home they can be made to occupy a small space. The complete horns come in neat pasteboard boxes. **Note dimensions and weight when boxed.**

Resonant. The acoustic properties of the new horn are pleasing and satisfying.

Highly Ornamental. The fibroid body is beautifully colored, adding to the attractiveness of the effect. The brightly polished bell enriches the ornamental features.

Transparent. This feature adds to the beauty and ornamental attractiveness of the horn.

Does Not Tarnish. The fibroid will retain its original lustre, and may be washed like glass.

The flexible elastic fibroid will yield to a blow and spring back into shape again where a **glass** horn <u>would</u> <u>break</u>.

LENGTH.	DIMENSIONS OF BOX.	Weight of Horn and Box	PRICE.
14-inch.	15 x 7 x 2 in.	9 oz.	$ .90
18-inch.	19 x 8 x 2 in.	16 oz.	1.50
24-inch.	24 x 11 x 3 in.	24 oz.	2.75
30-inch.	31 x 13 x 3 in.	31 oz.	4.00

Order from our [illegible]est Office.

COLUMBIA PHONOGRAPH COMPANY,

NEW YORK, 143-145 Broadway.
RETAIL BRANCH, 1155, [illegible], 1159 Broadway.
CHICAGO, 88 Wabash A[illegible]e.
ST. LOUIS, 720-722 Olive Street.
PHILADELPHIA, 1032 Chestnut Street.
BALTIMORE, 110 E. Baltimore Street.
CINCINNATI, 121 E. Fourth Street.
WASHINGTON, 919 Pennsylvania Avenue.
BUFFALO, 313 Main Street.
SAN FRANCISCO, 125 Geary Street.
LONDON, 122 Oxford Street.
PARIS, 34 Boulevard des Italiens.
BERLIN, 65 A Friedrichstrasse.

2-3. When the Columbia Phonograph Company purchased the rights to the Hogan collapsible celluloid horn, no time was lost in marketing what was thought to be a revolutionary product. Shipping costs, storage, ease of handling, and appearance all seemed to favor this new development. Why the celluloid horn never caught on is open to debate. When this 8 1/2" x 11 1/4" circular was distributed in 1900-1901, the future of the collapsible celluloid horn looked bright. *Courtesy of Ernest Carl Allen. (Value code: E)*

2-4. This 3" x 4 1/4" handbill advertised Lambert Indestructible Phonograph Records, introduced in 1900. Made of celluloid and dyed in various colors – notably pink – the Lambert records were years ahead of their time. Lambert's patent for molding in celluloid would eventually be acquired by Edison for his Blue Amberol cylinders introduced in 1912. *Courtesy of Ernest Carl Allen. (Value code: F)*

Special Price Reductions Montgomery Ward & Co., Chicago

Thornward Talking Machine Outfits at Special Prices

A complete $35.50 outfit for only $22.50 No such value was ever quoted at this price

Machine	$25.00
Dozen Records	5.00
Recorder	5.00
Record Case	.30
Oil Can	.10
Chip Brush	.10
	$35.50
Our special price	**$22.50**

The Thornward, shown in illustration, is the first and only talking machine ever offered at a reduced price. The retail value of all others is fixed by the manufacturers, and no dealer is allowed to change it. The Thornward is a new machine which we control absolutely, and, as we are anxious to get it well introduced, will sell a limited number of outfits at our special price of $22.50. Bear in mind this is not a "fake" advertisement, in which a cheap outfit is, by ambiguous wording, made to appear something more than it really is. On the other hand, we positively guarantee the machine to be equal in size, construction, durability and sound-reproducing qualities to those sold by dealers at $25.00; in fact, it is almost identical. Besides, our special price includes an outfit consisting of complete machine, an assortment of one dozen records, horn, record case, recorder, oil can and chip brush. It is made expressly for us by the best maker of sound-reproducing machines in the world. In construction it represents a combination of the best features taken from all other good machines using cylinder records. It possesses the exquisite reproducing qualities of the graphophone, together with the most notable improvements of the phonograph, as it is supplied with an adjustable end gate which supports the outer end of the mandrel, thus insuring a perfectly even and easy revolution of the record. Each is put up in a handsomely finished oak cabinet, with dimensions as follows: Length, 11½ inches; height, 10 inches; depth, 7¼ inches. It is supplied with a perfectly adjusted spring motor, well made, durable, and the upper works are all nickel plated and polished.

INDESTRUCTIBLE RECORDS—With our Thornward Outfit we will furnish, if preferred, the new patented INDESTRUCTIBLE RECORDS. This is a new line recently placed on the market which we can thoroughly recommend, not only on account of being, as the name implies, practically unbreakable, but the reproductions from them are louder than from the wax cylinders. The new records are made from a composition resembling celluloid, and may be handled or even dropped on the floor without injury. We will, however, furnish wax records if so requested. REMEMBER, we do not ask you to send cash with order, providing you prefer to have the Talking Machine sent C. O. D. on trial. In this case it is only necessary to order the outfit, stating your wishes and giving your name, postoffice address and shipping point. Upon receipt of same we will ship one of these splendid outfits, by freight, for your inspection. (The shipment is too heavy to ship by express, unless you live within 300 miles of Chicago, in which case the shipment can, if you prefer, be made by express C. O. D.) When the shipment arrives at your railroad station it will be necessary for you to deposit the price of same with your bank. The bill of lading will then be surrendered.

$22.50

and upon payment of freight charges you can take the outfit home. Give it five days' trial. If you are satisfied, after using it, that it is as we represent it, or, in other words, is a first-class Talking Machine, supplied with automatic reproducer, recorder, amplifying horn, one dozen best musical or talking records (your own selection), in a neat carrying case, oil can and brush, instruct your banker to send us the money. If it does not prove entirely satisfactory, return to station agent, obtain bill of lading showing it has been returned to us, present same to the bank, and the entire amount of deposit will be refunded; send us your expense bill, showing the amount of freight paid by you on the outfit, and this amount will be refunded.

Isn't This a Fair Offer? Do You Take any Risk?

Please note that when full amount of money is sent with order we allow our usual cash discount. If you wish to avail yourself of this special inducement enclose money order for price of outfit, $22.50, less our 2 per cent cash discount, or $22.05 net. We will then ship machine to you direct, with the understanding you are to have the privilege of five days' trial, and, if not satisfactory, to be returned at our expense, and your money will be returned to you, including any transportation charges paid by yourself. So thoroughly convinced are we of the fact that we have the latest, best and most meritorious talking machine on the market at anything like the price, that we are willing to take all the risk, as we feel sure that all purchasers will be more than pleased after a careful inspection and trial.

02235 Thornward Talking Machine Outfit, as described above, price, complete **$22.50**

2-5. This spring-summer 1901 Montgomery Ward catalog page (8" x 10 1/2") shows the "Thornward" – an odd hybrid incorporating an aging Type "N" Graphophone upper works coupled with a Type "A" motor. As might be surmised from its name, the "Thornward" was sold only through Montgomery Ward & Company. (Thorn was the name of the family that owned the firm.) *Courtesy of Ernest Carl Allen. (Value code: G, complete catalog)*

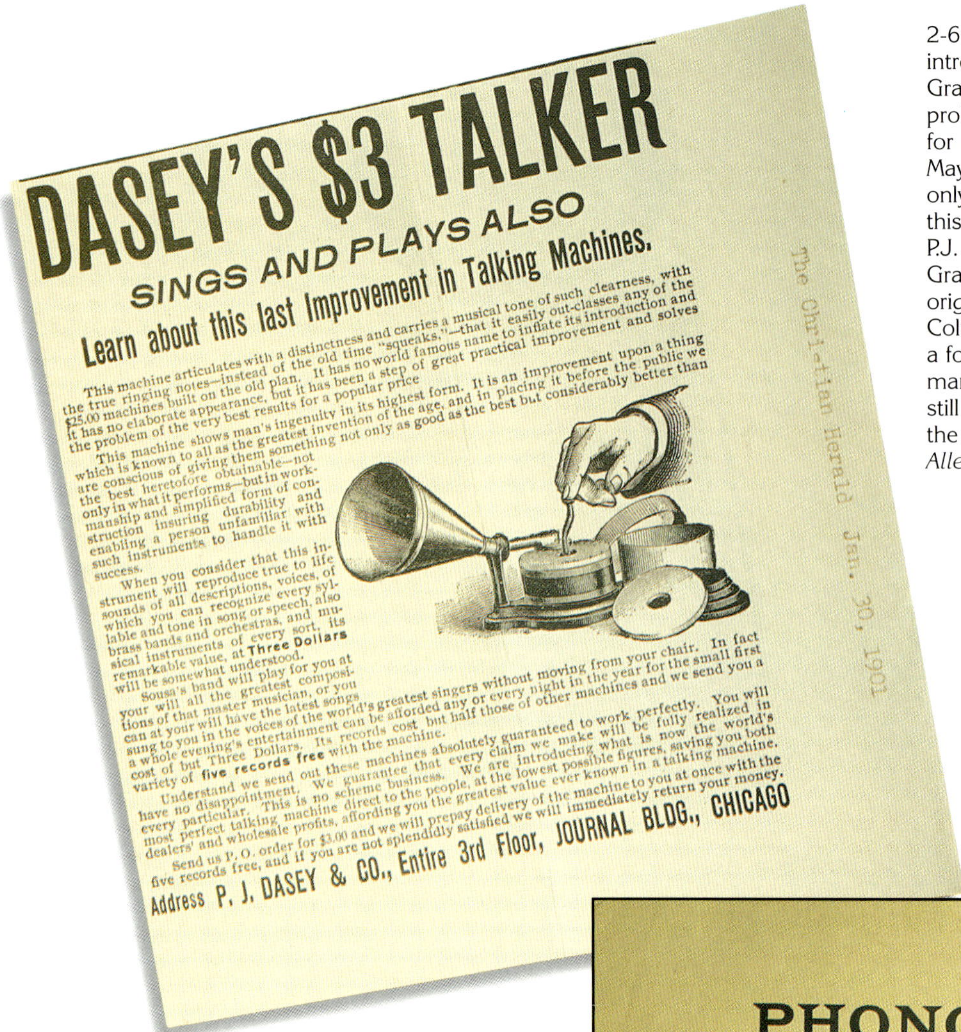

2-6. The Columbia "Toy Graphophone" was introduced in early 1900 at $3.00. It was the first Graphophone to play disc records, and it was probably intended to serve as a legal precedent for Columbia's entry into the disc market. By May 1900, Columbia was selling its "Toy" for only $1.50. It is therefore puzzling to discover this advertisement from January 1901, wherein P.J. Dasey was offering the same "Toy Graphophone," now on sale for a year, at the original $3.00 price. Perhaps he thought that Columbia had undervalued its product, and that a fortune awaited the man who could properly market the device! Most likely, however, he was still attempting to liquidate his initial purchase of the unpopular gadget. *Courtesy of Ernest Carl Allen. (Value code: H)*

2-7. Georges Carette & Company was a prolific maker of toys and metal novelties, including cylinder talking machines. Although the firm is most commonly associated with the humble "Puck" phonograph (a style of which is shown to the left), this flyer indicates that at least one "better" model was available. The "Puck" shown here is of the earliest design, featuring a straight flare horn, and a reproducer of black ebonite, based on the "gutta percha" Graphophone reproducer of the 1890s. Great Britain was a major market for German manufacturers such as Carette, though, as the text makes clear, cylinder talking machines had to be licensed through Edison Bell, which controlled the fundamental Phonograph and Graphophone patents in the British Isles. 8 1/4" x 10 1/4". *Courtesy of Ernest Carl Allen. (Value code: F)*

2-8. An unusual 1901 trade card (3" x 6") bearing a statement that would have pleased Edison, who placed great value on "the real work of thinking." The watch fob dates from 1906. In 1905, Edison bought the rights to a painting by Italian artist Pompeo Massani, showing two delighted old fossils being entertained by a European cylinder talking machine. The original title of the work, "Enrico!", was of no interest to Edison – he was not about to remind the public that operatic tenor Enrico Caruso was under exclusive contract to the Victor Talking Machine Company. An Edison Phonograph was inserted in the composition, and the image of the "old couple" (unimaginatively renamed "The Phonograph") was widely exploited. *Courtesy of Tom and Sandi McCarthy. (Value code: G, E)*

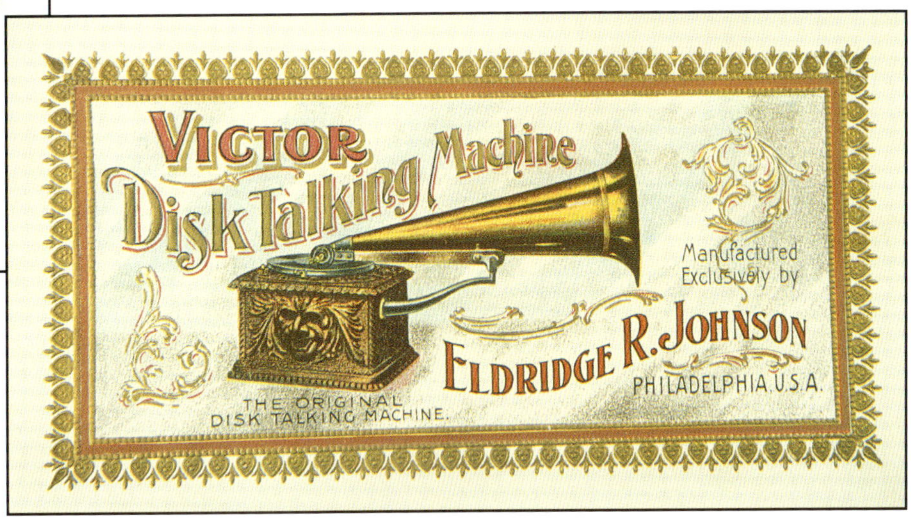

2-9. This 1901 Victor catalog cover bears a stylized rendering of a machine not known to have been manufactured – and all in color. *Courtesy of Ernest Carl Allen. (Value code: E)*

2-10. In July 1902, Victor recorded several Scots selections by the Kilties Band. In a bold break with tradition, Victor applied visually distinctive plaid labels to these issues (the first label Victor offered to the public in other than black and gold). Unfortunately, the tartan background interfered with the gold label lettering making it difficult to decipher. This deficiency, along with the limited appeal of the music, probably hastened the end of this short-lived experiment in eye-catching labels. *(Value code: F)*

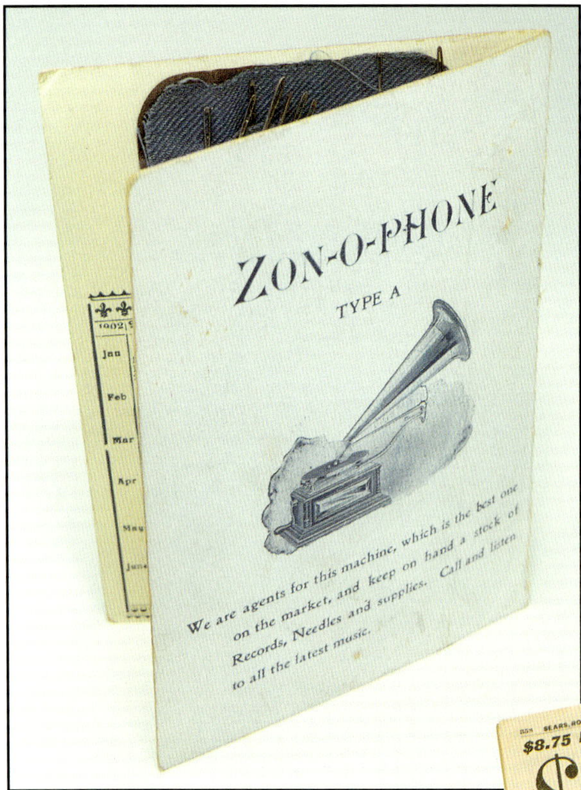

2-11. Any handy item was fair game for the touch of advertising. So, the supplier of this little folder of sewing needles reminded housewives of a fascinating modern contraption that could be found among his wares. The folder also bears a 1902 calendar. 3 3/4" x 5 1/4". *Courtesy of Tom Sideleau. (Value code: F)*

2-12. A spread from the 1903 Sears, Roebuck catalog of Graphophones and supplies. This shows both the early (left) and late model (right) "AK" Disc Graphophones being sold concurrently. It is also interesting to note that no "Grand" Graphophones (playing 5" diameter cylinders) appeared in Sears, Roebuck catalogs after 1902. *Courtesy of Ernest Carl Allen. (Value code: F)*

2-13. The raffle ticket at left, and a die-cut trade card in the form of a theatre ticket with envelope. As the text makes clear, an Edison Phonograph made live performances a thing of the past. The *trompe l'oeil* details are quite remarkable. *Courtesy of Tom and Sandi McCarthy. (Value code: G, F)*

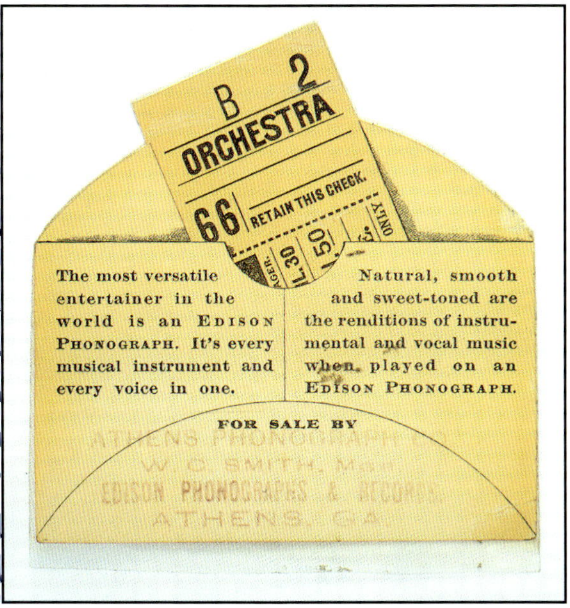

2-14. The reverse of the ticket trade card bears the stamp of the dealer whose business it promoted. Approximately 3 3/4" x 4 7/8". *Courtesy of the Charles Hummel collections.*

2-15. The image of the phonograph, always a reminder of its impact on society. A plate, 8 1/8" diameter, titled "Contrasts in Sound." On the left: a "monkey organ" entertains. On the right: "The phonograph – a whole opera in the voices of the Paris Opera singers. It's a marvel!" *Courtesy of Philippe Le Ray. (Value code: E)*

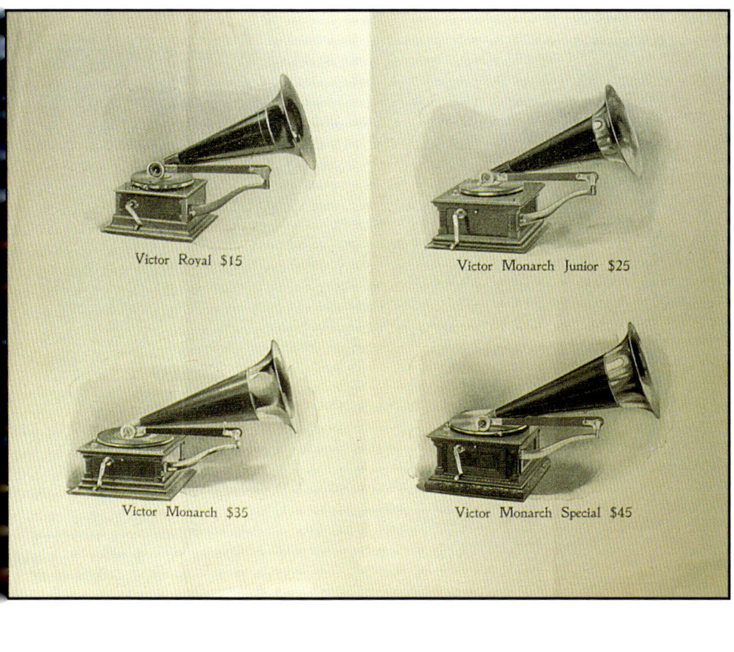

2-16. Upon this core of talking machines, the "Royal," "Monarch Junior," "Monarch," and "Monarch Special," Victor built a far-reaching empire. A catalog from 1903. *(Value code: E)*

Victor I $22 50

Rigid arm: more compact, better appearance, more substantial.
Cabinet: quartered-oak; metal corners.
Motor: one spring; plays nearly two 7-inch records, or one 10-inch, with one winding; and can be wound while playing.
Turn-table: 7-inch, for 7-inch records; can be used for 10-inch records.
Sound-box: Concert.
Horn: 16-inch black-japanned steel.
Needles: 200 with two-part box for new and worn needles.
A good machine for the money.

12

Victor
New Style
1903 Model
Just Out

13

2-17. Another 1903 Victor catalog showed the very first incarnation of the Victor "I" – actually a Victor "Royal" equipped with a back-mounted horn and tone arm (the non-tapering arm that Victor referred to as "Rigid"). Catalog measures 3 3/8" x 6". *Courtesy of Fred Guenther. (Value code: E)*

2-18. Dealer advertising affixed to talking machines was seldom as ostentatious as this: a nickel-plated brass nameplate from Ilsen & Company of Cincinnati, Ohio. This attribution is on the lid of an Edison "Home" Model "B" (introduced 1905) – unusually late for such elaborate markings. *Courtesy of Gregg Cline.*

2-19. The Victor Talking Machine Company of Camden, New Jersey was at the center of a commercial solar system of vast proportions. Through its affiliated companies in countries as far-flung as India and as accustomed as Europe or Canada, Victor disseminated semblances of the talking machines it sold in the United States. Sometimes, these instruments were identical to Victor's domestic lineup. Sometimes, the foreign-marketed machines incorporated cabinets, horns, or other elements drawn from local production. This handbill from the Compaignie Francaise du Gramophone illustrated important items from the French firm's catalog, 1903. One may easily recognize the association with Victor through the design of the products and the nomenclature ("Monarch"). Note the trusty "Trade Mark Berliner" Gramophone (so-called for its connection to the famous "His Master's Voice" logo) at the bottom, introduced in 1897 and still being flogged in foreign markets. The "Monarch" on its formidable stand with wrought-iron support for the colossal horn is unlike anything that appeared in the American catalog. *(Value code: F)*

2-20. This spectacular poster measures 29 3/4" x 41 1/2", and shows the renamed Victor line of mid-1904. The only difference between the "Monarch Jr." and the "Victor II" was the horn/arm configuration. The same was true of the "Monarch" and the "Victor III." *Courtesy of the Johnson Victrola Museum/Delaware State Museums. (Value code: VR)*

DIRECTIONS FOR UNPACKING AND ASSEMBLING

¶¶THE TALK-O-PHONE¶¶

After removing lid of the shipping box and excelsior on top of the machine, the various parts of the machine will be seen; look carefully through the packing as small parts might be contained therein, for the machine and parts are tested and counted carefully before being packed. Take the papers from the different packages, set the cabinet up, and on the top will be noticed a shaft protruding through the center; on this place the turn table; on the side of the cabinet where the transfer design is will be noticed two thumb screws in position; remove these and place bracket arm in position, replacing screws; on the extreme end of the bracket arm insert the horn carry arm, then in the small hole in the end, stick the horn support, the end turned toward the cabinet. Insert the sound box in the end of the horn carry arm, with the end pointing to your left when standing behind the machine; place the large end of the horn in position on horn support and insert end of sound box into the end of the metal elbow, pushing it back as far as it will go; have the needle arm pointing to the back of the machine and at an angle as shown in the above cut; see that winding crank is fitted securely, wind motor up until spring is nearly tight, but be very careful not to wind absolutely tight, as the best spring made will break on occasions like this, push the regulator lever to the right and turntable will revolve. Various records require different speed; same can be ganged to suit the individual taste; see that needle has been placed in the end of the needle arm and needle screw tightly clamped on the same, this will avoid any vibration; do not place the needle point on the record until the machine is running at its natural speed, then you get the full record, and after the record has been played use another needle, putting the old one in the end of the needle box furnished, as provided. By following these directions caref trouble will be experienced.

The Talk-o-Phone Company, Toledo, Oh

2-21. These directions accompanied each new "Talk-O-Phone" sold around 1905. Measures 6 1/2" x 9 1/2". *Courtesy of Freddie and Jeanette Blair. (Value code: F)*

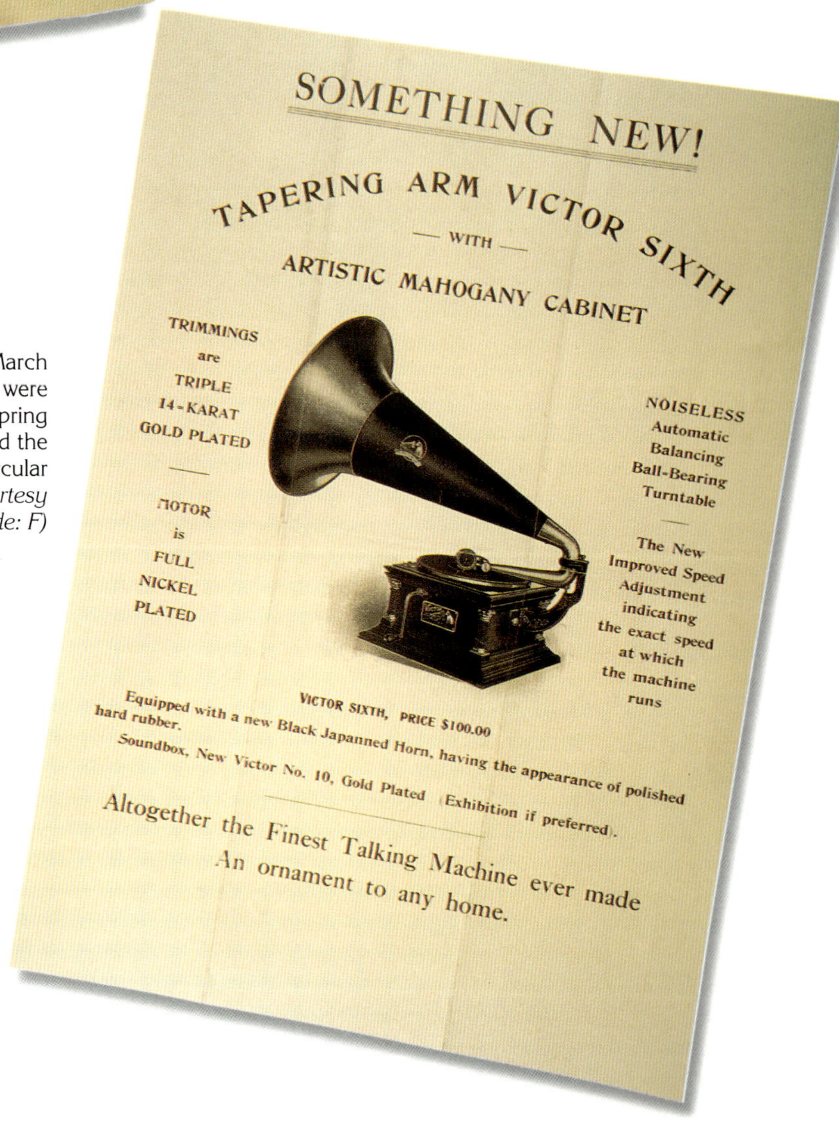

2-22. From November 1904 until March 1905, the very first Victor "VI" models were equipped with the earliest type of triple spring motor whose crank was located toward the front of the cabinet. This 8" x 11 1/4" circular dates from that four-month period. *Courtesy of Ernest Carl Allen. (Value code: F)*

2-23. The first version of the Victor "I" was simply a Victor "Royal" equipped with back-mounted horn and (so-called "Rigid") tone arm. That transitional model paved the way for the instrument announced in this 8" x 10" circular: the improved Victor "I", as first shipped in June 1905. *Courtesy of Ernest Carl Allen. (Value code: F)*

2-24. From a series of cards in which two adorable cherubs deconstruct a Columbia Disc Graphophone, Type "AK" (early version, 1902). *Courtesy of Steve and Ellie Saccente. (Value code: H)*

2-25. The motor casting of a Victor triple-spring motor, such as employed in the models "V" and "VI," was vulnerable to breaking under shock at its left extremity. To protect the motor in shipment, Victor used a specially-design wooden block to support the spring barrels. This notice, 4" x 5", advised customers to remove the block in order to free the barrels. *Courtesy of Charles McCarn. (Value code: H)*

2-26. For those of us who have needed to separate the winding gear from a set of Victor spring barrels, the specialized tool illustrated here would be a priceless boon, and result in considerably less vituperating the good name of Eldridge Johnson. "Instructions for Placing New Springs in Triple Barrel of the Victor Motor," 5 1/4" x 6 1/4". *Courtesy of Ernest Carl Allen. (Value code: G)*

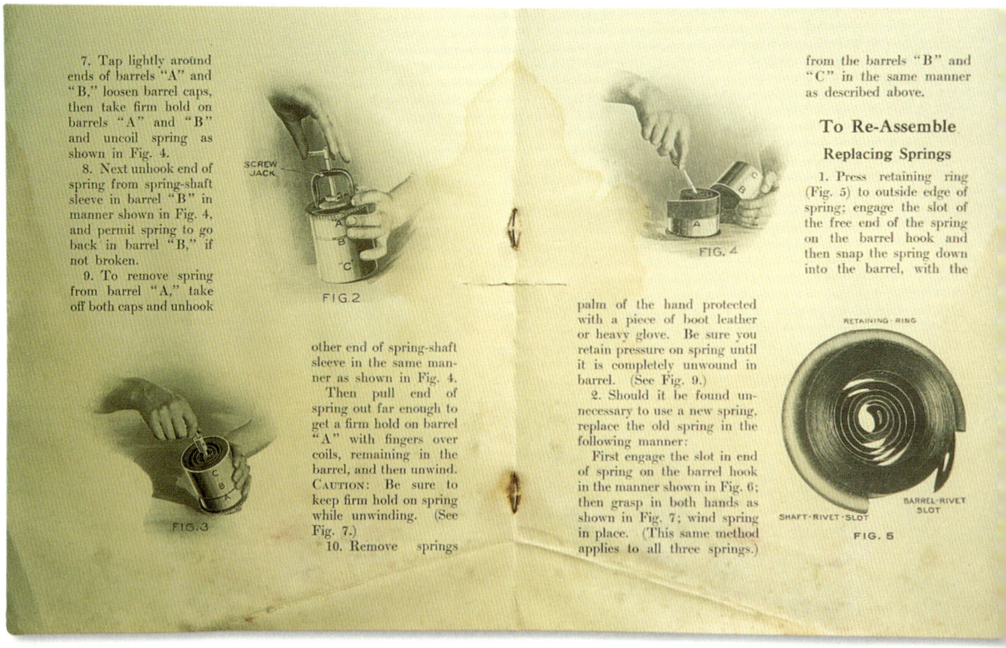

7. Tap lightly around ends of barrels "A" and "B," loosen barrel caps, then take firm hold on barrels "A" and "B" and uncoil spring as shown in Fig. 4.

8. Next unhook end of spring from spring-shaft sleeve in barrel "B" in manner shown in Fig. 4, and permit spring to go back in barrel "B," if not broken.

9. To remove spring from barrel "A," take off both caps and unhook other end of spring-shaft sleeve in the same manner as shown in Fig. 4. Then pull end of spring out far enough to get a firm hold on barrel "A" with fingers over coils, remaining in the barrel, and then unwind. CAUTION: Be sure to keep firm hold on spring while unwinding. (See Fig. 7.)

10. Remove springs from the barrels "B" and "C" in the same manner as described above.

SCREW JACK

FIG. 2

FIG. 3

FIG. 4

To Re-Assemble

Replacing Springs

1. Press retaining ring (Fig. 5) to outside edge of spring; engage the slot of the free end of the spring on the barrel hook and then snap the spring down into the barrel, with the palm of the hand protected with a piece of boot leather or heavy glove. Be sure you retain pressure on spring until it is completely unwound in barrel. (See Fig. 9.)

2. Should it be found unnecessary to use a new spring, replace the old spring in the following manner:

First engage the slot in end of spring on the barrel hook in the manner shown in Fig. 6; then grasp in both hands as shown in Fig. 7; wind spring in place. (This same method applies to all three springs.)

RETAINING RING

SHAFT-RIVET-SLOT

BARREL-RIVET SLOT

FIG. 5

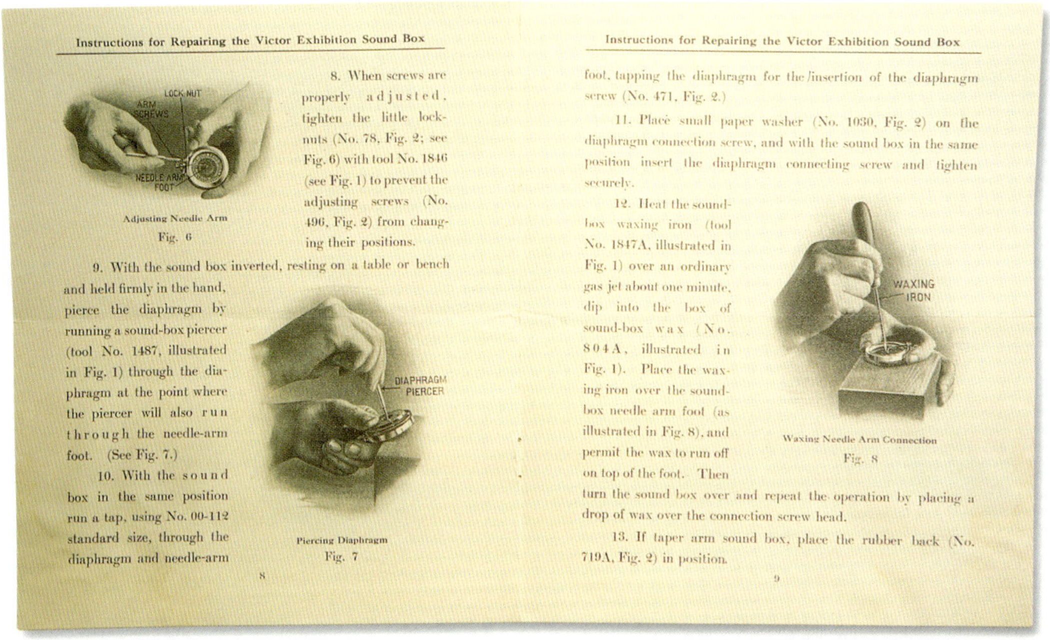

Instructions for Repairing the Victor Exhibition Sound Box

8. When screws are properly adjusted, tighten the little lock-nuts (No. 78, Fig. 2; see Fig. 6) with tool No. 1846 (see Fig. 1) to prevent the adjusting screws (No. 496, Fig. 2) from changing their positions.

LOCK-NUT

ARM SCREWS

NEEDLE ARM FOOT

Adjusting Needle Arm

Fig. 6

9. With the sound box inverted, resting on a table or bench and held firmly in the hand, pierce the diaphragm by running a sound-box piercer (tool No. 1487, illustrated in Fig. 1) through the diaphragm at the point where the piercer will also run through the needle-arm foot. (See Fig. 7.)

10. With the sound box in the same position run a tap, using No. 00-112 standard size, through the diaphragm and needle-arm

DIAPHRAGM PIERCER

Piercing Diaphragm

Fig. 7

8

Instructions for Repairing the Victor Exhibition Sound Box

foot, tapping the diaphragm for the insertion of the diaphragm screw (No. 471, Fig. 2.)

11. Place small paper washer (No. 1030, Fig. 2) on the diaphragm connection screw, and with the sound box in the same position insert the diaphragm connecting screw and tighten securely.

12. Heat the sound-box waxing iron (tool No. 1847A, illustrated in Fig. 1) over an ordinary gas jet about one minute, dip into the box of sound-box wax (No. 804A, illustrated in Fig. 1). Place the waxing iron over the sound-box needle arm foot (as illustrated in Fig. 8), and permit the wax to run off on top of the foot. Then turn the sound box over and repeat the operation by placing a drop of wax over the connection screw head.

WAXING IRON

Waxing Needle Arm Connection

Fig. 8

13. If taper arm sound box, place the rubber back (No. 719A, Fig. 2) in position.

9

2-27. "Instructions for Repairing the Victor Exhibition Sound Box," 5 1/4" x 6 1/4". *Courtesy of Ernest Carl Allen. (Value code: G)*

2-28. The "Automatic Needle Clamp" has been seen on Victor sound boxes of the 1903-1905 period, but little understood. Was it "standard equipment" or an option? Available on all models or only some? As this 3 1/2" x 5 3/4" brochure entitled "Victor Improvement on Exhibition Sound Box" makes clear, the clamp could be purchased for a dollar and appended to an existing device, or acquired complete with the sound box at the regular price of $5.00. *Courtesy of Ernest Carl Allen. (Value code: G)*

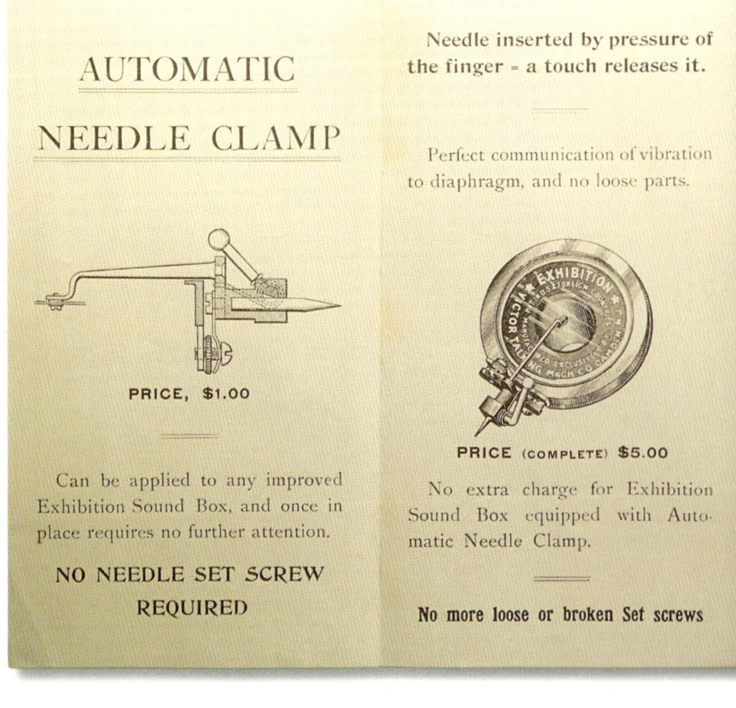

AUTOMATIC

NEEDLE CLAMP

PRICE, $1.00

Can be applied to any improved Exhibition Sound Box, and once in place requires no further attention.

NO NEEDLE SET SCREW REQUIRED

Needle inserted by pressure of the finger – a touch releases it.

Perfect communication of vibration to diaphragm, and no loose parts.

PRICE (COMPLETE) $5.00

No extra charge for Exhibition Sound Box equipped with Automatic Needle Clamp.

No more loose or broken Set screws

2-29. "The baby sends you flowery greetings." A charming girl with a Pathé phonograph. *(Value code: H)*

2-30. "Flemish Art," or pyrography, was popular as both a home craft and a decorating style during the first decade of the twentieth century. The purpose of this wooden box, ornamented with the "His Master's Voice" logo, is unknown. Was it commercially produced, or a handicraft project employing stock imagery? *(Value code: VR)*

2-31. A recurring theme: antipathy for the sound of the phonograph. *Courtesy of Ernest Carl Allen. (Value code: H)*

2-32. An enterprising retailer in France marked these wood-grained cardboard boxes of Pathé "Inter" (or "Intermediate," 3 1/2" diameter) cylinder records with the style of music contained therein. *Courtesy of Marie-Claude Stéger. (Value code: H, empty)*

2-33. A fanciful stock certificate for a French talking machine company. *Courtesy of the collection of Howard Hazelcorn. (Value code: F)*

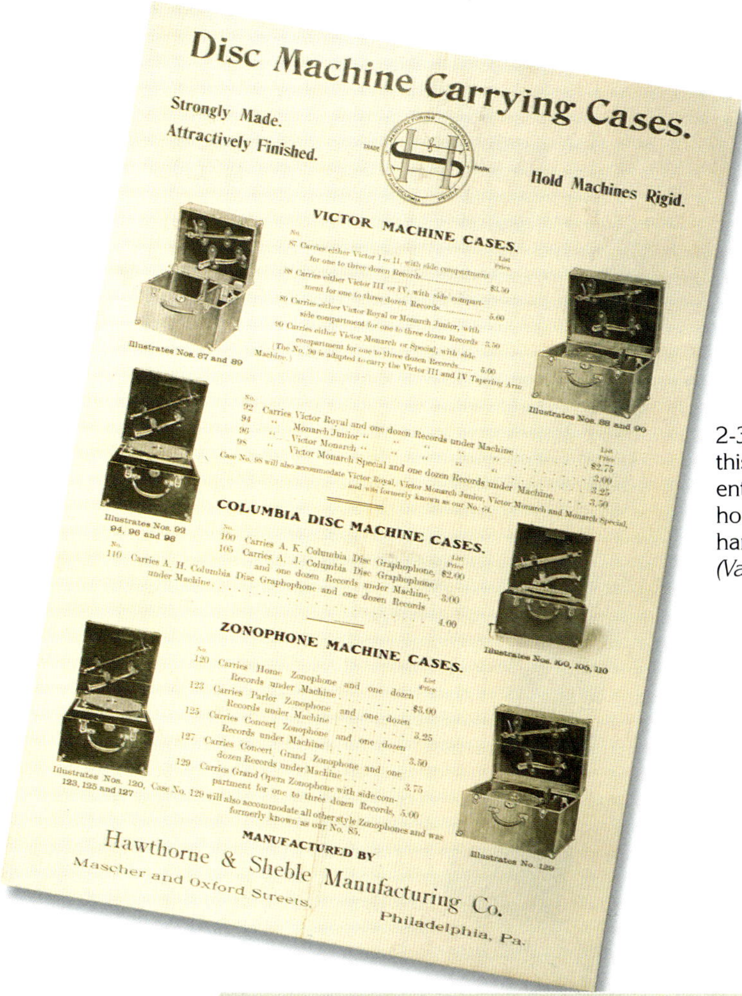

2-34. Carrying cases, such as the ones promoted in this Hawthorne & Sheble flyer accommodated the entire talking machine except the horn, which was housed separately in anything from a canvas bag to a hard-sided satchel. *Courtesy of Ernest Carl Allen. (Value code: F)*

2-35. This circular is mysteriously marked and numbered at the top. The nature of the contest/drawing is a mystery, but the requirement of the circular to redeem the prize was a clever maneuver to prevent it from being immediately discarded. The high number written at the upper right suggests that many contestants were involved, and that this document might have been distributed at a large exhibition. Measures 8 1/4" x 10 1/2". *Courtesy of Ernest Carl Allen. (Value code: VR)*

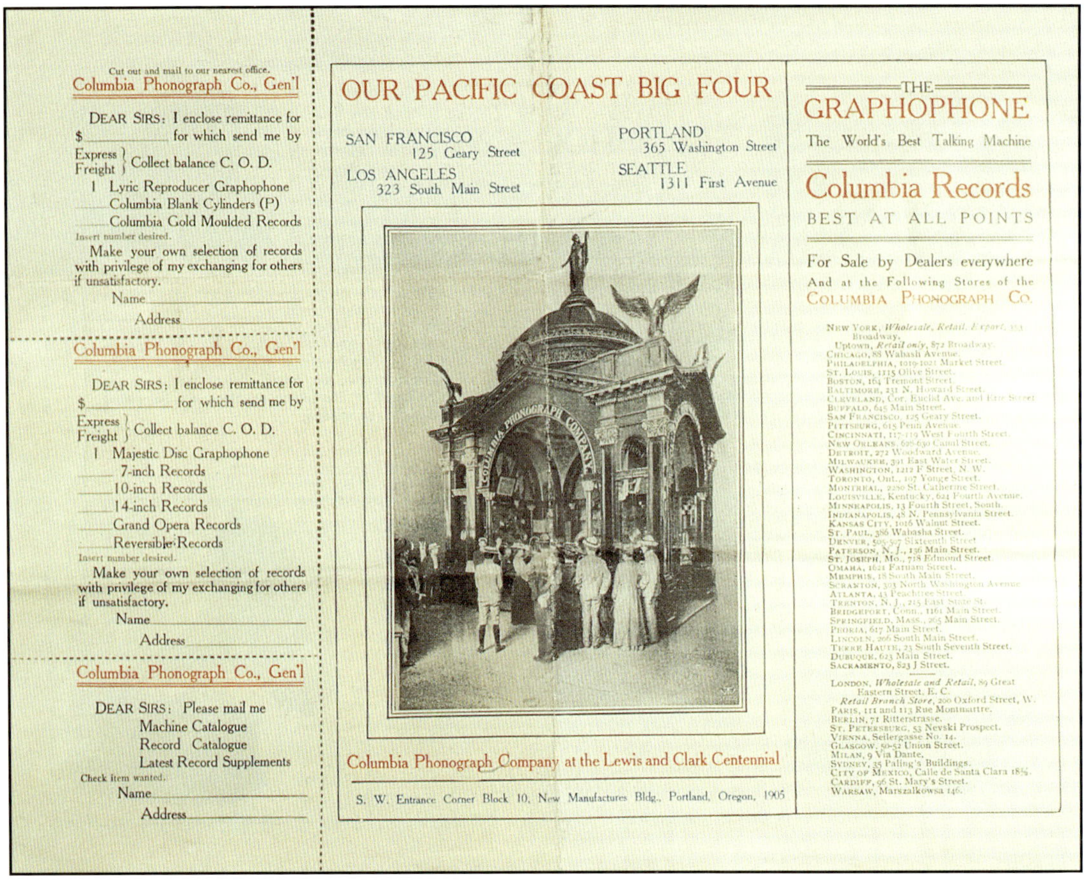

2-36. Examining the reverse side of this circular confirms that it was distributed at the Portland, Oregon Exhibition of 1905. Coupons at the left encouraged potential buyers to take the plunge. Of interest is the mention of 14" diameter disc records – a product Columbia manufactured only briefly. *Courtesy of Ernest Carl Allen.*

2-37. This little stamped metal advertising tray identified Kipp Brother as agents for Edison Phonographs as well as Regina music boxes. *(Value code: E)*

EDISON MAJESTIC
COIN-SLOT
PHONOGRAPH
Direct Current
Plays the Standard Size Record

Price $90. *Catalogue No. A 6501.* *Code Majestic*

SIZE:—Height, with low back, 54 inches; with high back, 72 inches; Base, 27x17. Weight:—Net, 150 lbs.; Gross, 217 lbs. Packed in one case, measuring 31x20x48 inches.

EQUIPMENT:—Flexible metal hearing tubes with ear-cups; Model B Reproducer; Model D Return Attachment (see page 8); wiping contact (see page 9); slot opening for one cent; cabinet in quartered oak, Flemish finish, highly polished, square top, fitted with bevel glass and beveled French plate glass mirror in back of interior; wires for connection with direct electric current.

OPTIONS:—24-inch horn and crane, or rubber hearing tubes with ear-cups, in place of flexible hearing tubes; coin-slot for any other coin than one cent.

THE mechanism of the Edison Conqueror Phonograph is used in the Edison Majestic. It is actuated by direct current. Keys are provided for lower part or door of cabinet; the top is secured by a latch or hook, which is accessible only when the door is open. The money box is securely attached to the cabinet and is equipped with a removable tin box, with which a reliable lock is furnished.

7

2-38. Of the Edison coin-operated Phonographs, the post-1906 models ("Excelsior," "Apex," and "Windsor") are the most commonly seen today. The "Majestic" (110 volt direct current) shown in this catalog, dated January 1, 1905, is an extreme rarity. 5 1/8" x 8". *Courtesy of the Tom and Joanna Broome. (Value code: E)*

2-39. The Tea Tray Company of Newark, New Jersey was a well-known manufacturer of horns and supporting cranes for talking machines. This catalog (circa 1906) offers a look at the variety of goods available to a dealer of the time. Measures 3 1/2" x 6 1/8". *Courtesy of Ernest Carl Allen. (Value code: VR)*

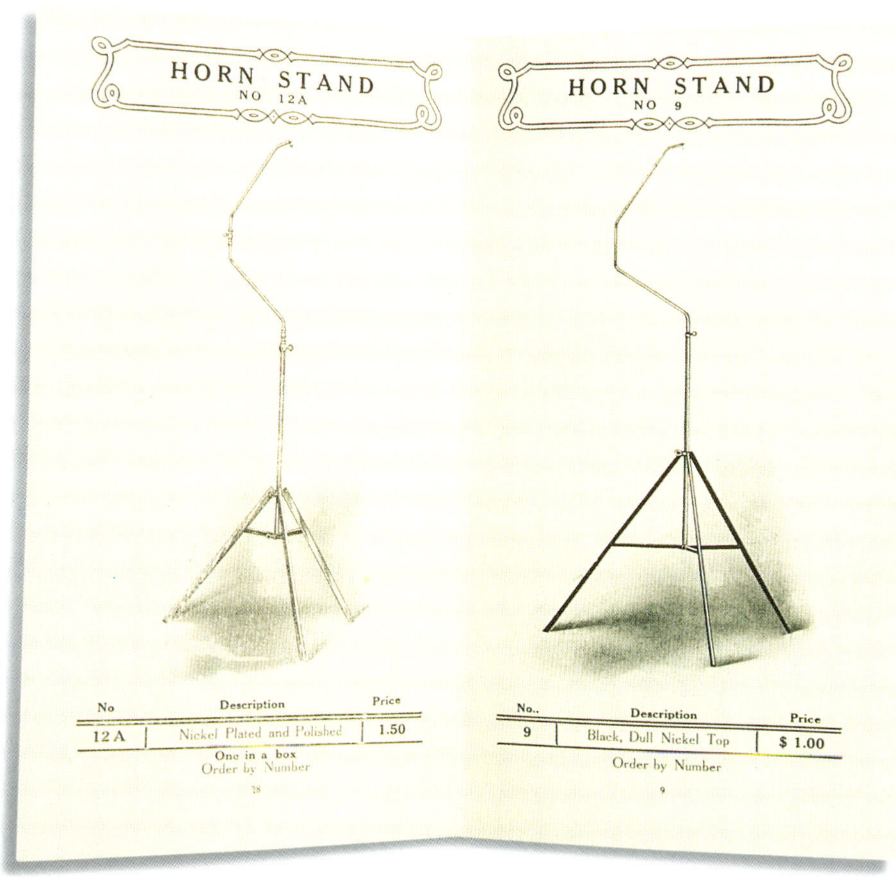

HORN STAND
NO 12A

No	Description	Price
12 A	Nickel Plated and Polished	1.50

One in a box
Order by Number

HORN STAND
NO 9

No.	Description	Price
9	Black, Dull Nickel Top	$ 1.00

Order by Number

9

2-40. In addition to its famous line of horns, the company also manufactured floor stands, a fact which is not commonly known. The two styles illustrated here are often seen by present-day collectors. *Courtesy of Ernest Carl Allen.*

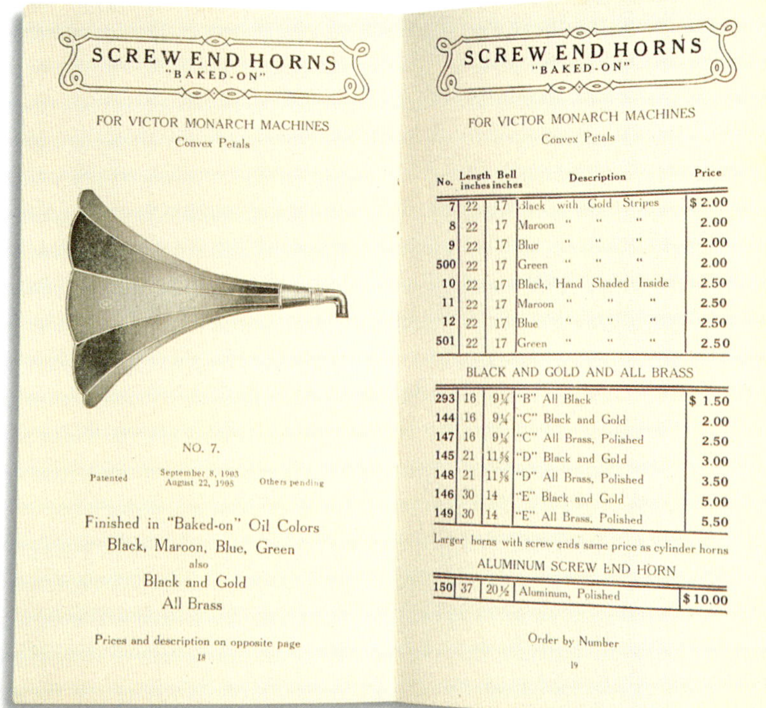

SCREW END HORNS
"BAKED-ON"

FOR VICTOR MONARCH MACHINES
Convex Petals

NO. 7.

Patented September 8, 1903 August 22, 1905 Others pending

Finished in "Baked-on" Oil Colors
Black, Maroon, Blue, Green
also
Black and Gold
All Brass

Prices and description on opposite page

18

SCREW END HORNS
"BAKED-ON"

FOR VICTOR MONARCH MACHINES
Convex Petals

No.	Length inches	Bell inches	Description	Price
7	22	17	Black with Gold Stripes	$ 2.00
8	22	17	Maroon " " "	2.00
9	22	17	Blue " " "	2.00
500	22	17	Green " " "	2.00
10	22	17	Black, Hand Shaded Inside	2.50
11	22	17	Maroon " " "	2.50
12	22	17	Blue " " "	2.50
501	22	17	Green " " "	2.50

BLACK AND GOLD AND ALL BRASS

No.	Length inches	Bell inches	Description	Price
293	16	9¼	"B" All Black	$ 1.50
144	16	9¼	"C" Black and Gold	2.00
147	16	9¼	"C" All Brass, Polished	2.50
145	21	11⅝	"D" Black and Gold	3.00
148	21	11⅝	"D" All Brass, Polished	3.50
146	30	14	"E" Black and Gold	5.00
149	30	14	"E" All Brass, Polished	5.50

Larger horns with screw ends same price as cylinder horns

ALUMINUM SCREW END HORN

No.	Length inches	Bell inches	Description	Price
150	37	20½	Aluminum, Polished	$ 10.00

Order by Number

19

2-41. Even horns for the early-style ("front-mounted") Victor "Monarch" machines could be ordered. Few knowledgeable collectors would consider a flower horn proper for a front-mounted Victor "Monarch," due to the extreme rarity of such combinations. Yet, this catalog proves otherwise. *Courtesy of Ernest Carl Allen.*

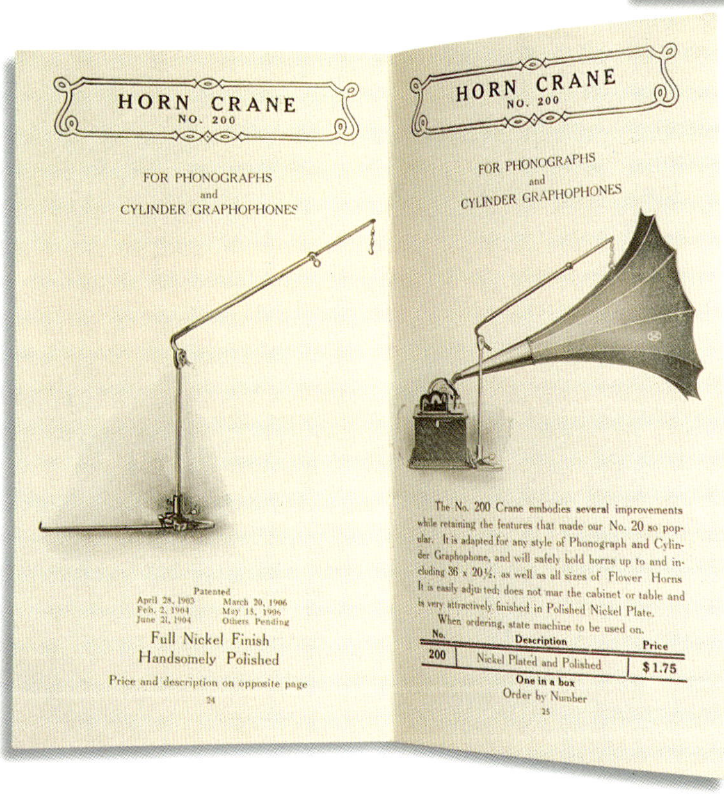

HORN CRANE
NO. 200

FOR PHONOGRAPHS
and
CYLINDER GRAPHOPHONES

Patented
April 28, 1903 March 20, 1906
Feb. 2, 1904 May 15, 1906
June 21, 1904 Others Pending

Full Nickel Finish
Handsomely Polished

Price and description on opposite page

24

HORN CRANE
NO. 200

FOR PHONOGRAPHS
and
CYLINDER GRAPHOPHONES

The No. 200 Crane embodies several improvements while retaining the features that made our No. 20 so popular. It is adapted for any style of Phonograph and Cylinder Graphophone, and will safely hold horns up to and including 36 x 20½, as well as all sizes of Flower Horns It is easily adjusted; does not mar the cabinet or table and is very attractively finished in Polished Nickel Plate.

When ordering, state machine to be used on.

No.	Description	Price
200	Nickel Plated and Polished	$1.75

One in a box
Order by Number

25

2-42. The Tea Tray Company made a clever and simple horn crane that clamped to the cabinet with no drilling or screws required. Multitudes of these devices must have been sold, to judge by their familiarity to collectors today. This horn support was best suited to Graphophones (as shown). *Courtesy of Ernest Carl Allen.*

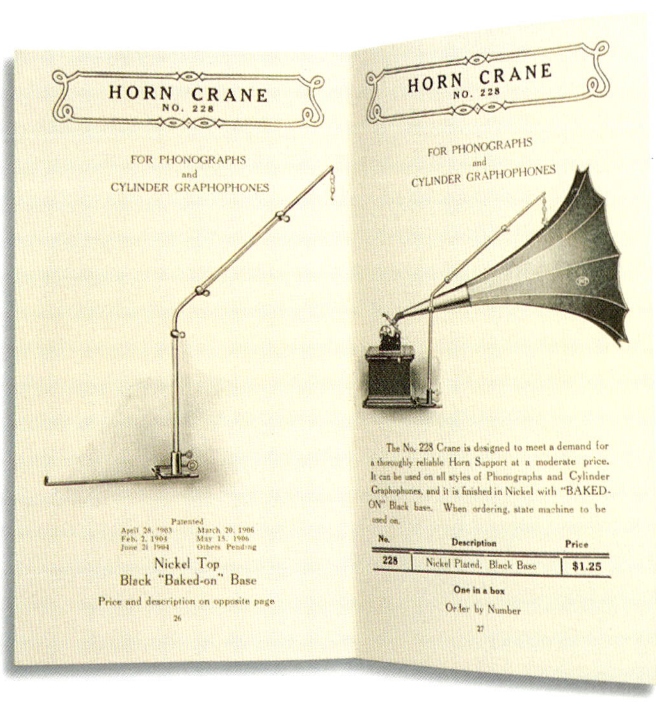

HORN CRANE
NO. 228

FOR PHONOGRAPHS
and
CYLINDER GRAPHOPHONES

Patented
April 28, 1903 March 20, 1906
Feb. 2, 1904 May 15, 1906
June 21 1904 Others Pending

Nickel Top
Black "Baked-on" Base

Price and description on opposite page

26

HORN CRANE
NO. 228

FOR PHONOGRAPHS
and
CYLINDER GRAPHOPHONES

The No. 228 Crane is designed to meet a demand for a thoroughly reliable Horn Support at a moderate price. It can be used on all styles of Phonographs and Cylinder Graphophones, and it is finished in Nickel with "BAKED-ON" Black base. When ordering, state machine to be used on.

No.	Description	Price
228	Nickel Plated, Black Base	$1.25

One in a box
Order by Number

27

2-43. A slightly less-costly horn crane was offered as well, another style that is commonly encountered today. *Courtesy of Ernest Carl Allen.*

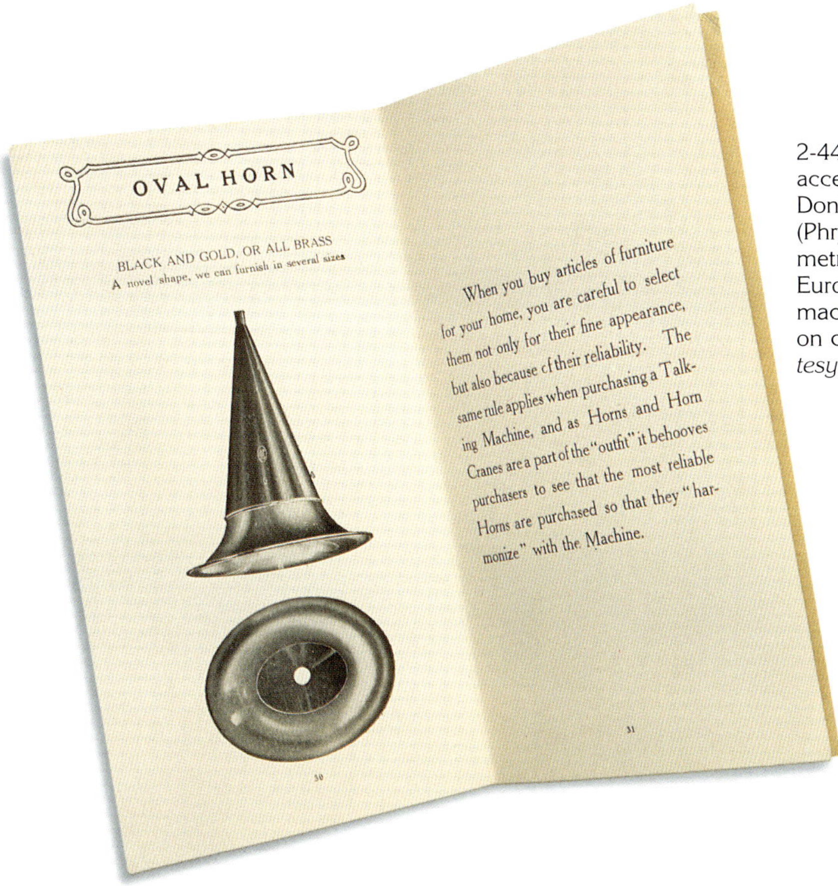

2-44. Finally, a never-seen accessory: the "Oval Horn." Don't blame the deliveryman! (Phrynis offered an asymmetrical flower horn in Europe, but all other talking machine horns were based on circular designs.) *Courtesy of Ernest Carl Allen.*

2-45. This 4 3/4" x 7 1/4" advertisement pictured the 1906 "Rectorphone" (which could be earned by selling 36 packages of bluing), and confidently stated, "Do not mistake this Motor Machine for the worthless toy disk affairs that turn by hand." This is a reference to cheap premium talking machines such as shown in *Discovering Antique Phonographs*, Fabrizio & Paul, Schiffer, 2000, illustrations 3-123 through 3-125. The "Rectorphone" was patented by Enoch J. Rector on August 15, 1905 (No. 797,020). It was originally intended to be powered by a treadle-driven sewing machine, but production models employed a motor as seen here. *Courtesy of Ernest Carl Allen. (Value code: F)*

2-46. A detail of a program from the Helena Theatre, Helena, Montana for a performance which took place on the last day of 1905. "Busy Bee" cylinder and disc talking machines were disseminated to the public through retail store promotions. Customers received coupons, commensurate with their cash purchases – a 25-cent ticket for each $5.00 spent, for example. These were accumulated against the "value" of a talking machine. Records were sold outright, and, in the instance of the "Busy Bee" machines, could constitute the real source of profit for the merchant. Both "Busy Bee" cylinder and disc instruments were designed to play only "Busy Bee" records. *(Value code: G)*

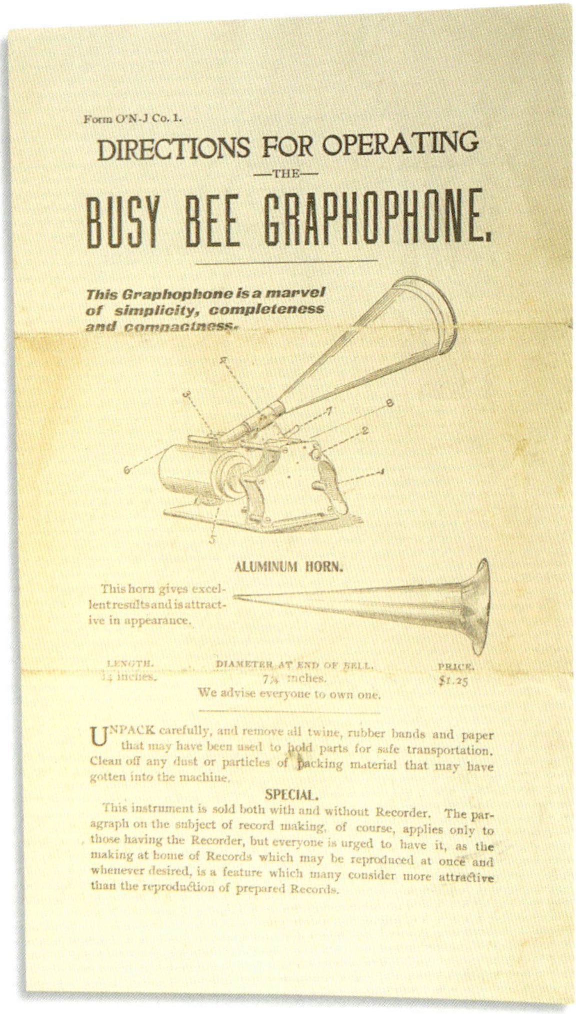
Form O'N-J Co. 1.

DIRECTIONS FOR OPERATING
—THE—
BUSY BEE GRAPHOPHONE.

This Graphophone is a marvel of simplicity, completeness and compactness.

ALUMINUM HORN.

This horn gives excellent results and is attractive in appearance.

LENGTH.	DIAMETER AT END OF BELL.	PRICE.
14 inches.	7¼ inches.	$1.25

We advise everyone to own one.

UNPACK carefully, and remove all twine, rubber bands and paper that may have been used to hold parts for safe transportation. Clean off any dust or particles of packing material that may have gotten into the machine.

SPECIAL.

This instrument is sold both with and without Recorder. The paragraph on the subject of record making, of course, applies only to those having the Recorder, but everyone is urged to have it, as the making at home of Records which may be reproduced at once and whenever desired, is a feature which many consider more attractive than the reproduction of prepared Records.

2-47. The profile of the "Busy Bee" Graphophone makes evident that it was manufactured by Columbia – virtually identical to Columbia's own Type "Q." The O'Neill-James Company of Chicago, which purchased these machines in huge quantities, to be circulated under its own "Busy Bee" brand, required one significant change in design. The mandrel of the "Busy Bee" was slightly enlarged to prohibit the playing of "ordinary" cylinder records. "Busy Bee" cylinders, recorded and produced by Columbia and distributed by O'Neill-James, were the only records that would fit. Instruction sheet, 5" x 8 1/2". *Courtesy of Ernest Carl Allen. (Value code: F)*

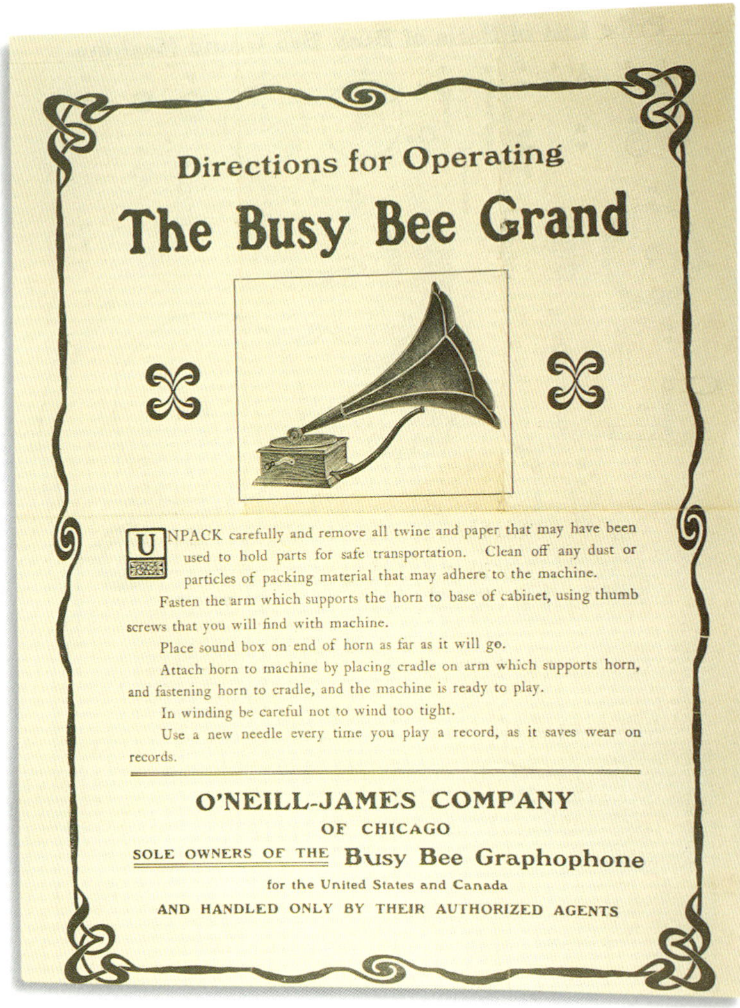
Directions for Operating

The Busy Bee Grand

UNPACK carefully and remove all twine and paper that may have been used to hold parts for safe transportation. Clean off any dust or particles of packing material that may adhere to the machine.

Fasten the arm which supports the horn to base of cabinet, using thumb screws that you will find with machine.

Place sound box on end of horn as far as it will go.

Attach horn to machine by placing cradle on arm which supports horn, and fastening horn to cradle, and the machine is ready to play.

In winding be careful not to wind too tight.

Use a new needle every time you play a record, as it saves wear on records.

O'NEILL-JAMES COMPANY
OF CHICAGO
SOLE OWNERS OF THE Busy Bee Graphophone
for the United States and Canada
AND HANDLED ONLY BY THEIR AUTHORIZED AGENTS

2-48. "Busy Bee Grand" talking machines were distributed in such substantial numbers that they are still a common sight to collectors today. The later version of the machine (most recognizable by a horn support which extends from the middle of a cabinet panel, as opposed to proceeding from the base molding as seen here) had a set of instructions pasted to the underside of the cabinet. Pasted-on instructions can be found under the earlier model, but those that lacked this feature came with this rare sheet of ephemera, 8 1/2" x 11", circa 1907. *Courtesy of Ernest Carl Allen. (Value code: F)*

CHRISTMAS GREETING

Order for Xmas or New Year's if You Wish to Buy on "EASY PAYMENTS"

My *Phono-Metro-Phonic Home-Course* solves the problem. I give you my personal instruction. You actually hear my voice—an accurate reproduction of it. My very accent and manner of speaking with nicety of articulation for overcoming words ordinarily difficult to the stammerer, are at your service whenever it suits your convenience. All that is best in the treatment of stammering and stuttering is embodied in my *Phono-Metro-Phonic Home-Course.*

GEORGE ANDREW LEWIS
35-41 Adelaide Street, DETROIT, MICHIGAN, U. S. A.

2-49. The potential of the phonograph for instruction and self-improvement was obvious. After the introduction of molded, mass-produced cylinders in 1901, home study became a practical reality. George Andrew Lewis was an American expert in speech-language pathology, specializing in stuttering. He sold specially-identified Columbia Type "Q" Graphophones, and specially-recorded Columbia two-minute cylinder records under his own name, circa 1905, before founding the Lewis Phono-metric Institute of Detroit. A complete home-cure course for stammerers was offered to the public by mail. A brochure, 10 1/2" x 14". *Courtesy of Ernest Carl Allen. (Value code: VR)*

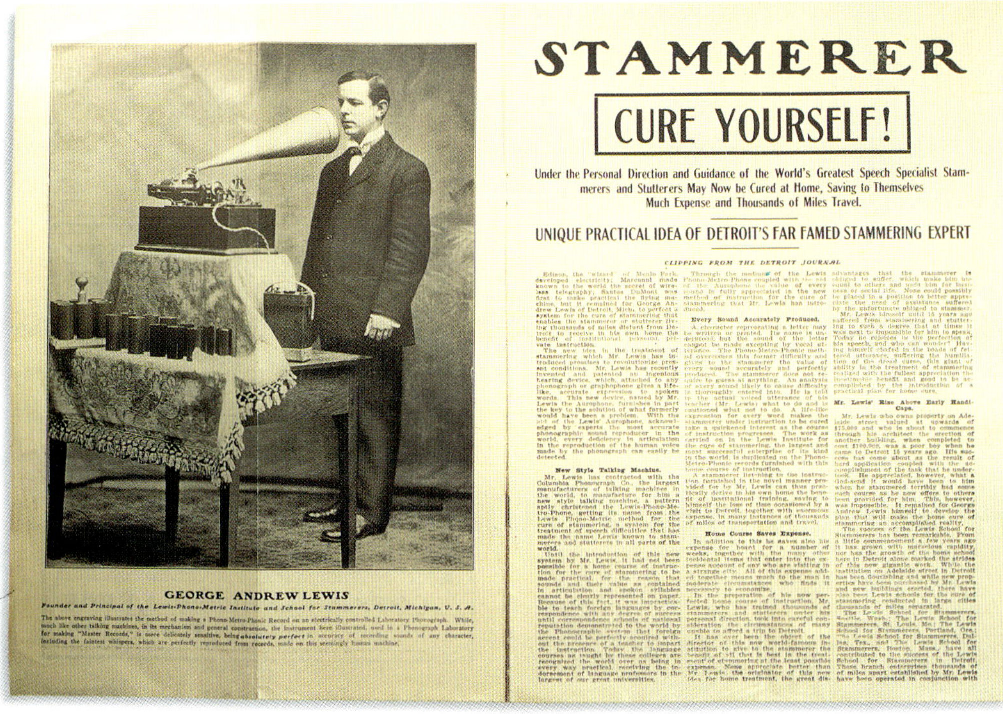

2-50. Mr. Lewis himself is shown in the interior of the brochure, recording with an Edison Class "M" electric Phonograph. A modern marketing specialist would advise Mr. Lewis to reduce the amount of text, in order to hold the reader's attention until the final section of the brochure (the money pitch) could be reached. *Courtesy of Ernest Carl Allen.*

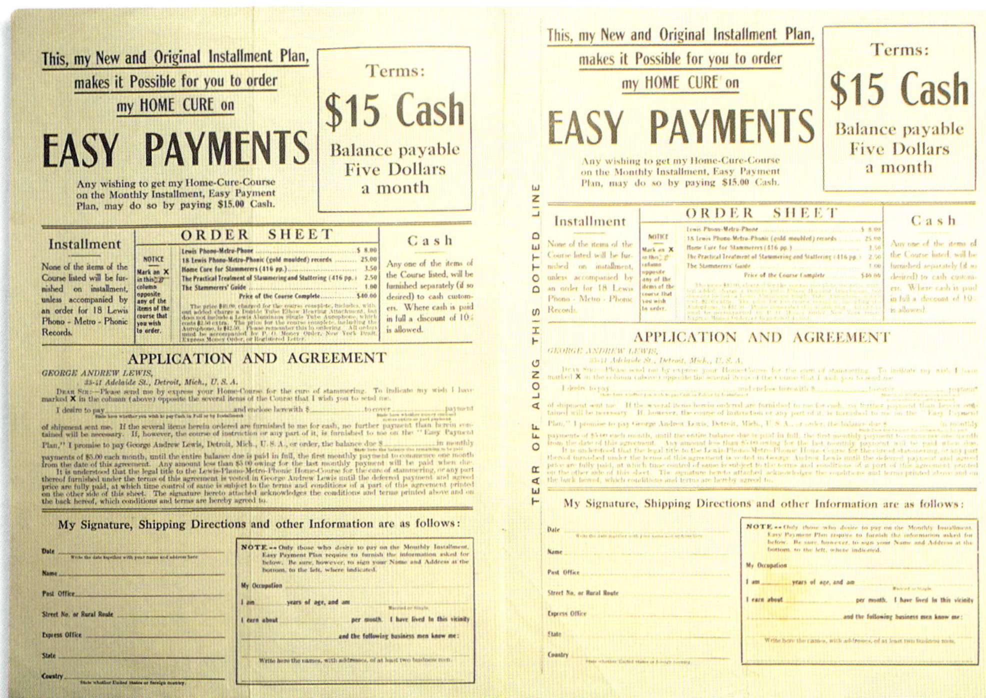

2-51. The proposition. $15.00 in cash needed to initiate the contract was a more considerable sum than might be expected, considering that the regular retail price of a Type "Q" Graphophone was only $7.50. The rarity of Lewis machines and records today could be attributed to a number of causes, not the least of which the inability of a stammerer to come up with the up-front money – about $300.00 in modern funds. *Courtesy of Ernest Carl Allen.*

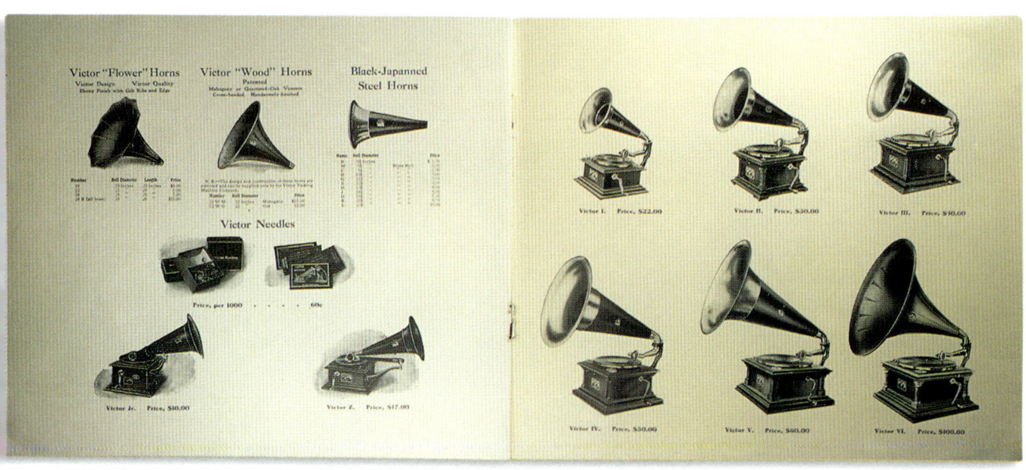

2-52. A 1906 Victor catalog showing, on the right, the company's complete line of talking machines. Stragglers from the firm's previous line of instruments, the "R" and "E," are shown at the left. These two models enjoyed tremendous popularity and continued to be made in an out-of-date style (with horn supported on an arm extending straight out from the instrument) even after an improved design was introduced that put the horn over the top of the cabinet. *(Value code: E)*

2-53. The Victor Talking Machine Company, in what was apparently an early attempt to promote records via sheet music, published this 7 1/4" x 9 1/2" example in 1906. Since this is the only such title we've encountered in the past 30 years, it is probable that the company was engaging in an experiment (short-lived) or this publication was prepared for a special event. *(Value code: VR)*

2-54. Two rare brochures promoting the "Indian" brand disc records of the American Record Company of Springfield, Massachusetts. Three talking machine entrepreneurs, Hawthorne, Sheble, and Prescott, conjoined to record and press a catalog of discs in unmistakable blue shellac, with a less-than-politically-correct label. The company enjoyed success 1905-1906, but succumbed to patent litigation. 3 3/8" x 6". *Courtesy of the Scott and Denise Corbett collection. (Value code: G, each)*

2-55. It was unusual for a retailer to use his own full-size label to cover the post-1901 Columbia or Edison cylinder record box label. (Earlier-style boxes had frequently received all-encompassing dealer labels.) Here is one example: a Columbia label completely enshrouded by that of Grubs Music Store. *Courtesy of the Gfell family collection. (Value code: F)*

Coin-operated talking machines appeared as early as 1889, and it wasn't long before exhibitors realized that their profits were limited by the machines' inability to offer more than one selection. To offer multiple titles simultaneously, it would be necessary to group a number of phonographs in one location: thus, the phonograph arcade was born. Half a century later, a single jukebox could provide dozens of selections, but for the first 25 years of its commercial life, coin-operated phonographs commonly were exhibited in clusters. Automatic instruments from that era are rare now, and even rarer are the show-card board signs used to advertise the available selections. These were sometimes worked up by the exhibitor, but often bought from professional firms such as the Hunt Show Card Company of 1024 Arch Street, Philadelphia, Pennsylvania. The following 20 original "marquees" are grouped much as they might have been found at an amusement park shortly after the turn of the twentieth century. *(Value code statement: The following coin-op signs are very rare, but occasionally sell in the "E" range, with some especially notable examples reaching into "D.")*

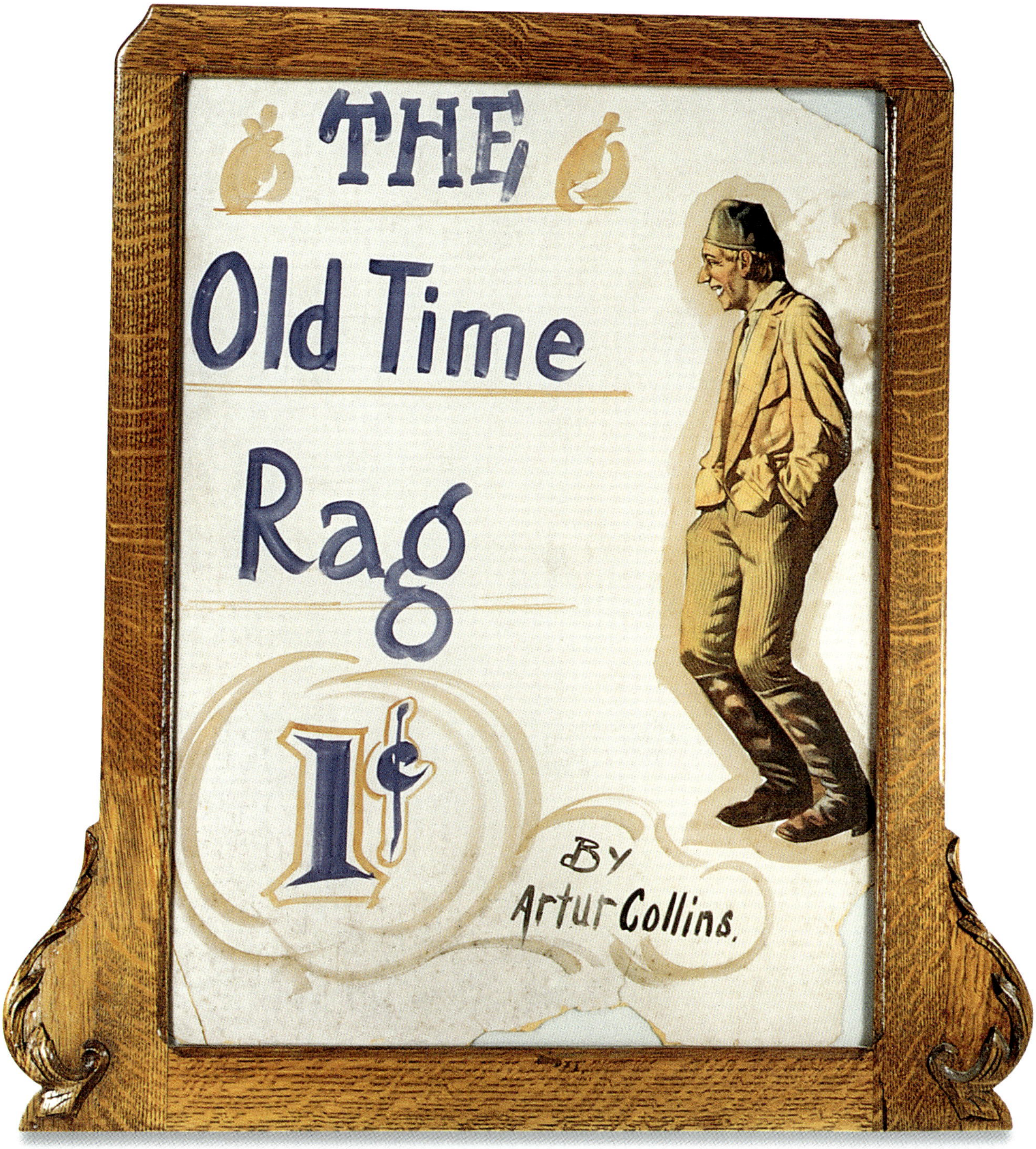

2-56 ...With apologies to Arthur Collins! *Courtesy of Bruce and Linda Peterson.*

2-57. "San Antonio" was one of a number of girl-meets-cowboy hits. Stamped "Hunt Show Card Co., 1024 Arch St., Phila., Pa." *Courtesy of Michael F. Khanchalian.*

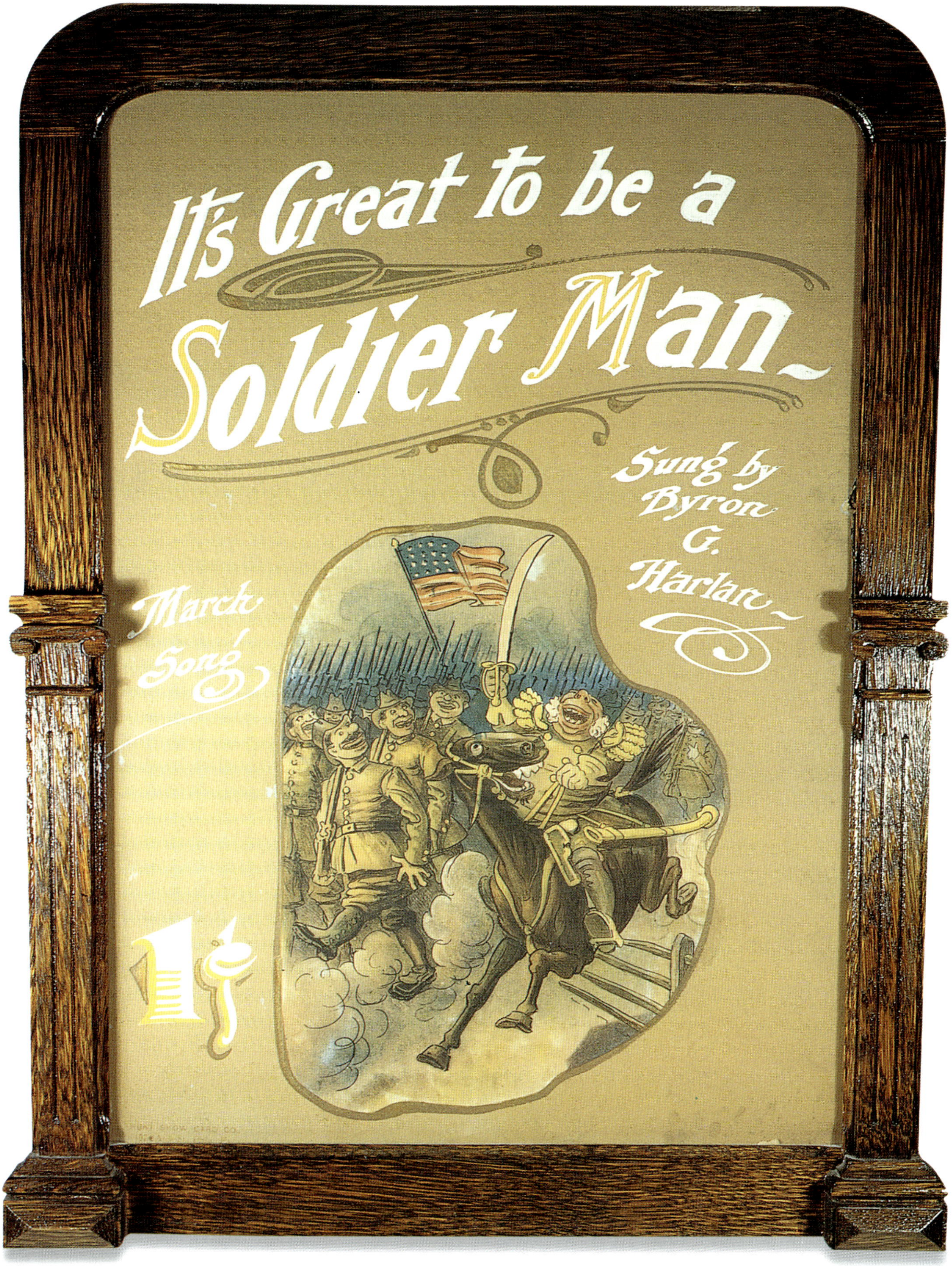

2-58. Not a well known song, but a delightfully rendered sign. Stamped "Hunt Show Card Co., 1024 Arch St., Phila., Pa." *Courtesy of Garry James.*

2-59. The stippled and dry-brushed background appears oddly modern. Measures 20" x 22 1/4".

2-60. A peculiarly melancholy creation for a "New March Song." Measures 18 1/4" x 22 1/4".

2-61. By using appliqué pictures, the sign painter could quickly create a finished piece. Hand-lettered, "Hunt, 227 N. 8th." Measures 17" x 24 1/2".

2-62. Adding a "vignette shadow" to an appliqué was a simple but effective way to create depth.

2-63. Here, a field of subtle color surrounds the applied figure, delineating it. Measures 17 1/4" x 23".

2-64. Or a strong outline might be used, in a complementary shade. Measures 16 1/4" x 23 3/4".

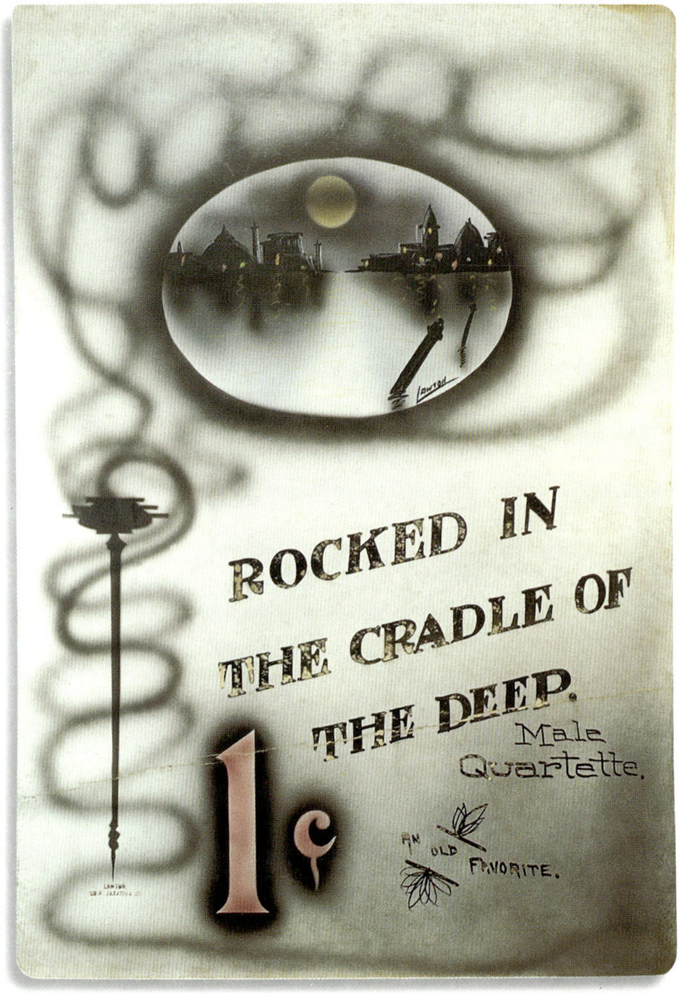

2-65. Here, the artist has dashed off a hand-painted night landscape, much the same as the inexpensively-produced paintings rendered during the nineteenth and early twentieth centuries for home decoration, mirrors, and the fronts of clocks. Signed, "Lawton." Measures 17 1/4" x 24 1/2".

2-66. It's likely that this cutout image of the "Yellow Kid" being pursued by a dog was chosen by happenstance, although the comic-page character it depicted was well renowned. In 1895, R.F. Outcault had created the wisecracking Irish tenement baby in the New York *World* (after a few mute appearances in *Truth* in 1894). For the next several years New York was Yellow Kid mad, and the character's fame spread countrywide. The Kid's popularity waned as the nineteenth century ended, however he did make appearances in Outcault's even more famous "Buster Brown" strip (begun 1902). Measures 17 1/2" x 24 1/2".

2-67. Mom looks pretty good for her age – but a more funereal song selection can hardly be imagined by those who have heard it! *Courtesy of the Domenic DiBernardo collection.*

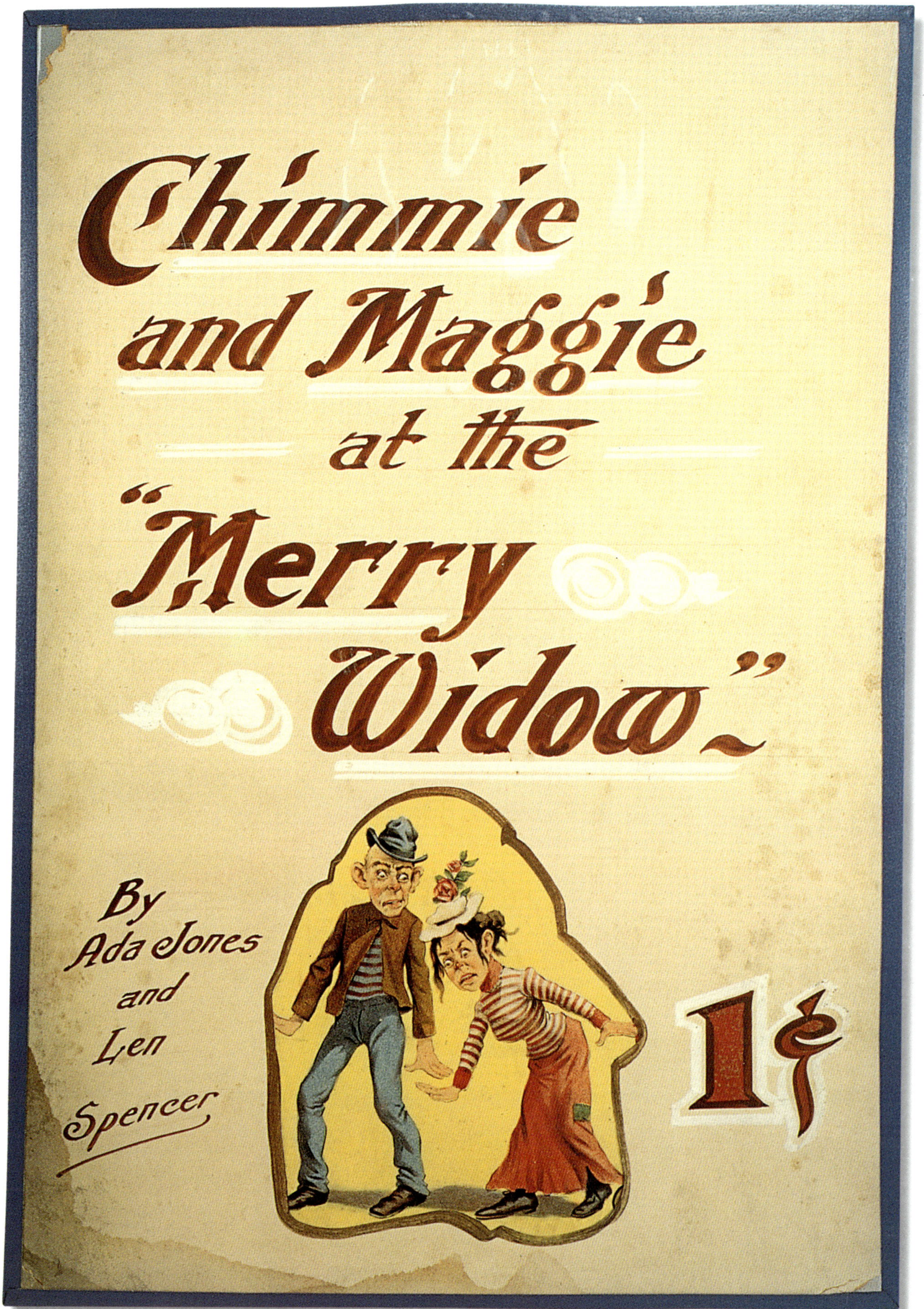

2-68. Rough and uneducated tenement dwellers were the frequent butt of turn-of-the-twentieth-century humor (as in the "Yellow Kid"). Hully gee, wot moit be a bigger laff den Chimmie an his gal Maggie at de high-toned "Merry Widder" on Broadway? (*Courtesy of the Domenic DiBernardo collection.*

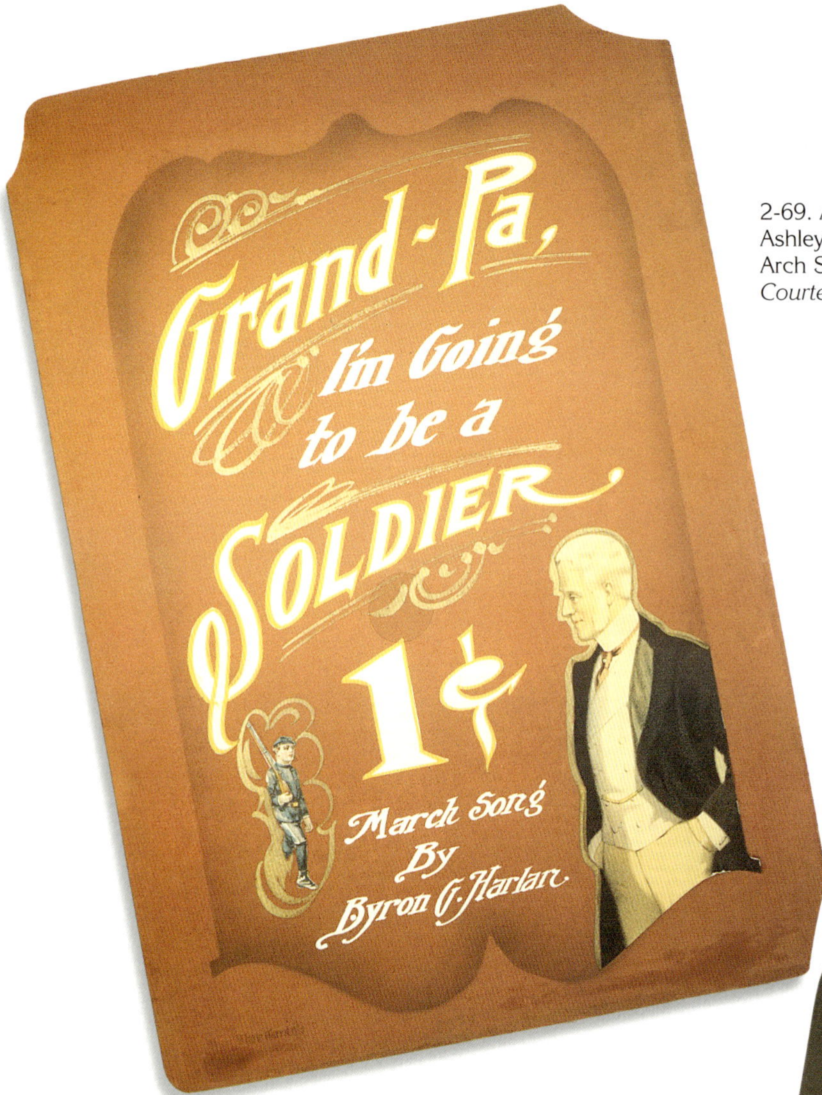

2-69. A cadet parading, and who's that looking on – Ashley Wilkes? Stamped "Hunt Show Card Co., 1024 Arch St., Phila., Pa." Measures 17 1/4" x 24 1/4". *Courtesy of the Domenic DiBernardo collection*.

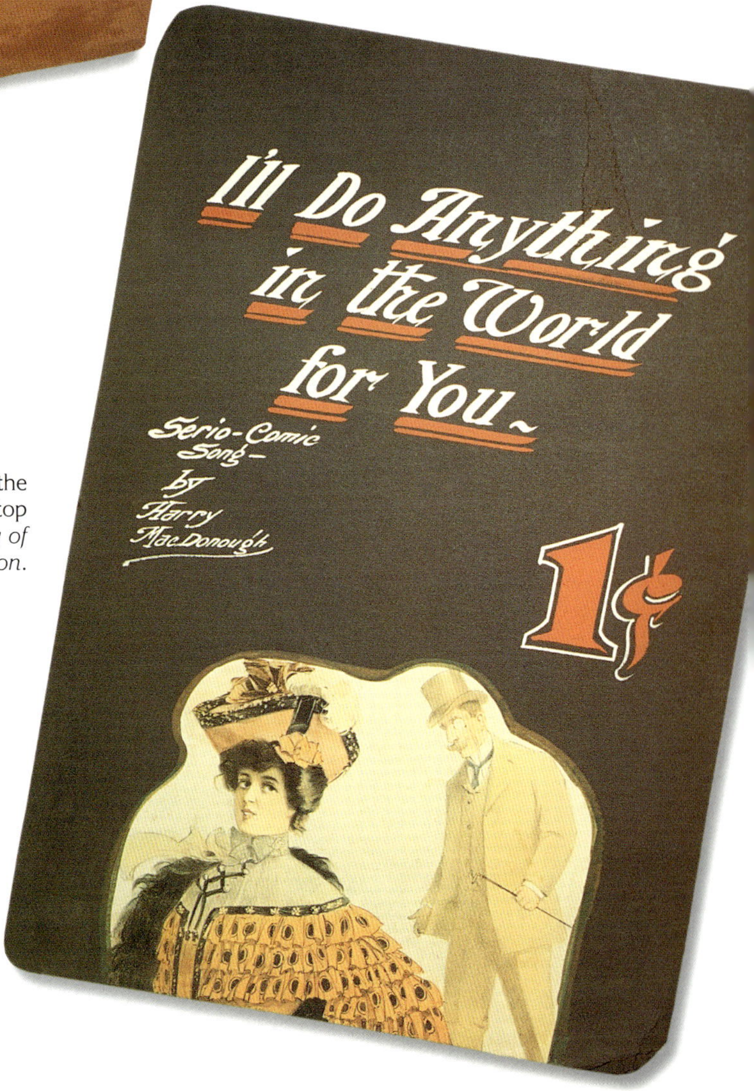

2-70. A rather peculiar layout, where it seems the dapper gentleman leans to avoid bumping his top hat on the periphery of the "balloon." *Courtesy of the Domenic DiBernardo collection*.

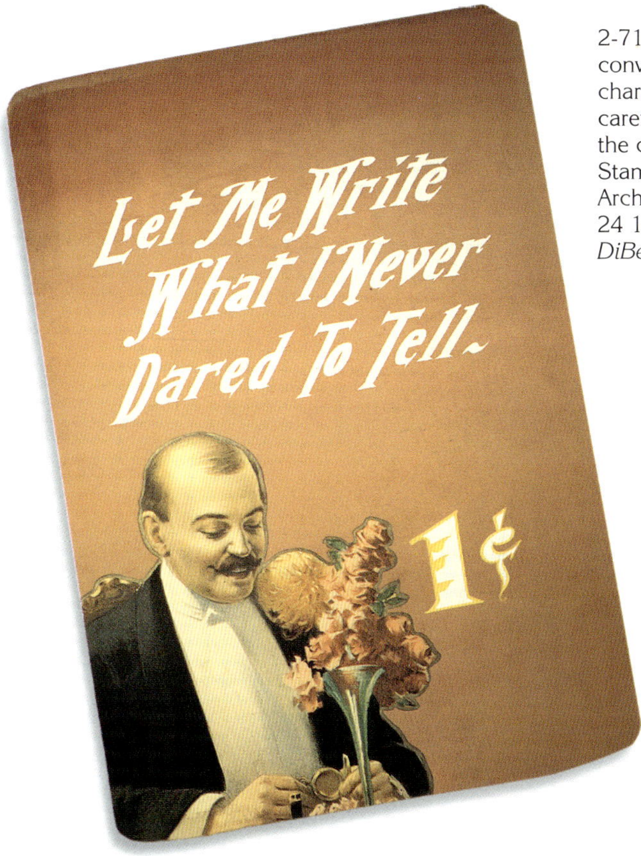

2-71. This discreet-looking gentleman convinces us he would be tongue-tied. A charming composition was created by careful positioning of the elements and the choice of a background color. Stamped "Hunt Show Card Co., 1024 Arch St., Phila., Pa." Measures 17 1/4" x 24 1/4". *Courtesy of the Domenic DiBernardo collection.*

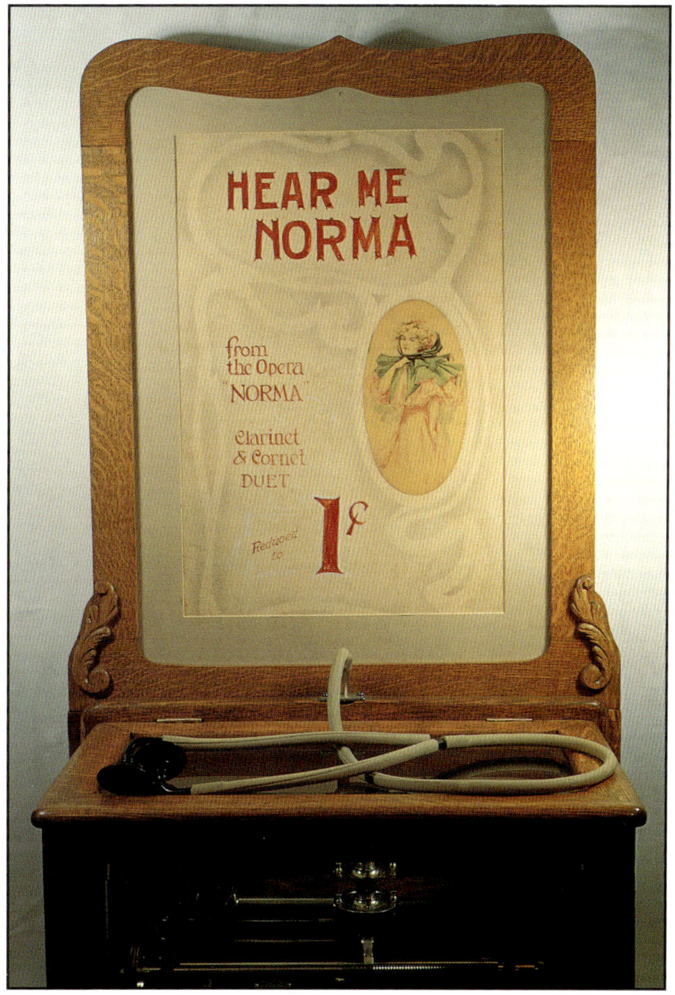

2-72. A more elaborate design than commonly seen, for a very popular selection. *Courtesy of Nina and Dave Heitz.*

2-73. The grinning political nabob shown here makes it clear that the song's setting is not the Pacific Northwest! Stamped "Hunt Show Card Co., 1024 Arch St., Phila., Pa.". Measures 16 1/2" x 22 1/2". *Courtesy of Nina and Dave Heitz.*

2-74. Coin-op signs seldom editorialized, though there was plenty of space to do so. This one unconventionally identifies a "Big Hit." Measures 16 1/2" x 22 1/2". *Courtesy of Nina and Dave Heitz.*

2-75. A sophisticated composition. Stamped "Hunt Show Card Co., 1024 Arch St., Phila., Pa.".

2-76. In actual use, this window decal from the 1905-1907 period would not have been framed, but affixed to a dealer's window. An identically designed twin, lettered "EDISON RECORDS," was also available. Made by Palm Fechteler & Company, New York City and Chicago. Measures 11 3/4" x 32". *Courtesy of Nina and Dave Heitz. (Value code: E)*

2-77. The English Columbia Phonograph Company, General, supplied this record box label in 1906, illustrating the "Lyric" reproducer. *Courtesy of George Glastris. (Value code: F)*

2-78. When Victor introduced the Victrola in late 1906, the talking machine industry was stood on its ear. After a decade of associating horns with recorded entertainment, the public would embrace the concept of a phonograph that didn't look like a phonograph. Of course, it took several years before the internal-horn talking machine made significant inroads, but Victor's design coup was clearly a path to the future. Caught flat-footed, Columbia marketed the piano-inspired "Symphony Grand" Graphophone in 1907. Pamphlet measures 4 7/8" x 8". *Courtesy of Ernest Carl Allen. (Value code: VR)*

2-79. The interior pages showed an open view of the "Symphony Grand," and released a barrage of flowery prose extolling its benefits. On another page, the writer gushed, "…[T]he Symphony Grand Graphophone is in every sense the fulfillment of a promise that was almost beyond hope." Note that this instrument was not called a "Grafonola" yet – and indeed, its components and their configuration resemble an external-horn Graphophone more than an internal-horn Victrola. *Courtesy of Ernest Carl Allen.*

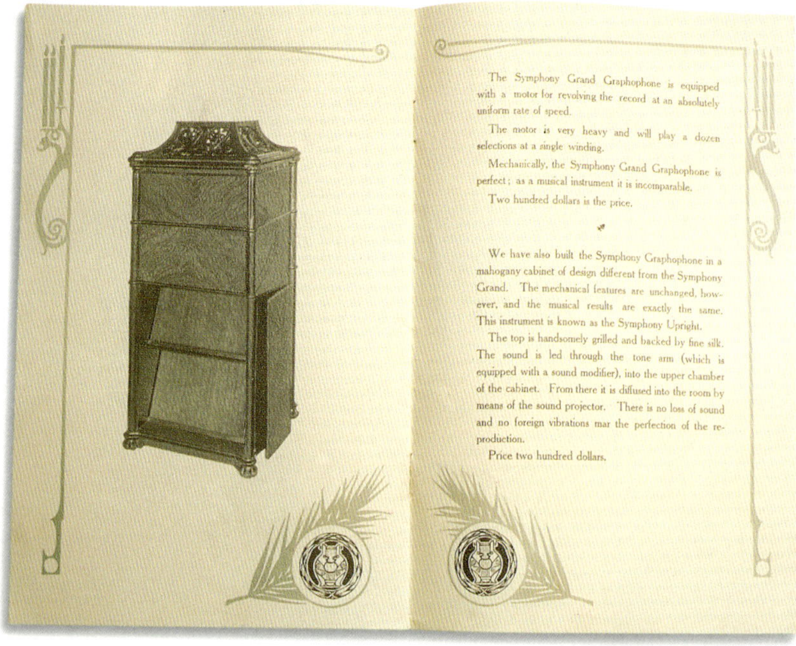

The Symphony Grand Graphophone is equipped with a motor for revolving the record at an absolutely uniform rate of speed.

The motor is very heavy and will play a dozen selections at a single winding.

Mechanically, the Symphony Grand Graphophone is perfect; as a musical instrument it is incomparable.

Two hundred dollars is the price.

We have also built the Symphony Graphophone in a mahogany cabinet of design different from the Symphony Grand. The mechanical features are unchanged, however, and the musical results are exactly the same. This instrument is known as the Symphony Upright.

The top is handsomely grilled and backed by fine silk. The sound is led through the tone arm (which is equipped with a sound modifier), into the upper chamber of the cabinet. From there it is diffused into the room by means of the sound projector. There is no loss of sound and no foreign vibrations mar the perfection of the reproduction.

Price two hundred dollars.

2-80. For those who preferred a more "Victrola-like" instrument, Columbia thoughtfully provided, "...the Symphony Graphophone in a mahogany cabinet design different from the Symphony Grand... This instrument is known as the Symphony Upright." Customers could take their pick of either Symphony Graphophone for $200.00. *Courtesy of Ernest Carl Allen.*

2-81. Around 1907, the Universal Talking Machine Manufacturing Company adopted a trademark consisting of a small child listening to a Zonophone. It was titled, "On Speaking Terms." The inspiration of "His Master's Voice" is inescapable, and several other companies attempted (albeit unsuccessfully) to develop a trademark with the appeal of Nipper and the Gramophone. Interestingly, Zonophone was, at this time, a wholly owned subsidiary of the Victor Talking Machine Company, the Nipper trademark's proud owner. Advertising cards with the Victor logo were widely distributed, but this framed Zonophone example is a rarity. Approximate size, 8" x 10". *Courtesy of Norm and Janyne Smith. (Value code: VR)*

2-82. This 3 1/2" x 5 7/8" flyer touted a "Sound Modifier" for any type of Victor talking machine. The Victor Distributing & Export Company was a prominent jobber and (less well-known) retailer of Victor products. *Courtesy of the Johnson Victrola Museum/ Delaware State Museums. (Value code: F)*

2-83. Pages from the British trade journal *The Talking Machine News*, 1907-1908, bound, 8 1/4" x 10 3/8". In this double-page spread, the "SIMP" talking machines were grandly promoted. The brand name may seem rather unpleasant to us today, conjuring up "simper, pimple, and wimp." Yet this remains proof that dreaming up stupid names for commercial objects has not been limited to the present day automobile industry. The technical system employed here (two horns, with a dedicated reproducing diaphragm for each horn) was being exploited in the United States by the Duplex Phonograph Company. *Courtesy of Ernest Carl Allen. (Value code: bound volumes VR)*

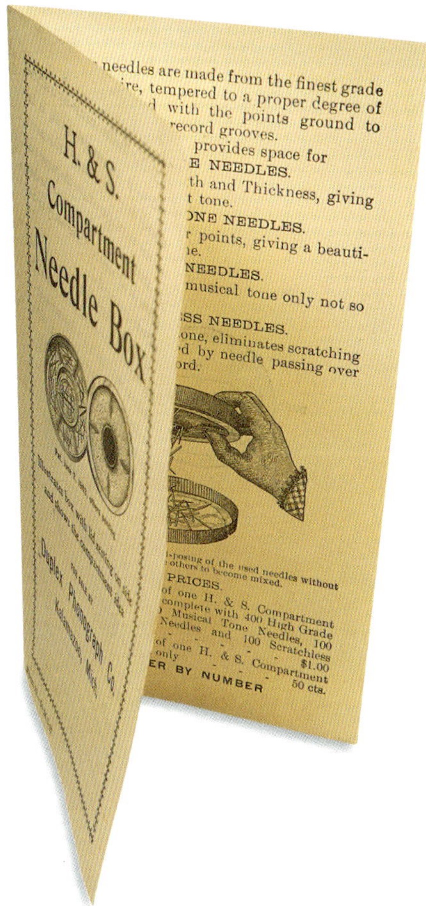

2-84. The Duplex Phonograph Company of Kalamazoo, Michigan distributed this February 1907 flyer for the H&S Compartment Needle Box, manufactured by Hawthorne & Sheble of Philadelphia. Note the blank area on the lower front where a dealer's imprint could be added, as seen here. Measures 3 3/8" x 6 1/4". *Courtesy of Ernest Carl Allen. (Value code: F)*

2-85. A 4" x 6" catalog cover of enduring appeal. Teddy Roosevelt was midway through his second term, and his well-known exclamation – "De-lighted!" – found its way into the advertising of various firms. *Courtesy of Ernest Carl Allen. (Value code: E)*

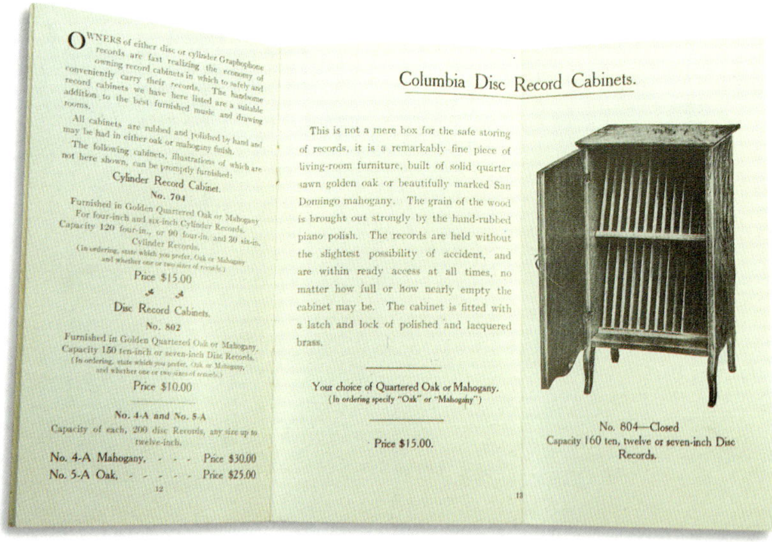

OWNERS of either disc or cylinder Graphophone records are fast realizing the economy of owning record cabinets in which to safely and conveniently carry their records. The handsome record cabinets we have here listed are a suitable addition to the best furnished music and drawing rooms.

All cabinets are rubbed and polished by hand and may be had in either oak or mahogany finish.

The following cabinets, illustrations of which are not here shown, can be promptly furnished:

Cylinder Record Cabinet.
No. 704
Furnished in Golden Quartered Oak or Mahogany
For four-inch and six-inch Cylinder Records.
Capacity 120 four-in., or 90 four-in. and 30 six-in. Cylinder Records.
(In ordering, state which you prefer, Oak or Mahogany and whether one or two sizes of records.)
Price $15.00

Disc Record Cabinets.
No. 802
Furnished in Golden Quartered Oak or Mahogany.
Capacity 150 ten-inch or seven-inch Disc Records.
(In ordering, state which you prefer, Oak or Mahogany, and whether one or two sizes of records.)
Price $10.00

No. 4-A and No. 5-A
Capacity of each, 200 disc Records, any size up to twelve-inch.
No. 4-A Mahogany, - - - Price $30.00
No. 5-A Oak, - - - - - Price $25.00

12

Columbia Disc Record Cabinets.

This is not a mere box for the safe storing of records, it is a remarkably fine piece of living-room furniture, built of solid quarter sawn golden oak or beautifully marked San Domingo mahogany. The grain of the wood is brought out strongly by the hand-rubbed piano polish. The records are held without the slightest possibility of accident, and are within ready access at all times, no matter how full or how nearly empty the cabinet may be. The cabinet is fitted with a latch and lock of polished and lacquered brass.

Your choice of Quartered Oak or Mahogany.
(In ordering specify "Oak" or "Mahogany")

Price $15.00.

No. 804—Closed
Capacity 160 ten, twelve or seven-inch Disc Records.

13

2-86. This 1907 Columbia catalog of "Talking Machine Supplies" features cylinder and disc record storage cabinets manufactured by the Herzog Art Furniture Company of Saginaw, Michigan. *Courtesy of Tom Sideleau. (Value Code: E)*

Columbia Disc Record Cabinets.

No. 816. Cabinet for holding 200 12-in. or 10-in. Disc Records.

It is impossible to show the beauty of this elegant cabinet in an illustration. Possibly the most distinctive and attractive feature is the [illegible] marking of the selected quarter-sawn oak or rich mahogany of the gracefully carved door. But even in a picture [illegible] the quiet, artistic design of this cabinet would give it distinction. The wood used in making this Cabinet has been selected with great care in order to secure the most striking markings. It is finished with genuine hand-carved ornamentation and "shaped" legs. The cabinet has a most brilliant hand-rubbed finish.

Your choice in Quartered Oak or Mahogany. Price $35.00.
(In ordering, specify "Oak" or "Mahogany")

14

Columbia Disc Record Cabinets.

No. 832. Cabinet for holding 200 12-in. or 10-in. Disc Records.

The quiet elegance and simplicity of the design of this cabinet could hardly be surpassed. It is an example of cabinet-making handicraft—fit for a place of honor in anybody's music-room. Made in either quarter-sawn golden oak or clear grain selected West Indian mahogany. It is without a fault in the design or in the way the design has been carried out by the cabinet-maker. The diagonally marked wood of the double doors and of the pillars gives the simplicity of the design a most striking and effective richness.

Price $40.00.
(In ordering specify "Oak" or "Mahogany")

15

2-87. *Courtesy of Tom Sideleau.*

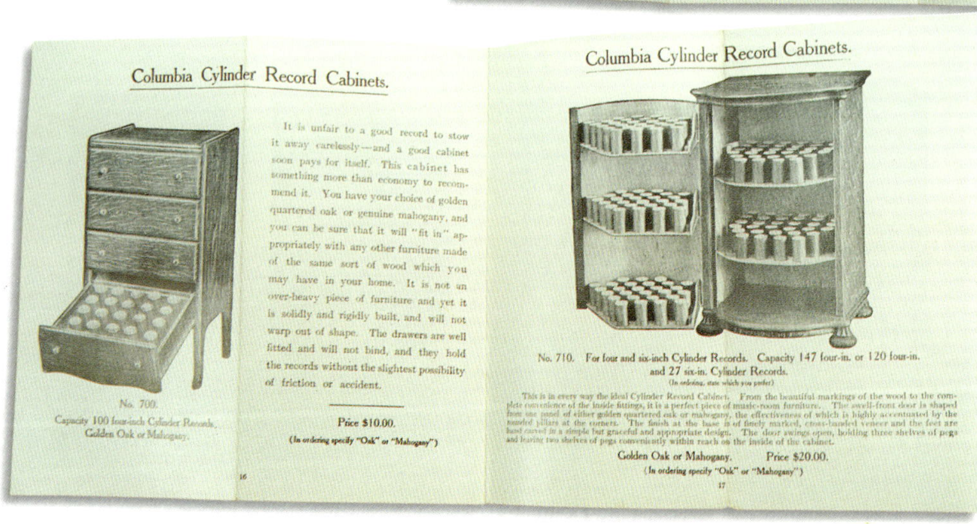

Columbia Cylinder Record Cabinets.

It is unfair to a good record to stow it away carelessly—and a good cabinet soon pays for itself. This cabinet has something more than economy to recommend it. You have your choice of golden quartered oak or genuine mahogany, and you can be sure that it will "fit in" appropriately with any other furniture made of the same sort of wood which you may have in your home. It is not an over-heavy piece of furniture and yet it is solidly and rigidly built, and will not warp out of shape. The drawers are well fitted and will not bind, and they hold the records without the slightest possibility of friction or accident.

No. 709.
Capacity 100 four-inch Cylinder Records.
Golden Oak or Mahogany.

Price $10.00.
(In ordering specify "Oak" or "Mahogany")

16

Columbia Cylinder Record Cabinets.

No. 710. For four and six-inch Cylinder Records. Capacity 147 four-in. or 120 four-in. and 27 six-in. Cylinder Records.
(In ordering, state which you prefer)

This is in every way the ideal Cylinder Record Cabinet. From the beautiful markings of the wood to the complete convenience of the inside fittings, it is a perfect piece of music-room furniture. The swell-front door is shaped from one panel of either golden quartered oak or mahogany, the effectiveness of which is highly accentuated by the rounded pillars at the corners. The finish at the base is of finely marked, cross-banded veneer and the feet are hand carved in a simple but graceful and appropriate design. The door swings open, holding three shelves of pegs and leaving two shelves of pegs conveniently within reach on the inside of the cabinet.

Golden Oak or Mahogany. Price $20.00.
(In ordering specify "Oak" or "Mahogany")

17

2-88. Another page of the 1907 Columbia catalog illustrates, on the right, the barrel-fronted Herzog cabinet which is enormously popular with collectors today. *Courtesy of Tom Sideleau.*

Columbia Cylinder Record Cabinets.

No. 714—Closed. Capacity 180 four-in., or 108 four-in and 72 six-in. Cylinder Records.
(In ordering, state which you desire)

The simplicity and [illegible]

Golden Oak or Mahogany. Price $20.00.
(In ordering specify "Oak" or "Mahogany")

18

Ask for complete list of Columbia Disc and Cylinder Records.

THE difference between Columbia Records and ordinary process records is much too plain to miss.

Popular songs by popular singers, concerts by bands and orchestras, character sketches in dialect, instrumental solos, the greatest triumphs of the stars of opera—no matter what the selection may be, the reproduction is nothing less than perfect. Every vibrant note of the living voice pours out full and round, clear and mellow, with no flatness, no scraping sound, no artificial quality of tone.

Columbia Records are made in the largest factory of its kind in the world, under the original patents covering the recording of musical sounds, and by the men who developed the Graphophone to its present place as a musical instrument.

2-89. *Courtesy of Tom Sideleau.*

2-90. This Yankee Doodle Dandy enlivened the cover of Victor's catalog of new records for July 1907. *Courtesy of Tom and Sandi McCarthy. (Value code: F)*

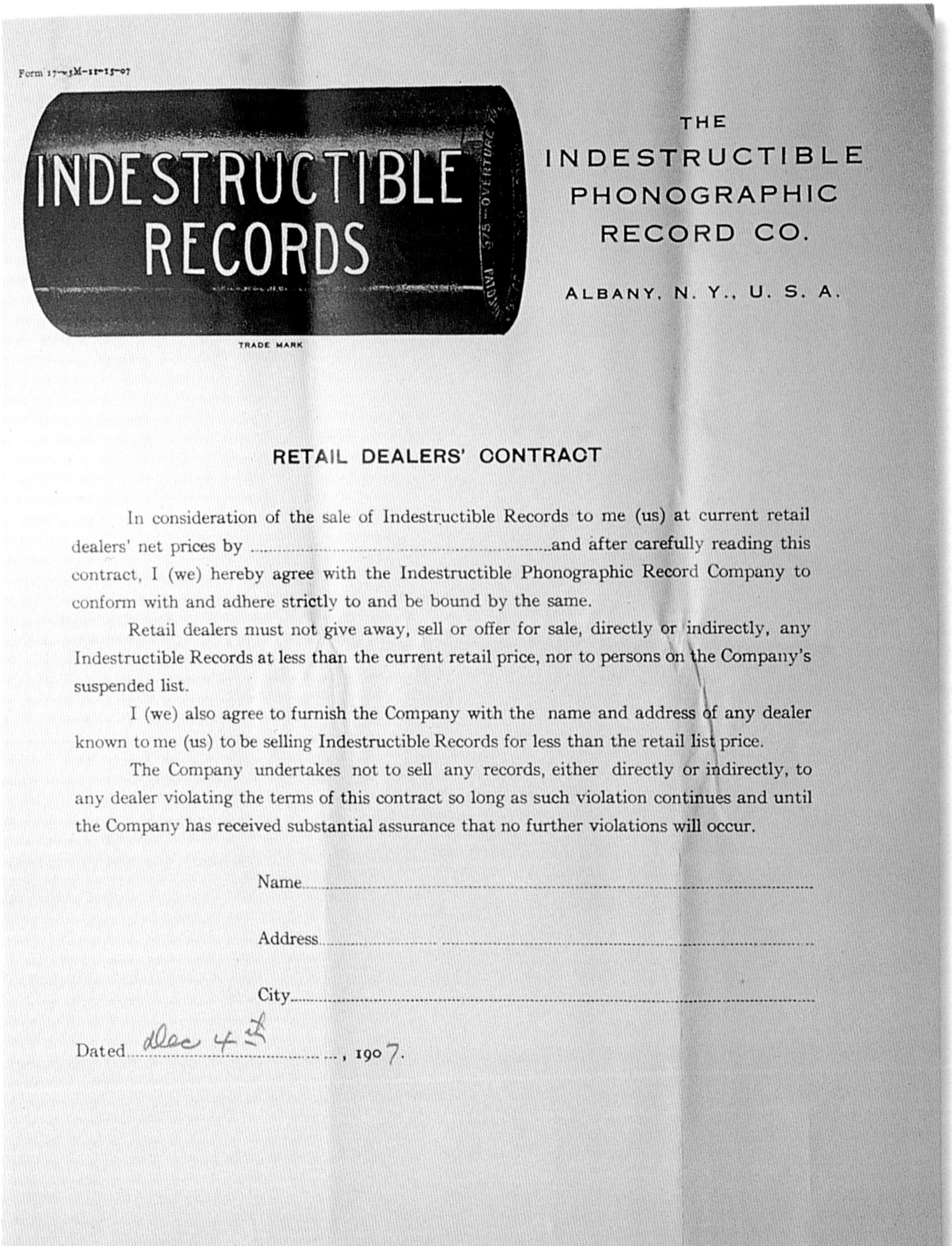

Form 17-5M-11-15-07

INDESTRUCTIBLE RECORDS

TRADE MARK

THE INDESTRUCTIBLE PHONOGRAPHIC RECORD CO.

ALBANY, N. Y., U. S. A.

RETAIL DEALERS' CONTRACT

In consideration of the sale of Indestructible Records to me (us) at current retail dealers' net prices by ....and after carefully reading this contract, I (we) hereby agree with the Indestructible Phonographic Record Company to conform with and adhere strictly to and be bound by the same.

Retail dealers must not give away, sell or offer for sale, directly or indirectly, any Indestructible Records at less than the current retail price, nor to persons on the Company's suspended list.

I (we) also agree to furnish the Company with the name and address of any dealer known to me (us) to be selling Indestructible Records for less than the retail list price.

The Company undertakes not to sell any records, either directly or indirectly, to any dealer violating the terms of this contract so long as such violation continues and until the Company has received substantial assurance that no further violations will occur.

Name..

Address..

City..

Dated Dec 4th ...., 1907.

2-91. A profound rarity: a dealer's contract from the Indestructible Record Company of Albany, New York. Indestructible was the direct heir to the Lambert monopoly on celluloid cylinder records in the United States. Lambert had begun manufacturing cylinders of durable, molded celluloid in 1900, but lack of capital and legal battles kept the company from developing. Consequently, most American cylinder records continued to be molded of fragile "wax" (actually, waxlike soap) until Indestructible arrived upon the scene in 1907. The company would have a long history. In 1908 the firm's entire output began to be sold under the Columbia brand, thus saving Columbia the necessity of producing its own celluloid records. After 1910 the Columbia contract ended and Indestructible continued to sell its records (both two and four minute varieties) under the name "Everlasting Indestructible" until the factory was destroyed by fire (and what a conflagration *that* must have been, considering the explosive nature of cellulose nitrate) in 1922. *Courtesy of the Tom and Joanna Broome. (Value code: VR)*

2-92. The Talking Machine Company of Chicago offered a line of disc record storage cabinets that were designed to fit a particular talking machine. Shown here from a 3 1/4" x 6 1/4" catalog entitled "Six Victor Outfits," circa 1908, "The Cabinet that Matches" a Victor "III." *Courtesy of Ernest Carl Allen. (Value code: E)*

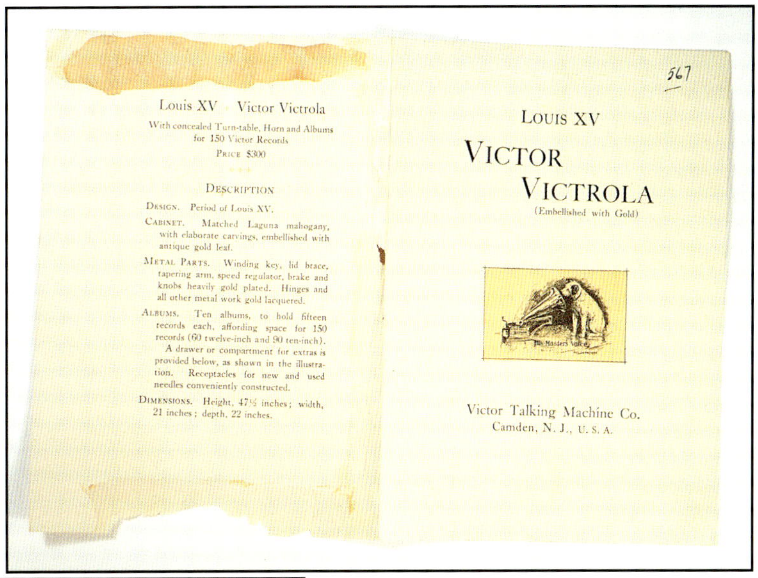

2-93. A brochure for the "Louis XV" Victrola, "(Embellished with Gold)." The $300.00 price offers a general (but not a decisive) clue to the model's better-known identity as revealed in the next caption. Brochure measures 7" x 10" opened. *(Value code: VR)*

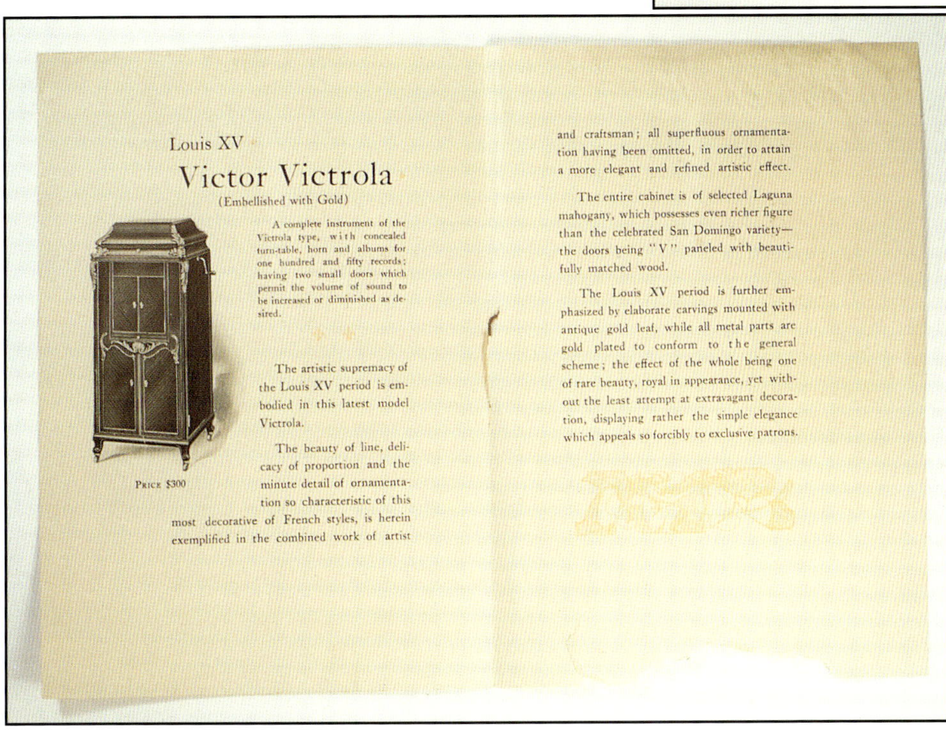

2-94. The interior of the brochure shown in the previous illustration pictures a model commonly known to collectors as the "XX" (Twenty). Introduced March 1,1908, and produced for only one year, the price of this uncommon machine was reduced to $250.00 in December. At that time, 275 of these Victrolas remained unsold in Victor's inventory. Surviving examples today are among the rarest of the rare.

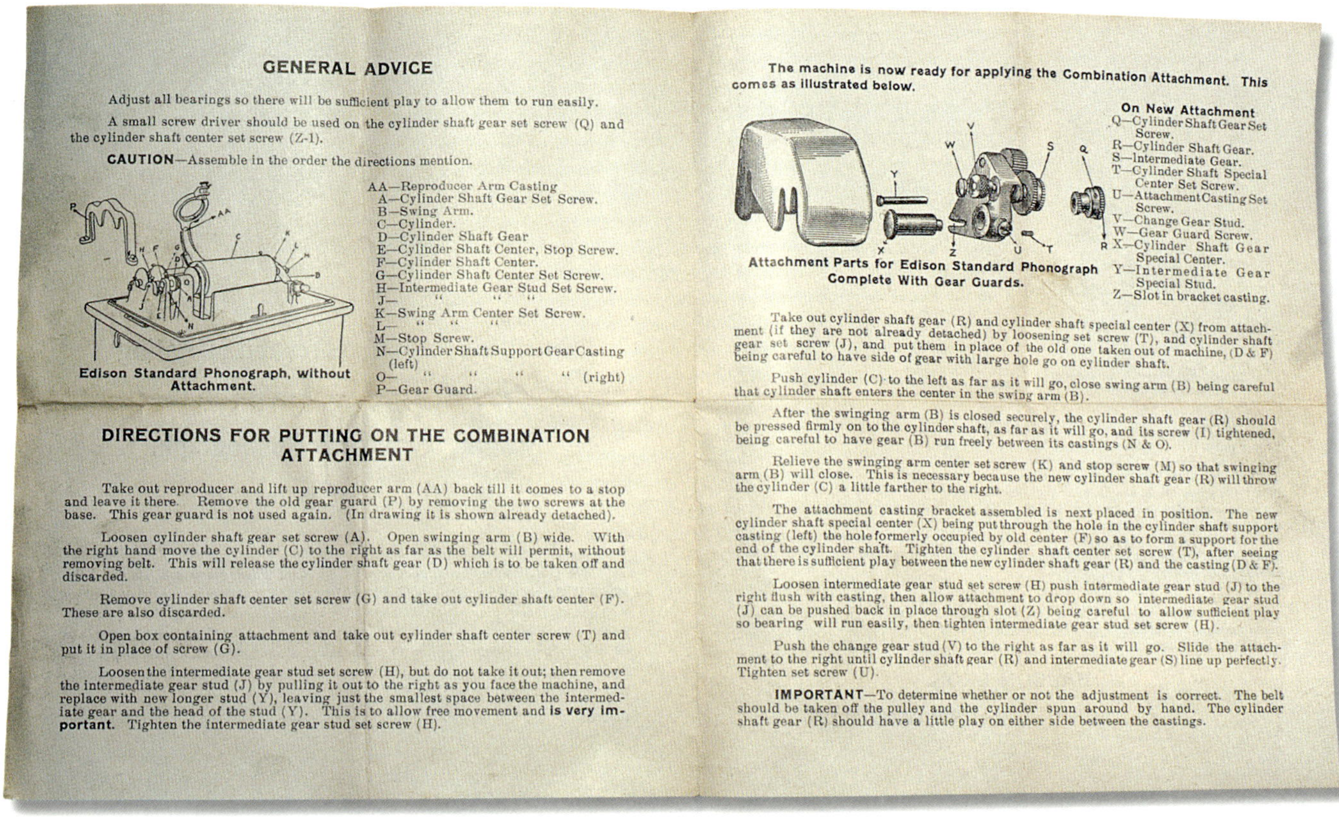

GENERAL ADVICE

Adjust all bearings so there will be sufficient play to allow them to run easily.

A small screw driver should be used on the cylinder shaft gear set screw (Q) and the cylinder shaft center set screw (Z-1).

CAUTION—Assemble in the order the directions mention.

AA—Reproducer Arm Casting
A—Cylinder Shaft Gear Set Screw.
B—Swing Arm.
C—Cylinder.
D—Cylinder Shaft Gear
E—Cylinder Shaft Center, Stop Screw.
F—Cylinder Shaft Center.
G—Cylinder Shaft Center Set Screw.
H—Intermediate Gear Stud Set Screw.
J— " " "
K—Swing Arm Center Set Screw.
L— " " "
M—Stop Screw.
N—Cylinder Shaft Support Gear Casting (left)
O— " " " " (right)
P—Gear Guard.

Edison Standard Phonograph, without Attachment.

DIRECTIONS FOR PUTTING ON THE COMBINATION ATTACHMENT

Take out reproducer and lift up reproducer arm (AA) back till it comes to a stop and leave it there. Remove the old gear guard (P) by removing the two screws at the base. This gear guard is not used again. (In drawing it is shown already detached).

Loosen cylinder shaft gear set screw (A). Open swinging arm (B) wide. With the right hand move the cylinder (C) to the right as far as the belt will permit, without removing belt. This will release the cylinder shaft gear (D) which is to be taken off and discarded.

Remove cylinder shaft center set screw (G) and take out cylinder shaft center (F). These are also discarded.

Open box containing attachment and take out cylinder shaft center screw (T) and put it in place of screw (G).

Loosen the intermediate gear stud set screw (H), but do not take it out; then remove the intermediate gear stud (J) by pulling it out to the right as you face the machine, and replace with new longer stud (Y), leaving just the smallest space between the intermediate gear and the head of the stud (Y). This is to allow free movement and **is very important.** Tighten the intermediate gear stud set screw (H).

The machine is now ready for applying the Combination Attachment. This comes as illustrated below.

Attachment Parts for Edison Standard Phonograph Complete With Gear Guards.

On New Attachment
Q—Cylinder Shaft Gear Set Screw.
R—Cylinder Shaft Gear.
S—Intermediate Gear.
T—Cylinder Shaft Special Center Set Screw.
U—Attachment Casting Set Screw.
V—Change Gear Stud.
W—Gear Guard Screw.
X—Cylinder Shaft Gear Special Center.
Y—Intermediate Gear Special Stud.
Z—Slot in bracket casting.

Take out cylinder shaft gear (R) and cylinder shaft special center (X) from attachment (if they are not already detached) by loosening set screw (T), and cylinder shaft gear set screw (J), and put them in place of the old one taken out of machine, (D & F) being careful to have side of gear with large hole go on cylinder shaft.

Push cylinder (C) to the left as far as it will go, close swing arm (B) being careful that cylinder shaft enters the center in the swing arm (B).

After the swinging arm (B) is closed securely, the cylinder shaft gear (R) should be pressed firmly on to the cylinder shaft, as far as it will go, and its screw (I) tightened, being careful to have gear (B) run freely between its castings (N & O).

Relieve the swinging arm center set screw (K) and stop screw (M) so that swinging arm (B) will close. This is necessary because the new cylinder shaft gear (R) will throw the cylinder (C) a little farther to the right.

The attachment casting bracket assembled is next placed in position. The new cylinder shaft special center (X) being put through the hole in the cylinder shaft support casting (left) the hole formerly occupied by old center (F) so as to form a support for the end of the cylinder shaft. Tighten the cylinder shaft center set screw (T), after seeing that there is sufficient play between the new cylinder shaft gear (R) and the casting (D & F).

Loosen intermediate gear stud set screw (H) push intermediate gear stud (J) to the right flush with casting, then allow attachment to drop down so intermediate gear stud (J) can be pushed back in place through slot (Z) being careful to allow sufficient play so bearing will run easily, then tighten intermediate gear stud set screw (H).

Push the change gear stud (V) to the right as far as it will go. Slide the attachment to the right until cylinder shaft gear (R) and intermediate gear (S) line up perfectly. Tighten set screw (U).

IMPORTANT—To determine whether or not the adjustment is correct. The belt should be taken off the pulley and the cylinder spun around by hand. The cylinder shaft gear (R) should have a little play on either side between the castings.

2-95. These 1908 directions enabled the owner of an Edison "Standard" Phonograph Model "B" to convert his machine to play both two- and four-minute cylinder records, employing a $7.50 "Combination Attachment." The kit consisted of the parts shown on the upper right-hand page, plus a four-minute Model "H" reproducer. *(Value code: H)*

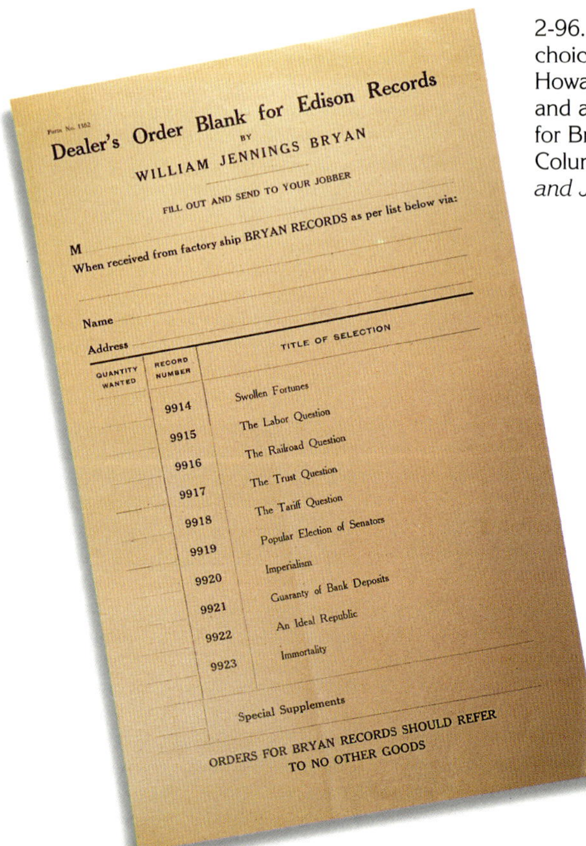

Form No. 1362

Dealer's Order Blank for Edison Records

BY

WILLIAM JENNINGS BRYAN

FILL OUT AND SEND TO YOUR JOBBER

M

When received from factory ship BRYAN RECORDS as per list below via:

Name

Address

QUANTITY WANTED	RECORD NUMBER	TITLE OF SELECTION
	9914	Swollen Fortunes
	9915	The Labor Question
	9916	The Railroad Question
	9917	The Trust Question
	9918	The Tariff Question
	9919	Popular Election of Senators
	9920	Imperialism
	9921	Guaranty of Bank Deposits
	9922	An Ideal Republic
	9923	Immortality
		Special Supplements

ORDERS FOR BRYAN RECORDS SHOULD REFER TO NO OTHER GOODS

2-96. Long before broadcast political conventions, voters were aided in their choice by recorded speeches of the Presidential candidates of 1908, William Howard Taft (Republican) and William Jennings Bryan (Democrat, in his third and again unsuccessful shot at the White House). Shown here is an order blank for Bryan's speeches on Edison two-minute cylinder records, 6" x 9 1/2". Columbia and Victor also issued records by both candidates. *Courtesy of Tom and Joanna Broome. (Value code: VR)*

2-97. The widespread recognition of recording artists generated by the talking machine was a factor of increasing importance as the industry evolved. Manuel Romain, a singer for Edison during the first decade of the twentieth century, here lent his name to a British publication, circa 1908: a 5" x 7 1/4" compilation of the words to ballads, which were Romain's specialty. *Courtesy of Michael F. Khanchalian. (Value code: G)*

2-98. A prize steer. Have you guessed the connection? *(Value code: H)*

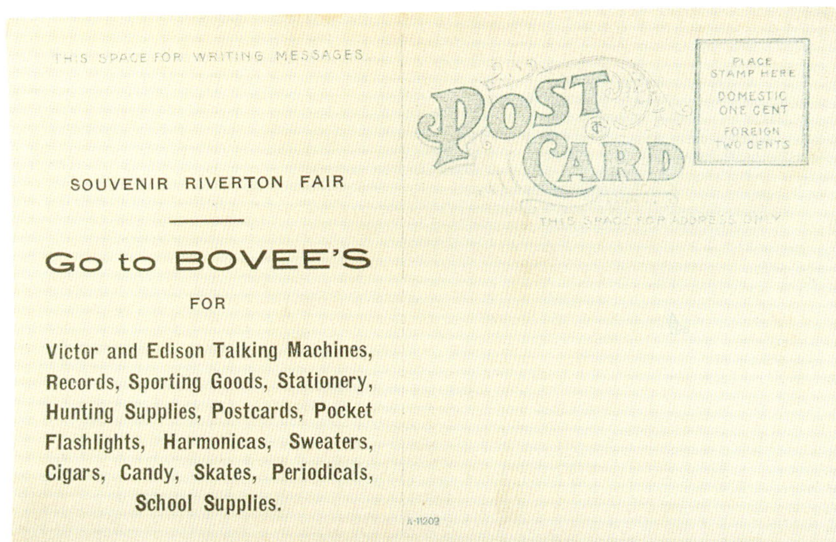

2-99. There is none. Just a way of getting your attention for the real message.

2-100. From *The Talking Machine News* (British), November 1907: "Symphonion" Disc Machines. This was a Swiss brand (akin to the "Symphonion" music boxes of the period) which was being distributed in Great Britain by the venerable house of John G. Murdoch, well-known for importing talking machines. One of the more interesting members of the line was the "Nickel King No. 3," top left, which was completely fashioned of metal except for a low wooden base. *Courtesy of Ernest Carl Allen. (Value code: bound volumes VR)*

2-101. From the French periodical *L'Illustration*, October 3, 1908. The American Presidential race was in full swing, with Democrat Bryan battling Republican Taft in a war of words – brought to you by your friendly talking machine record. At the "Automatic Vaudeville" in New York City, a "virtual" debate was enacted with coin-slot phonographs holding forth in the real voices of the candidates. Note the horn issuing from behind "Bryan," emitting either a catchy tune or the recorded words of the orator himself. Note also that Mr. Taft appeared to have lost 150 pounds! The French caption reads, "The phonographs recite, in payment of a *sou*, the latest speeches of the presidential candidates." *(Value code: H)*

The Voice of the Victor
Is the Voice of the Masters
The Trade Journal of the Victor Talking Machine Co., Camden, N. J.
Vol. I. MARCH, 1906. No. 1

All Mankind Now Awake to the Voice of the Victor

2-102. *The Voice of the Victor* was the house organ of the Victor Talking Machine Company. If you worked for Victor and somehow had failed to notice how wonderful the company was, *The Voice of the Victor* would set you straight. Here, from March 1906, is Volume 1, No. 1 (9" x 12"). *Courtesy of the Charles Hummel collections. (Value code: F)*

2-103. The Edison Phonograph Monthly was a well-known dealer publication of the National Phonograph Company (Edison's distributing organization), and after a 1910 reorganization, Thomas A. Edison, Incorporated. Over the years of its publication (1903-1917), the cover underwent design changes as shown in these issues from 1905, 1908, and 1909. *Courtesy of Tom and Joanna Broome. (Value code: F)*

2-104. Each of the major talking machine companies had its dealer magazine. Edison published "The Edison Phonograph Monthly." Victor's was called "The Voice of the Victor." Perhaps the least imaginatively titled publication was Columbia's "The Columbia Record." These latter are rarely seen. Shown are issues from November 1910 and May 1909. Both measure 6 3/4" x 10". *Courtesy of Ernest Carl Allen. (Value code: F)*

2-105. Almost a century ago, some lucky ticket holder won a Victor Talking Machine and records on February 27, 1909. Experienced collectors might assume that the prize was a Victor "II." Yet, just three days earlier, the price of the Victor "II" had been raised from $30.00 to $32.50. The price increase doubtless came into effect after the raffle began, and in any event it is likely the winner received a Victor "I" ($25.00) and a quantity of records, rather than a "II" (similar to that shown in a 1904 catalog). *Courtesy of the Johnson Victrola Museum/Delaware State Museums. (Value code: F)*

2-106. The instrument illustrated here is most commonly recognized as a "Standard Disc Talking Machine" Type "X2." More obscurely, it was available as an "Imperial," a "Harmony," or under the name of the firm which actually manufactured it: Columbia. The Hamilton Silver Company was offering its dealers a premium scheme, and though these initial retailer propositions were purposely vague (the humdrum details came later), one may infer an important point. "...and no installments to bother with" suggests an element that was basic to talking machine premium promotions: the customer paid *cash* to qualify. *Courtesy of Ernest Carl Allen. (Value code: G)*

2-107. The eye-catching cover of this brochure promoting the Edison "Fireside" Phonograph (introduced in Great Britain, August 1909) is a literal interpretation of the instrument's name. The table supporting the machine is a little closer to the roaring hearth than most collectors would care to entertain. *Courtesy of George Glastris. (Value code: F)*

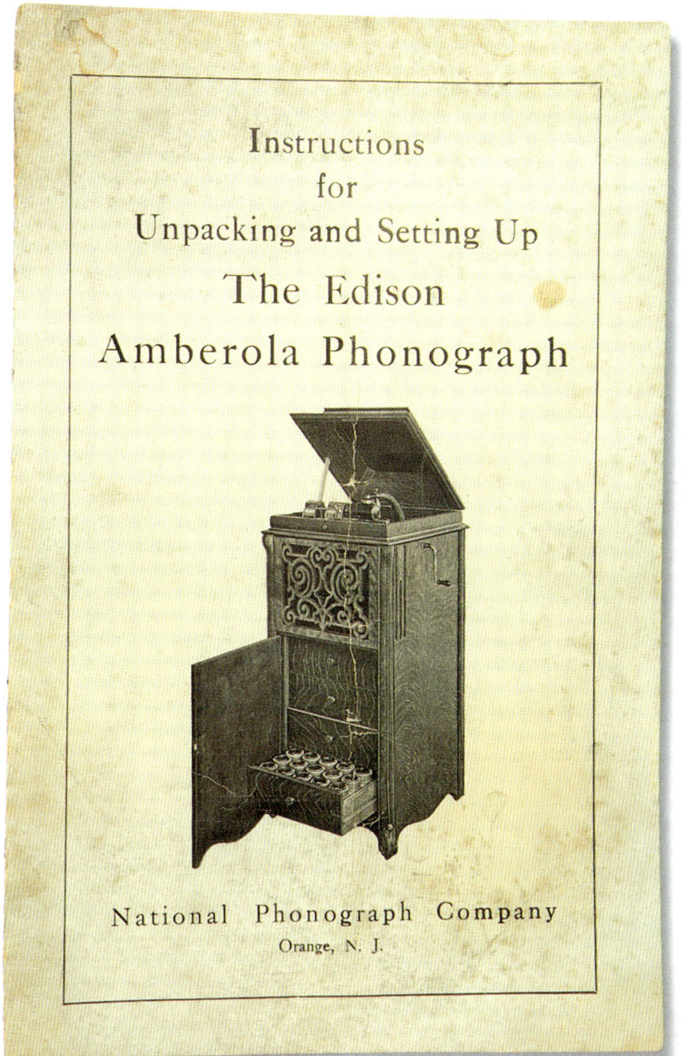

2-108. The 1906 introduction of the concealed-horn "Victrola" by the Victor Talking Machine Company had been the ambuscade of the decade for other manufacturers in the phonograph industry. Columbia, Edison, Pathé, Hawthorne & Sheble all realized that they would have to enter the internal-horn talking machine field in order to compete. Edison's first contribution to the genre was the "Amberola" "1(A)" of 1909. Nobody at West Orange was acknowledging the familiar use of the suffix "ola"… as in "Victrola." Shown here is a rare instruction booklet for setup and operation, 5 1/2" x 8 1/2". *Courtesy of the Babcock House Museum, John H. Perschbacher. (Value code: F)*

2-109. Talking machine manufacturers would be labeled in today's pop psychological terms as "control freaks." Edison held the monocratic notion that his dealers existed for the good of the National Phonograph Company, and they were lucky to get the work. Therefore, dealers and jobbers were rigidly regulated by all manner of company decrees. Dealers didn't want to be stuck with out-of-date or discontinued Edison cylinders, and squawked until the front office issued its "Continuous Exchange Proposition" in May 1909, whereby old records could be returned against new orders. The scolds at West Orange, however, couldn't avoid the opportunity to complain: from the April 1910 *Edison Phonograph Monthly*, "If we could only discover some method of impressing upon our Dealers the absolute necessity of following the provisions of the Continuous Exchange Proposition *literally* what a relief it would be!" What a world, what a world… *Courtesy of Tom and Joanna Broome. (Value code: F)*

2-110. A 6" x 7 1/2" plaque depicting the "His Master's Voice" trademark, pressed in the same material from which disc records were made (shellac). From Deutshen Grammophon Gesellschaft, the German branch of the worldwide network of Gramophone companies. *Courtesy of the Charles Hummel collections. (Value code: VR)*

2-111. Many manufacturers, especially those in possession of important patents, were obsessed with the ultimate control of their products in the hands of their dealers and even after purchased by consumers. Antitrust and price-related legislation would eventually curtail company claims of sovereignty over goods that had left the factory. Back in the days of patriarchal meddling, however, Edison was at the front of the pack. Even the common practice of "bird-dogging," whereby a former store customer received an "incentive" for encouraging another retail sale, required a ponderously-worded document. The dealer certified that he had given, as an honorarium for services rendered, six "Special Amberol" cylinder records (which could not under any circumstances be sold!) to a bona fide, registered Edison Phonograph owner. And heaven help anyone who fibbed! Circa 1910. *Courtesy of the Charles Hummel collections. (Value code: G)*

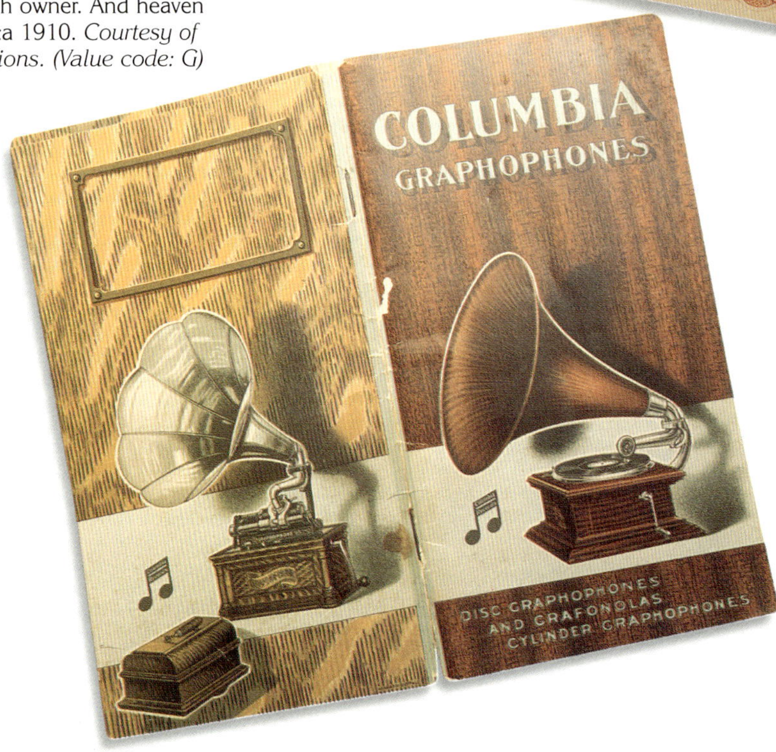

2-112. Columbia – alone among the "Big Three" dominant talking machine companies – offered both cylinder and disc merchandise during the first decade of the twentieth century. By 1910, the end was in sight for Columbia's cylinder line. That year's Columbia Graphophone catalog featured a posh mahogany background and a matching Disc Graphophone on the front cover. However, lowlier oak and a cylinder machine (a "BET") had been relegated to the back cover. *Courtesy of Ernest Carl Allen. (Value code: E)*

2-113. This undated 6" x 8" photograph shows the local baseball team sponsored by the St. Louis Talking Machine Company – obviously a dealer in Victor goods! Circa 1910. *Courtesy of Ernest Carl Allen. (Value code: F)*

2-114. A suitably mystical cover adorns an Edison record supplement of Masonic Hymns on Amberol cylinders. *(Value code: H)*

2-115. As an alternative to wooden record storage boxes, Victor inaugurated cardstock record boxes, and finally, bound albums. This 18" x 23 1/2" poster would have been displayed in a dealer's showroom in 1909/1910 to promote the new product. *Courtesy of Tom and Sandi McCarthy. (Value code: F)*

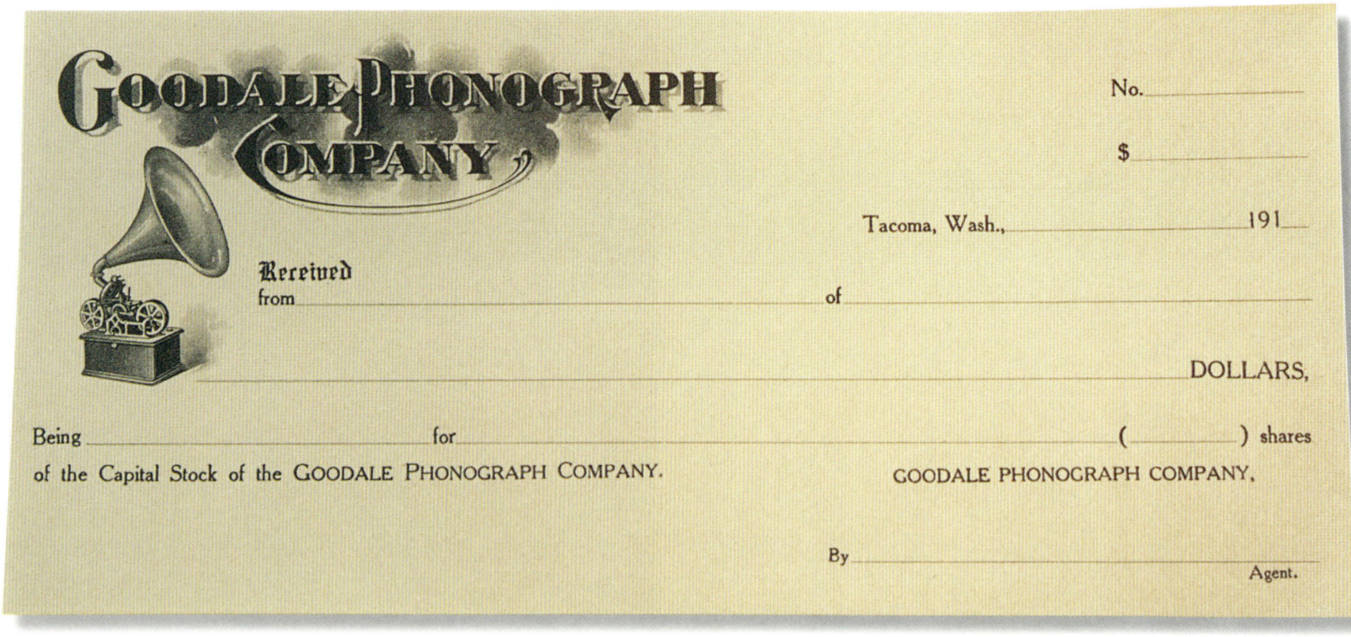

GOODALE PHONOGRAPH COMPANY

No.________

$________

Tacoma, Wash.,________191__

Received from________of________

________DOLLARS,

Being________for________(________) shares

of the Capital Stock of the GOODALE PHONOGRAPH COMPANY.

GOODALE PHONOGRAPH COMPANY,

By________

Agent.

2-116. Franklin C. Goodale, of Tacoma, Washington, designed a most unusual talking machine in 1908, which employed celluloid film as its recording medium. Mr. Goodale applied for a patent which was granted December 28, 1909 (No. 944,608), and he supposedly capitalized his company at $5,000,000. This receipt was to be issued to a Goodale stock purchaser. A similar method of reproduction, employing celluloid film stock, would be employed in the (German) Hiller Talking Clock (1911). Could Goodale's invention have been applied to motion picture sound synchronization as well as home entertainment? One wonders how a movie projector might have been incorporated into the complicated design. Moreover, it is doubtful that the venture advanced beyond the prototype stage. One example of Goodale's phonograph is known to exist. *Courtesy of Tom and Sandi McCarthy. (Value code: VR)*

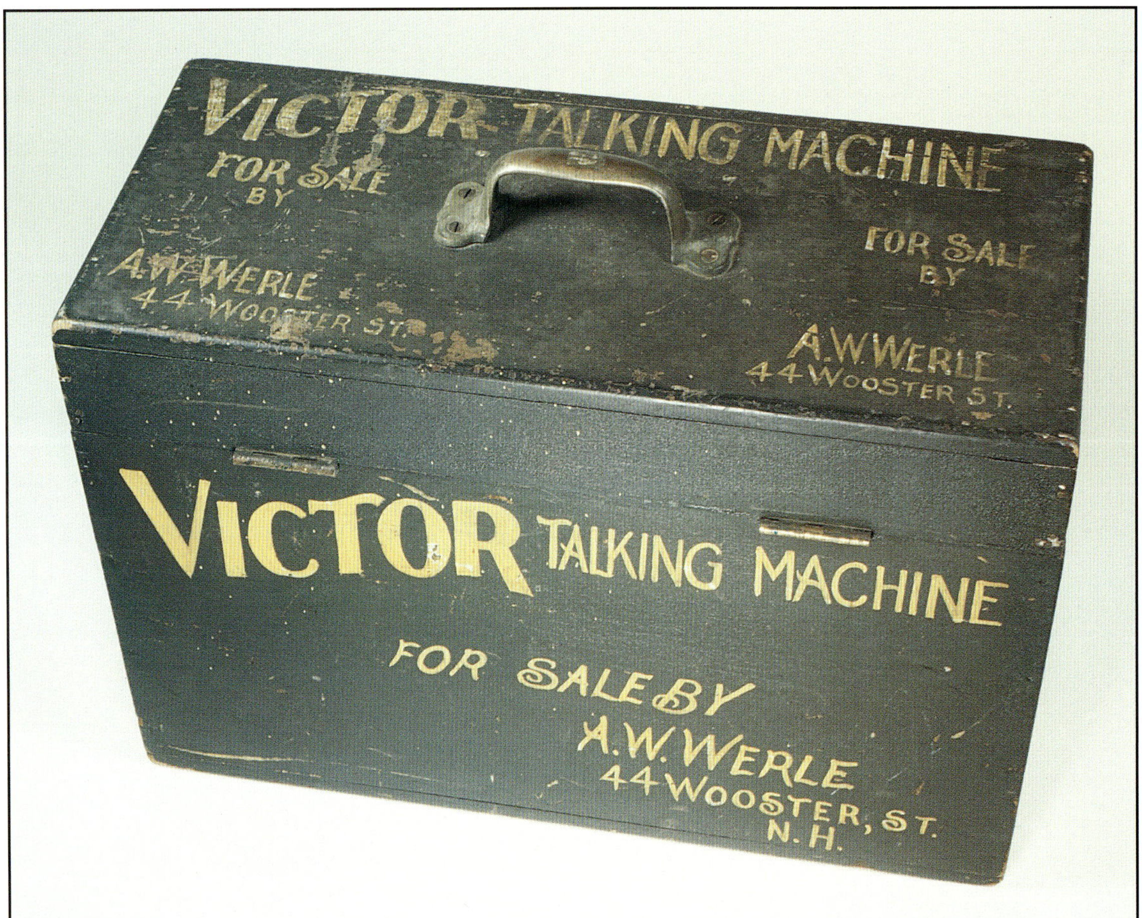

2-117. The repairman working for Victor dealer A.W. Werle used this toolbox to keep customers' machines in good order. Measures 19" wide x 12 1/2" high x 8 3/4" deep. *Courtesy of the Johnson Victrola Museum/ Delaware State Museums. (Value code: VR)*

Chapter Three:

1911-1920

3-1. An unusually large (6 1/2" x 13 3/4") and attention-getting trade card for the Reading Phonograph Company. Circa 1910. *Courtesy of Tom and Sandi McCarthy. (Value code: E)*

3-2. This 4" x 9" blotter for Lamb's shows the 1910 table model Victrola "XI." *Courtesy of Tom and Sandi McCarthy. (Value code: F)*

3-3. This smoking set features a small music box beneath the gramophone, and a "Nipper" effigy that wags its head as the music plays. Probably manufactured in Switzerland, the originally monochromatic dog was painted mottled white – possibly to resemble a family pet. He doesn't look much like Nipper! Base measures 8 1/2" x 12". *Courtesy of Tom and Sandi McCarthy. (Value code: C)*

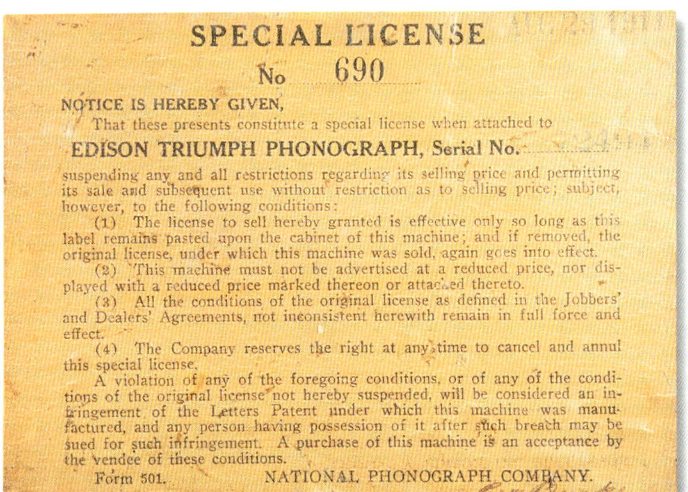

SPECIAL LICENSE

No 690

NOTICE IS HEREBY GIVEN,

That these presents constitute a special license when attached to

EDISON TRIUMPH PHONOGRAPH, Serial No.

suspending any and all restrictions regarding its selling price and permitting its sale and subsequent use without restriction as to selling price; subject, however, to the following conditions:

(1) The license to sell hereby granted is effective only so long as this label remains pasted upon the cabinet of this machine; and if removed, the original license, under which this machine was sold, again goes into effect.

(2) This machine must not be advertised at a reduced price, nor displayed with a reduced price marked thereon or attached thereto.

(3) All the conditions of the original license as defined in the Jobbers' and Dealers' Agreements, not inconsistent herewith remain in full force and effect.

(4) The Company reserves the right at any time to cancel and annul this special license.

A violation of any of the foregoing conditions, or of any of the conditions of the original license not hereby suspended, will be considered an infringement of the Letters Patent under which this machine was manufactured, and any person having possession of it after such breach may be sued for such infringement. A purchase of this machine is an acceptance by the vendee of these conditions.

Form 501. NATIONAL PHONOGRAPH COMPANY.

3-4. The book of rules and regulations at Thomas A. Edison, Incorporated was fuller than a picnic basket at a Sunday Social. Retail prices could not be reduced, secondhand Phonographs could not be sold, and so forth. Occasionally, however, if one caught the company on a good day, a retailer could obtain a "Special License" (stamped, dated and signed, of course). This particular document was glued by a merchant to the bottom of an Edison "Triumph" Phonograph Model "B," allowing him to reduce the retail price for clearance. By 1911, the machine had remained in unsold inventory for approximately four years. *Courtesy of Terry Paul Baer.*

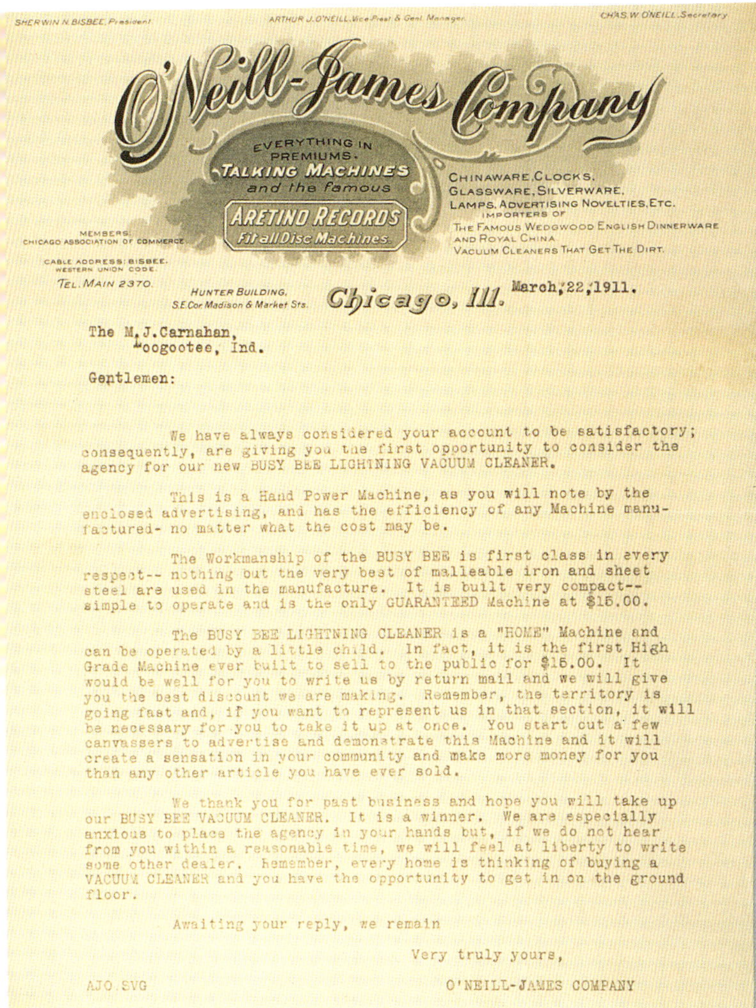

SHERWIN N. BISBEE, President — ARTHUR J. O'NEILL, Vice Pres't & Genl. Manager — CHAS. W. O'NEILL, Secretary

O'Neill-James Company

EVERYTHING IN PREMIUMS.
TALKING MACHINES and the famous ARETINO RECORDS Fit all Disc Machines.

CHINAWARE, CLOCKS, GLASSWARE, SILVERWARE, LAMPS, ADVERTISING NOVELTIES, ETC.
IMPORTERS OF THE FAMOUS WEDGWOOD ENGLISH DINNERWARE AND ROYAL CHINA.
VACUUM CLEANERS THAT GET THE DIRT.

MEMBERS: CHICAGO ASSOCIATION OF COMMERCE.
CABLE ADDRESS: BISBEE. WESTERN UNION CODE.
TEL. MAIN 2370.
HUNTER BUILDING, S.E. Cor. Madison & Market Sts.

Chicago, Ill. March, 22, 1911.

The M. J. Carnahan,
Loogootee, Ind.

Gentlemen:

We have always considered your account to be satisfactory; consequently, are giving you the first opportunity to consider the agency for our new BUSY BEE LIGHTNING VACUUM CLEANER.

This is a Hand Power Machine, as you will note by the enclosed advertising, and has the efficiency of any Machine manufactured- no matter what the cost may be.

The Workmanship of the BUSY BEE is first class in every respect-- nothing but the very best of malleable iron and sheet steel are used in the manufacture. It is built very compact-- simple to operate and is the only GUARANTEED Machine at $15.00.

The BUSY BEE LIGHTNING CLEANER is a "HOME" Machine and can be operated by a little child. In fact, it is the first High Grade Machine ever built to sell to the public for $15.00. It would be well for you to write us by return mail and we will give you the best discount we are making. Remember, the territory is going fast and, if you want to represent us in that section, it will be necessary for you to take it up at once. You start out a few canvassers to advertise and demonstrate this Machine and it will create a sensation in your community and make more money for you than any other article you have ever sold.

We thank you for past business and hope you will take up our BUSY BEE VACUUM CLEANER. It is a winner. We are especially anxious to place the agency in your hands but, if we do not hear from you within a reasonable time, we will feel at liberty to write some other dealer. Remember, every home is thinking of buying a VACUUM CLEANER and you have the opportunity to get in on the ground floor.

Awaiting your reply, we remain

Very truly yours,

AJO.SVG

O'NEILL-JAMES COMPANY

3-5. No one could accuse Arthur J. O'Neill of being a pill. He was a live wire from the word go! O'Neill, the son of Irish immigrants, hailed from Wisconsin, and began his business career as a commercial traveler. A turn-of-the-twentieth-century Chicago newspaper entry: "Arthur J. O'Neill was a caller at the office of Chicago Dairy Produce Tuesday. Mr. O'Neill was at one time perhaps the best known solicitor [salesman] traveling the west in the interests of the Chicago [dairy product] market, but has for some years now been connected with the Robert Johns' Co. dealers in China, this city." Robert Johns was soon to play an important role in the Standard Talking Machine Company of Chicago, which was a direct outgrowth of the chinaware business. O'Neill followed the lead of his former employer and got into talking machines, too. O'Neill's brands were "Busy Bee," "Yankee Prince," and "Aretino." Yet, this letterhead from 1911 makes it clear that china still figured high on O'Neill's list of merchandise. The latest addition was the "Busy Bee Lightning Vacuum Cleaner," borrowing the name from the talking machines that by this point were no longer being sold (though O'Neill continued to offer the "Aretino" brand). Note the reference to "a Hand Power Machine" – the intrepid housewife (or maid for the fortunate) obtained a surfeit of aerobic activity pumping the gadget while trying to negotiate the floors and rugs. *Courtesy of Ernest Carl Allen. (Value code: G)*

3-6. A descriptive handbill offering a Standard Talking Machine as a premium for $25.00 worth of clothing from the Outlet Clothing House of Salem, Massachusetts. A period photograph showing these flyers in actual use may be found on page 55 of *Antique Phonograph Advertising: An Illustrated History*, Fabrizio & Paul, Schiffer, 2002. Handbill measures 7" x 10 1/2". *(Value code: F)*

3-7. The Keen-O-Phone Company, brainchild of Morris Keen, talking machine inventor and entrepreneur, had a factory on Orthodox Street, in an industrial area of Philadelphia. The firm's salesrooms were located at 227 S. Broad Street, Philadelphia, not far from the Victor Talking Machine Company's store. In this 5" x 8" 1911 catalog, Keen offered his initial line of disc instruments, among the most imaginatively designed and superbly constructed talking machines to be produced anywhere, any time. On the cover, El Dorado: a Model "XXX" in quarter-sawn oak. No more than a handful of oak Keen-O-Phones could have been produced, with the exception of the firm's less-expensive external horn models. This catalog offered various wood finishes, but it's likely oak was withdrawn in favor of mahogany-finish only, which was consistent with contemporary tastes in home furnishing. *Courtesy of Ernest Carl Allen. (Value code: VR)*

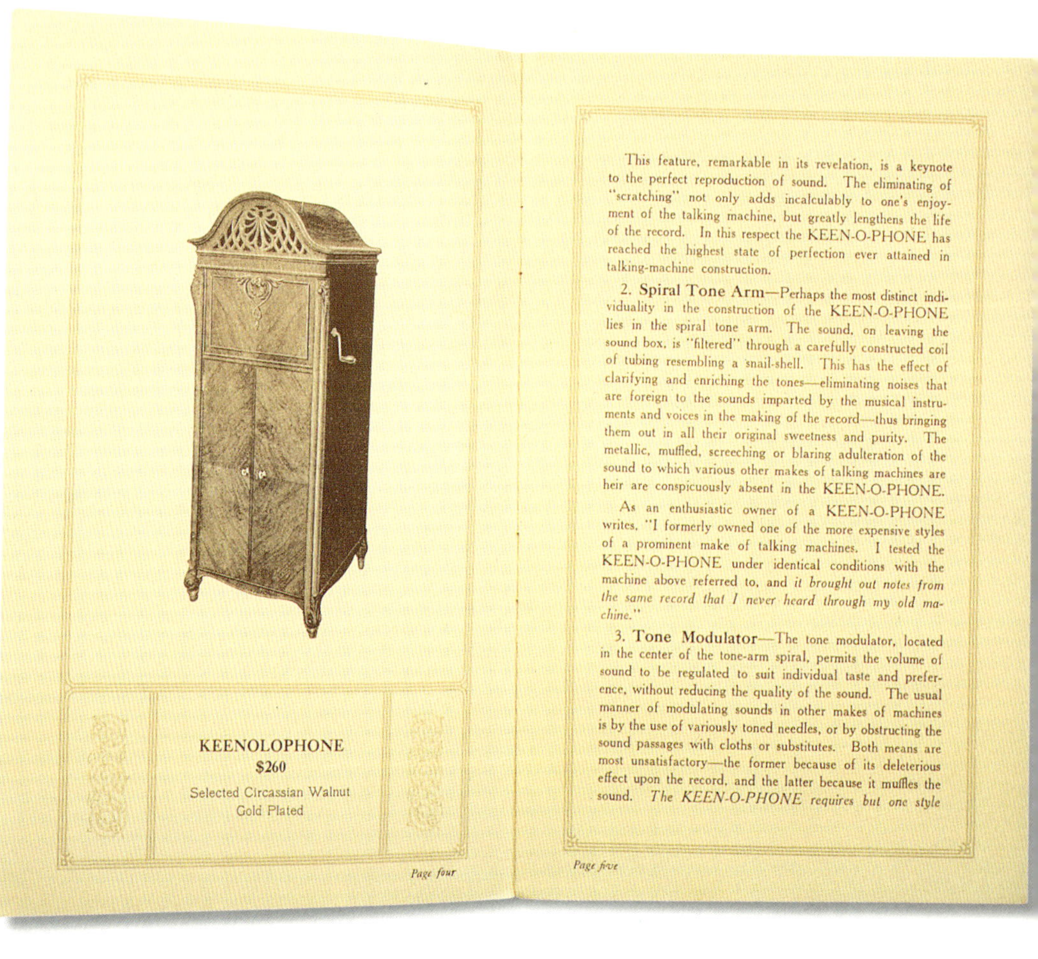

KEENOLOPHONE

$260

Selected Circassian Walnut
Gold Plated

Page four

This feature, remarkable in its revelation, is a keynote to the perfect reproduction of sound. The eliminating of "scratching" not only adds incalculably to one's enjoyment of the talking machine, but greatly lengthens the life of the record. In this respect the KEEN-O-PHONE has reached the highest state of perfection ever attained in talking-machine construction.

2. **Spiral Tone Arm**—Perhaps the most distinct individuality in the construction of the KEEN-O-PHONE lies in the spiral tone arm. The sound, on leaving the sound box, is "filtered" through a carefully constructed coil of tubing resembling a snail-shell. This has the effect of clarifying and enriching the tones—eliminating noises that are foreign to the sounds imparted by the musical instruments and voices in the making of the record—thus bringing them out in all their original sweetness and purity. The metallic, muffled, screeching or blaring adulteration of the sound to which various other makes of talking machines are heir are conspicuously absent in the KEEN-O-PHONE.

As an enthusiastic owner of a KEEN-O-PHONE writes, "I formerly owned one of the more expensive styles of a prominent make of talking machines. I tested the KEEN-O-PHONE under identical conditions with the machine above referred to, and *it brought out notes from the same record that I never heard through my old machine.*"

3. **Tone Modulator**—The tone modulator, located in the center of the tone-arm spiral, permits the volume of sound to be regulated to suit individual taste and preference, without reducing the quality of the sound. The usual manner of modulating sounds in other makes of machines is by the use of variously toned needles, or by obstructing the sound passages with cloths or substitutes. Both means are most unsatisfactory—the former because of its deleterious effect upon the record, and the latter because it muffles the sound. *The KEEN-O-PHONE requires but one style*

Page five

3-8. Nirvana: a Circassian walnut Model "XXX." This instrument debuted at $225.00 in mahogany (rapidly reduced to $210.00 and then $200.00), but an additional $25.00 would transform it into the exalted object shown here. That was a serious bargain, considering that Victor charged $50.00 for the same privilege. It's unknown if any Keen-O-Phone in Circassian walnut was produced or has survived. Mechanic and inventor Thomas Kraemer, formerly associated with Philadelphia's Hawthorne & Sheble, patented a mechanical feed motor that was used in Keen-O-Phones made during 1911 (the "Berliner patent," which the motor was designed to circumvent, officially expired in early 1912). Coupled with Morris Keen's peculiar "Spiral Tone Arm" (Patent No. 907,814, December 29, 1908), the Keen-O-Phone evidenced a high degree of technical sophistication. *Courtesy of Ernest Carl Allen.*

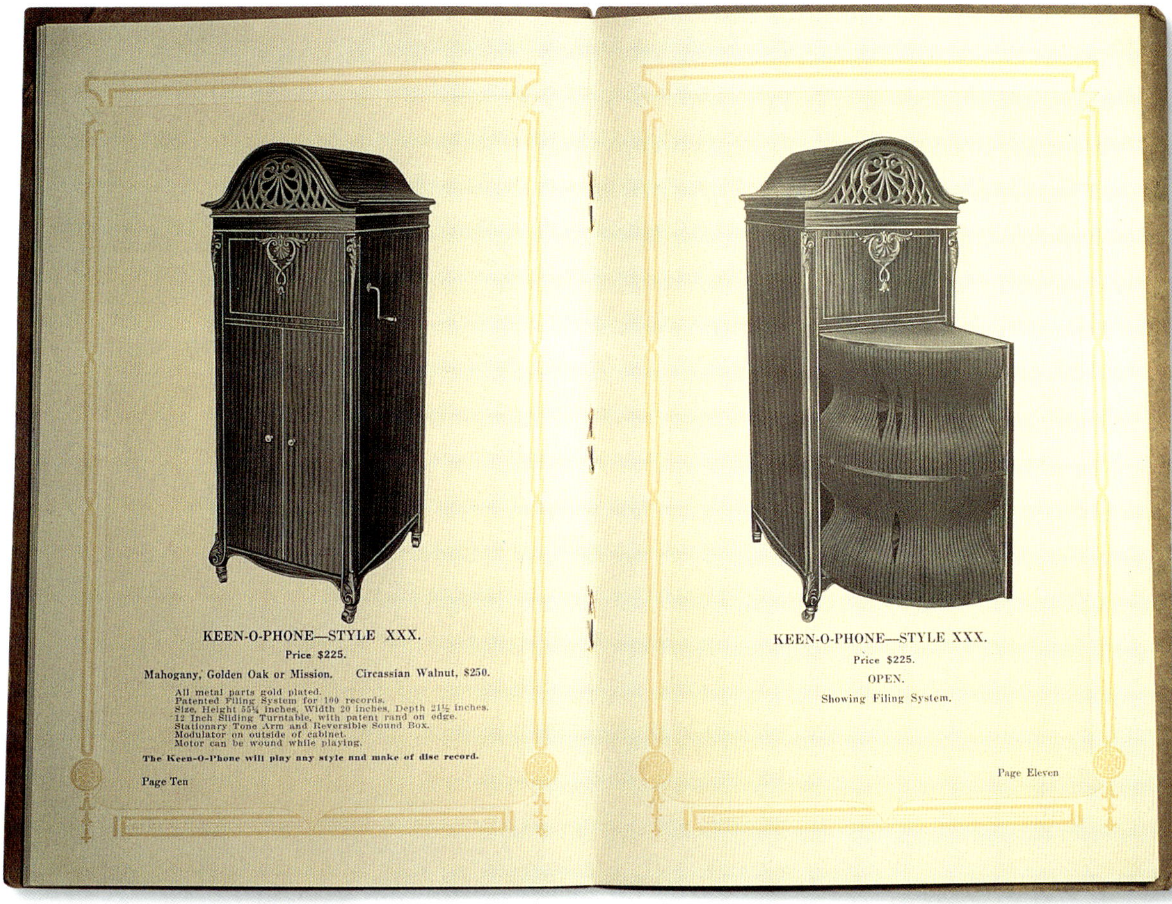

3-9. Model "XXX" with its voluptuous-looking record storage. Keen-O-Phone cabinets were made by the Pooley Furniture Company of Philadelphia, maker of some of the finest cabinetry in the talking machine industry. *Courtesy of Ernest Carl Allen.*

3-10. Models "XXV" and "XX". The cupboard of the style on the left was customarily fitted with a Pooley automatic selecting record storage system. The two models shown here switched number designations in an advertisement in the March 1913 trade journal *Talking Machine World*. We wondered if this was intentional, but it appears to have been a typographical error. However, if we consider the fact that the cabinet lid/horn of one would fit the other, and were interchanged according to the whim of the customer – there is ample reason to confuse the two. *Courtesy of Ernest Carl Allen.*

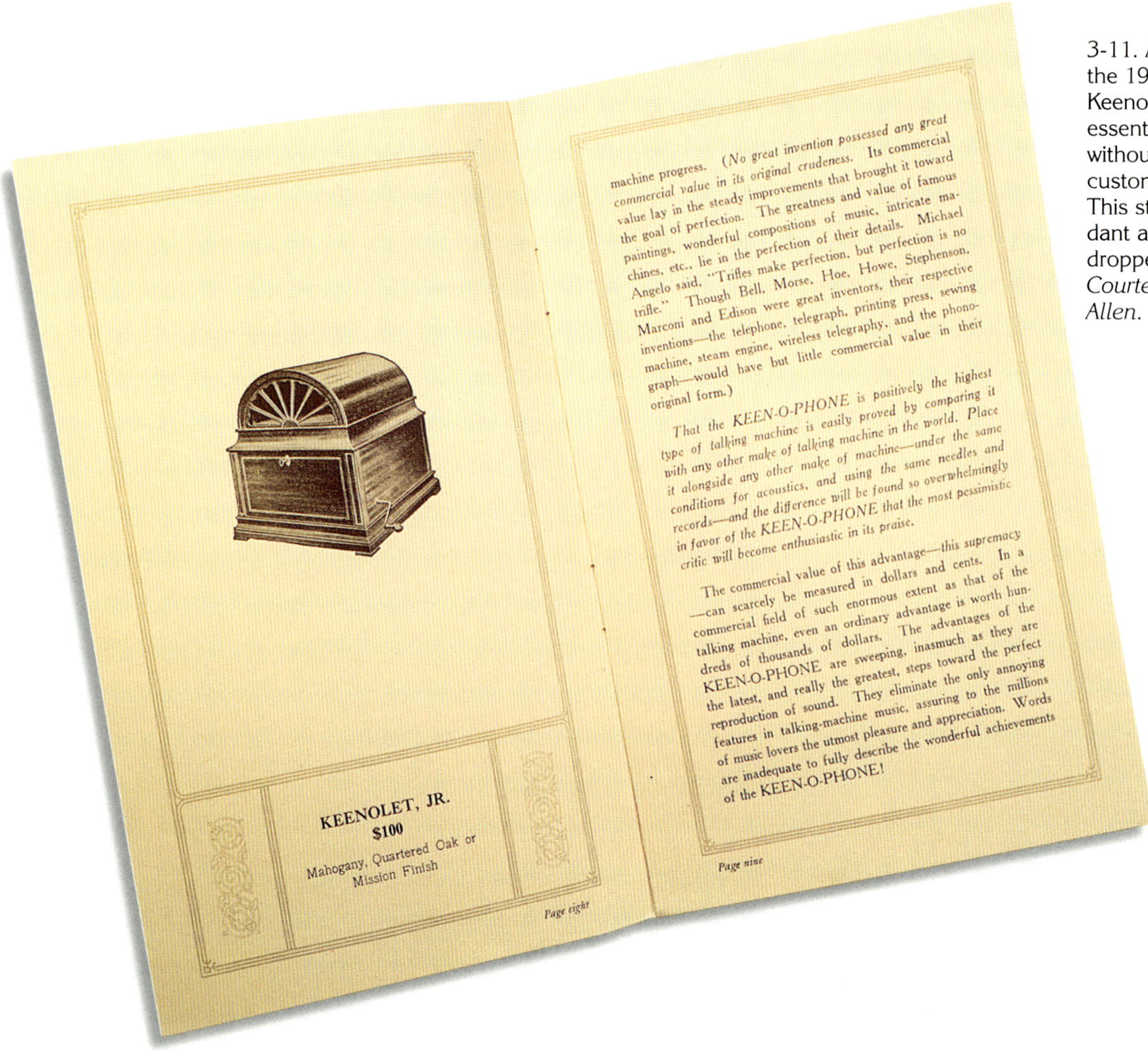
KEENOLET, JR.
$100
Mahogany, Quartered Oak or Mission Finish

Page eight

machine progress. (*No great invention possessed any great commercial value in its original crudeness.* Its commercial value lay in the steady improvements that brought it toward the goal of perfection. The greatness and value of famous paintings, wonderful compositions of music, intricate machines, etc., lie in the perfection of their details. Michael Angelo said, "Trifles make perfection, but perfection is no trifle." Though Bell, Morse, Hoe, Howe, Stephenson, Marconi and Edison were great inventors, their respective inventions—the telephone, telegraph, printing press, sewing machine, steam engine, wireless telegraphy, and the phonograph—would have but little commercial value in their original form.)

That the KEEN-O-PHONE is positively the highest type of talking machine is easily proved by comparing it with any other make of talking machine in the world. Place it alongside any other make of machine—under the same conditions for acoustics, and using the same needles and records—and the difference will be found so overwhelmingly in favor of the KEEN-O-PHONE that the most pessimistic critic will become enthusiastic in its praise.

The commercial value of this advantage—*this supremacy*—can scarcely be measured in dollars and cents. In a commercial field of such enormous extent as that of the talking machine, even an ordinary advantage is worth hundreds of thousands of dollars. The advantages of the KEEN-O-PHONE are sweeping, inasmuch as they are the latest, and really the greatest, steps toward the perfect reproduction of sound. They eliminate the only annoying features in talking-machine music, assuring to the millions of music lovers the utmost pleasure and appreciation. Words are inadequate to fully describe the wonderful achievements of the KEEN-O-PHONE!

Page nine

3-11. Another mystery of the 1911 catalog: the Keenolet, Jr. It was, essentially, a Model "XX" without the high legs that customarily supported it. This style proved redundant and was quickly dropped from the line. *Courtesy of Ernest Carl Allen.*

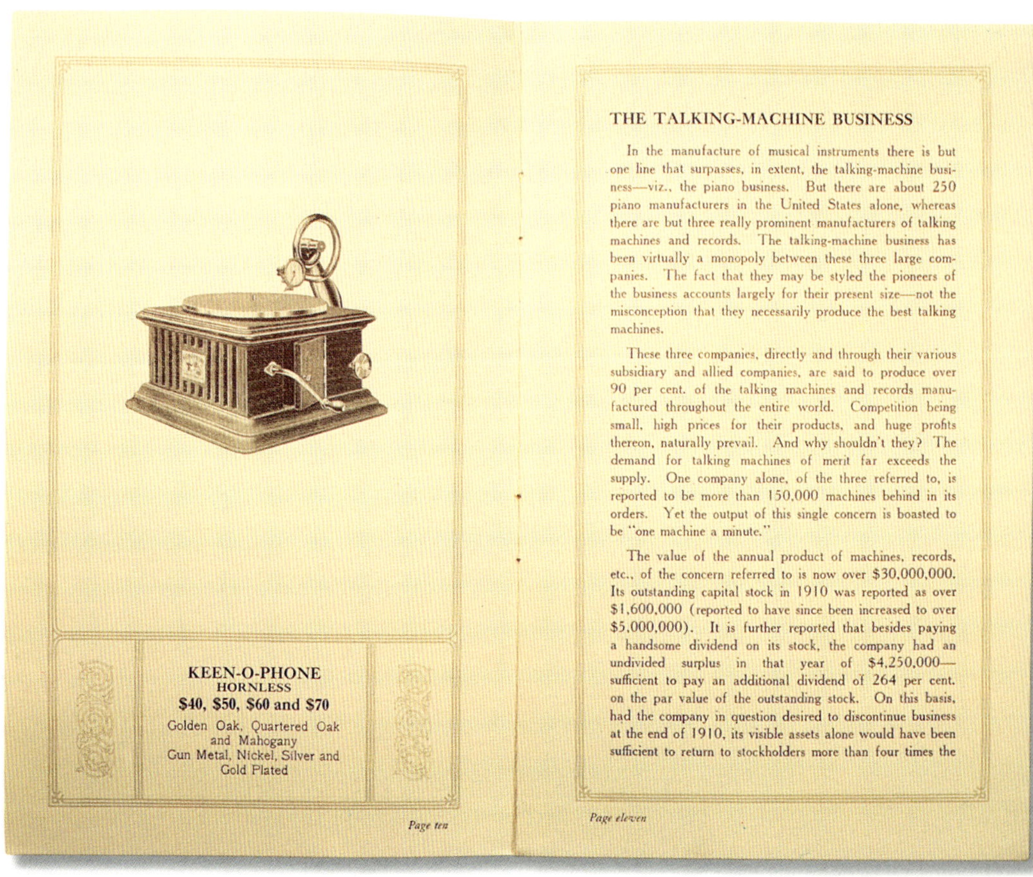
KEEN-O-PHONE
HORNLESS
$40, $50, $60 and $70
Golden Oak, Quartered Oak and Mahogany
Gun Metal, Nickel, Silver and Gold Plated

Page ten

THE TALKING-MACHINE BUSINESS

In the manufacture of musical instruments there is but one line that surpasses, in extent, the talking-machine business—viz., the piano business. But there are about 250 piano manufacturers in the United States alone, whereas there are but three really prominent manufacturers of talking machines and records. The talking-machine business has been virtually a monopoly between these three large companies. The fact that they may be styled the pioneers of the business accounts largely for their present size—not the misconception that they necessarily produce the best talking machines.

These three companies, directly and through their various subsidiary and allied companies, are said to produce over 90 per cent. of the talking machines and records manufactured throughout the entire world. Competition being small, high prices for their products, and huge profits thereon, naturally prevail. And why shouldn't they? The demand for talking machines of merit far exceeds the supply. One company alone, of the three referred to, is reported to be more than 150,000 machines behind in its orders. Yet the output of this single concern is boasted to be "one machine a minute."

The value of the annual product of machines, records, etc., of the concern referred to is now over $30,000,000. Its outstanding capital stock in 1910 was reported as over $1,600,000 (reported to have since been increased to over $5,000,000). It is further reported that besides paying a handsome dividend on its stock, the company had an undivided surplus in that year of $4,250,000—sufficient to pay an additional dividend of 264 per cent. on the par value of the outstanding stock. On this basis, had the company in question desired to discontinue business at the end of 1910, its visible assets alone would have been sufficient to return to stockholders more than four times the

Page eleven

3-12. Curiouser and curiouser. This instrument emitted its sound through a conventional grille-enclosed front panel, but what could be the purpose of the "hatch" on the side? A hypothesis. The Keen-O-Phone motor was very large and it is unlikely that there was enough room in this cabinet (the same cabinet as the external horn Model "IV") for a dedicated internal horn structure. Most likely, the works of the motor "swam" in the middle of an open chamber that sufficed for an internal horn. The early Victrola "IV" was designed the same way. Amplification of the sound would be partially interrupted by the intrusion of the motor, so perhaps the "hatch" was there to allow some of the diffused sound to escape? Notice the considerable list of finishes for cabinet and hardware. Were any silver-plated examples made? Who would care to keep up the maintenance of easy-tarnishing silver? *Courtesy of Ernest Carl Allen.*

3-13. The customers be hanged! Morris Keen was determined to give them silver-plated hardware whether they wanted it not. The Model "II" provided no other option. *Courtesy of Ernest Carl Allen.*

SOUND BLENDING

MODULATOR

KEEN-O-PHONE II
$60
Golden Quartered Oak
Silver Plated

Page fourteen

any other manufacturing business of equal extent and possibilities.

It is not the purpose of the issuers of this booklet to dilate at length upon statistics pertaining to the profits that may be made in the business. "Comparisons (of figures) are odious" and can be shaped into almost any form to suit the fancy, flatter the imagination and convince the most skeptical mind. But if—

Merit of and demand for product,
Field for business,
Producing and selling ability,
Mechanical ingenuity, and
Capable management

stand for anything—and they are unquestionably the "five senses" of the manufacturing business—then the KEEN-O-PHONE COMPANY faces one of the greatest opportunities of the day.

A well-appointed and well-regulated factory of modest size—one capable of turning out, say, 100 average-priced machines per day—should produce a clear profit of over $300,000 annually. (The three large talking-machine companies previously referred to turn out about 5500 machines per day.) Add to this the profits made from the manufacture of records, and the figures mount to almost incredible size.

THE KEEN-O-PHONE COMPANY

The KEEN-O-PHONE is the invention of Morris Keen, an experienced scientific engineer and talented musician. Mr. Keen's experiments, which have resulted in the wonderful machine that bears his name, date back more than ten years.

Page fifteen

3-14. Keen-O-Phones were of superior quality, but pricey. Roughly comparable to a Victor "III," the Model "I" cost ten dollars more. Kraemer's mechanical-feed motor was complicated and expensive to produce. *Courtesy of Ernest Carl Allen.*

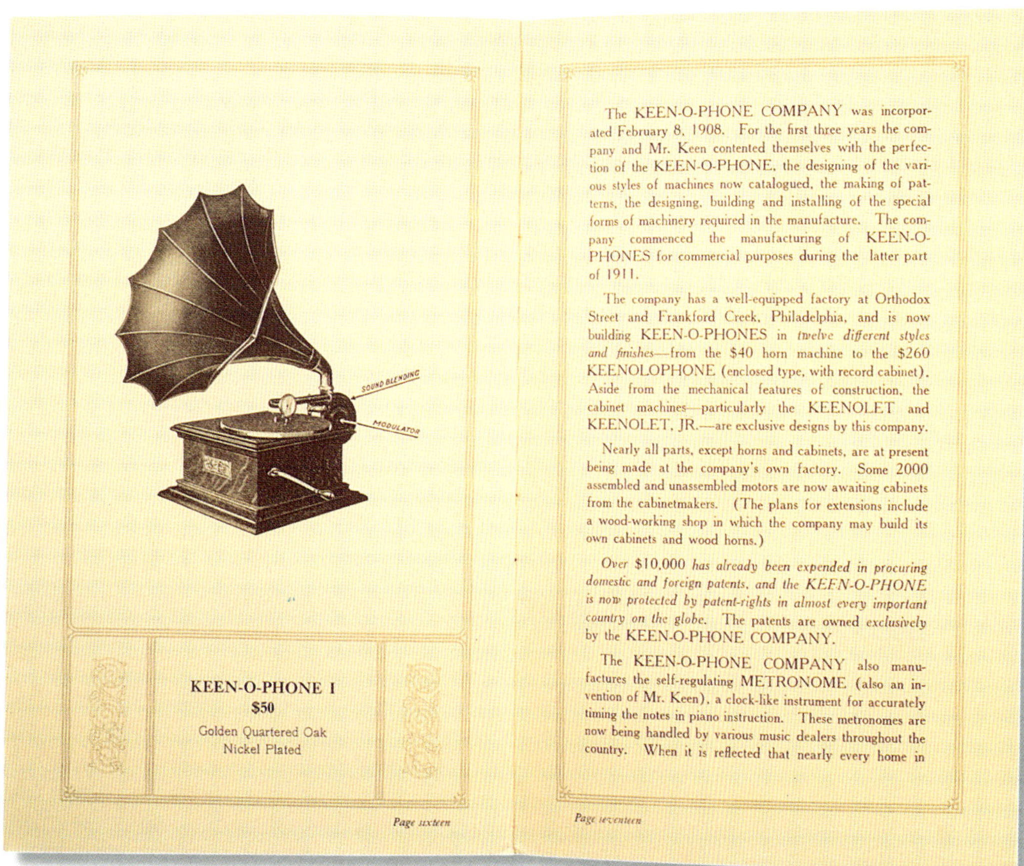
SOUND BLENDING

MODULATOR

KEEN-O-PHONE I
$50
Golden Quartered Oak
Nickel Plated

Page sixteen

The KEEN-O-PHONE COMPANY was incorporated February 8, 1908. For the first three years the company and Mr. Keen contented themselves with the perfection of the KEEN-O-PHONE, the designing of the various styles of machines now catalogued, the making of patterns, the designing, building and installing of the special forms of machinery required in the manufacture. The company commenced the manufacturing of KEEN-O-PHONES for commercial purposes during the latter part of 1911.

The company has a well-equipped factory at Orthodox Street and Frankford Creek, Philadelphia, and is now building KEEN-O-PHONES in *twelve different styles and finishes*—from the $40 horn machine to the $260 KEENOLOPHONE (enclosed type, with record cabinet). Aside from the mechanical features of construction, the cabinet machines—particularly the KEENOLET and KEENOLET, JR.—are exclusive designs by this company.

Nearly all parts, except horns and cabinets, are at present being made at the company's own factory. Some 2000 assembled and unassembled motors are now awaiting cabinets from the cabinetmakers. (The plans for extensions include a wood-working shop in which the company may build its own cabinets and wood horns.)

Over $10,000 *has already been expended in procuring domestic and foreign patents, and the* KEEN-O-PHONE *is now protected by patent-rights in almost every important country on the globe.* The patents are owned *exclusively* by the KEEN-O-PHONE COMPANY.

The KEEN-O-PHONE COMPANY also manufactures the self-regulating METRONOME (also an invention of Mr. Keen), a clock-like instrument for accurately timing the notes in piano instruction. These metronomes are now being handled by various music dealers throughout the country. When it is reflected that nearly every home in

Page seventeen

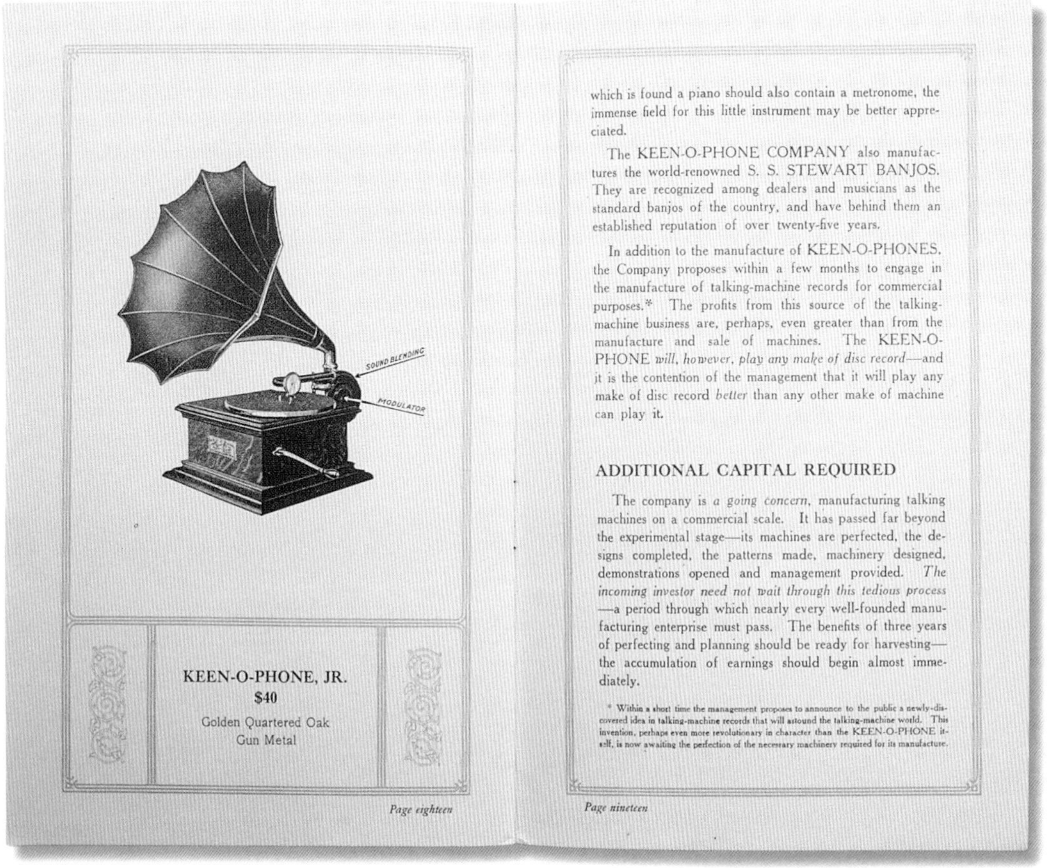

SOUND BLENDING

MODULATOR

KEEN-O-PHONE, JR.
$40
Golden Quartered Oak
Gun Metal

Page eighteen

which is found a piano should also contain a metronome, the immense field for this little instrument may be better appreciated.

The KEEN-O-PHONE COMPANY also manufactures the world-renowned S. S. STEWART BANJOS. They are recognized among dealers and musicians as the standard banjos of the country, and have behind them an established reputation of over twenty-five years.

In addition to the manufacture of KEEN-O-PHONES, the Company proposes within a few months to engage in the manufacture of talking-machine records for commercial purposes.* The profits from this source of the talking-machine business are, perhaps, even greater than from the manufacture and sale of machines. The KEEN-O-PHONE *will, however, play any make of disc record*—and it is the contention of the management that it will play any make of disc record *better* than any other make of machine can play it.

ADDITIONAL CAPITAL REQUIRED

The company is *a going concern*, manufacturing talking machines on a commercial scale. It has passed far beyond the experimental stage—its machines are perfected, the designs completed, the patterns made, machinery designed, demonstrations opened and management provided. *The incoming investor need not wait through this tedious process*—a period through which nearly every well-founded manufacturing enterprise must pass. The benefits of three years of perfecting and planning should be ready for harvesting—the accumulation of earnings should begin almost immediately.

* Within a short time the management proposes to announce to the public a newly-discovered idea in talking-machine records that will astound the talking-machine world. This invention, perhaps even more revolutionary in character than the KEEN-O-PHONE itself, is now awaiting the perfection of the necessary machinery required for its manufacture.

Page nineteen

3-15. It must have become evident that the catalog was too extensive for a new company to support (note Keen's plea for investors on the right hand page). All table model Keen-O-Phones were discontinued following the expiration of the "Berliner Patent," when the firm eliminated the costly mechanical feed motor. *Courtesy of Ernest Carl Allen.*

Oxford Concertolas
High Grade Hornless Disc Talking Machines

These talking machines mark the culmination of inventive genius in the science of sound reproduction. They are the last word in talking machines, the final result of costly experimental work extending over years, the crowning triumph of mechanical art in the reproduction of musical sound.

The sound amplifying compartment, which in these machines replaces the well known external horn, is contained wholly within the cabinet, and the sound waves are conducted from the reproducer to this concealed amplifying compartment or sound chamber by tapering tone arm, which affords a perfectly smooth and unobstructed passage for the sound waves. The wonderfully sensitive sound analyzing reproducer, in combination with the scientifically calculated taper of the tone arm, and the acoustically perfect sound amplifying compartment, results in the most marvelously true to life reproduction of the human voice and the most perfect reproduction of the tones of all musical instruments.

No. 1 Oxford Concertola $9.90
No. 20V5091
Polished oak cabinet, 11½ inches square at base, 6½ inches high. Turntable, 10 inches in diameter. Single spring noiseless motor. Plays one 10-inch record with one winding and may be wound while running. Extra sensitive sound analyzing reproducer. Tapering tone arm.
Shipping weight, 27 pounds.

No. 2A Oxford Concertola $17.75
No. 20V5094
Polished oak cabinet, 14¾ inches square at base, 8 inches high. Turntable, 10 inches in diameter. Double spring noiseless motor. Plays three 10-inch records with one winding and may be wound while running. Extra sensitive sound analyzing reproducer. Tapering tone arm.
Shipping weight, 41 pounds.

Style IX Keen-O-Phone Talking Machine
Manufacturer's Price, $85.00
Our Price, $38.00
No. 20V5098
Beautifully finished mahogany cabinet, standing 40 inches high and 18 inches wide by 19 inches deep.
Turntable, 12 inches in diameter.
Extra heavy double spring motor. Plays three 10-inch or two 12-inch records with one winding. Can be wound while playing.
Adjustable tone arm and reversible sound box.
Will play any standard record, including Oxford, Columbia, Victor, Edison and Pathe.
The lower compartment of the cabinet is equipped with the Pooley filing device, which accommodates twenty-six 12-inch and twenty-four 10-inch records, or fifty 12-inch records. Any record can be instantly located and always be returned to its proper place.
The Style IX Keen-O-Phone is a beautifully finished machine of exceptional merit in every way. It is sold by the manufacturer at $85.00. Through a very fortunate purchase, we secured a large number of these machines, which we are able to offer at $38.00, less than one-half the regular price.
Shipping weight, 112 lbs.

No. 5 Oxford Concertola $28.95
No. 20V5099
Beautifully finished quarter sawed oak cabinet with cover. 17½ inches square at base, 13 inches high. Turntable, 11¾ inches in diameter. Triple spring noiseless motor. Plays four 10-inch records with one winding. Can be wound while running. New model, extra sensitive sound analyzing reproducer. Tapering tone arm.
Shipping weight, 67 pounds.

SEARS, ROEBUCK AND CO., CHICAGO, ILLINOIS. 729

3

3-16. The Keen-O-Phone Company ceased operation in 1914, when problems with distribution and sales failed to support the excellence of the products and the considerable expense of creating them. It appears that the Pooley Furniture Company was a collateral victim, declaring bankruptcy (though later reorganized) because of bad credit extended to Keen-O-Phone. The stock of unsold instruments was hustled out at profoundly reduced prices. Here, from a 1915 Sears, Roebuck & Company catalog, the $85.00 Model "IX" was selling for a mere $38.00. *Courtesy of Ernest Carl Allen. (Value code: H)*

3-17. The Keen-O-Phone story is not quite over, folks, so please stay in your seats. The McKinley Music Company, well known as a publisher of sheet music, offered "Perfect Perpetual Satisfaction" in the object of the "McKinley Talking Machine." At $15.00 for the lesser model, absolute bliss was cheaper to obtain than the cost of Zen Buddhism classes. However, we draw your attention to the instruments illustrated here, which were actually manufactured by the Rex Talking Machine Corporation and were the direct descendents of the Keen-O-Phone. Early in 1914, the Keen-O-Phone Company had halted operations. Yet, among the firm's assets were a fully equipped factory on Orthodox Street in Philadelphia and a downtown vertical cut recording studio (the only "sapphire ball" record facility in the United States, though Pathé was in the process of setting up in Brooklyn). The Rex Talking Machine Corporation was formed in Philadelphia with the intention of taking over Keen-O-Phone's factory and record business. According to an article in the February 1914 *Talking Machine World*: "...Thomas Kraemer is superintendent of the factory, while the recording end of the business is under the charge of Frederick W. Hager and Charles L. Hibbard." Rex planned to profit by its predecessor's mistakes: costly, superbly constructed instruments were out; affordable, serviceably built machines were in. This philosophy may be witnessed in the Rexes shown in this 8 3/8" x 10 5/8" flyer from 1915: they lacked even the amenity of a decorative grille over the opening of the internal horn. "McKinley Velvet Records" were Rex vertically-recorded discs. A flyer, 8 3/8" x 10 5/8". *Courtesy of Ernest Carl Allen. (Value code: G)*

3-18. This oversized folding postcard was mailed to Victor dealers in 1911 to advertise the full-sized show card for October. The 22" x 29" show card could be ordered from jobbers/distributors for 50 cents each, or directly from Victor for 70 cents. *Courtesy of the Johnson Victrola Museum/Delaware State Museums. (Value code: G)*

3-19. For Christmas 1911, Victor dealers could purchase the "Ready Made Window Display No. 26." Flanking the fireplace are a Victrola "XVI" and "XIV." A Victor "II" and "III" guard the corners of the hearth, while Nipper listens to a Victor "Junior." The new Victrola "IV" (with early metal grille) is also on display, and is the only model to feature a display card with its price. *Courtesy of the Johnson Victrola Museum/Delaware State Museums. (Value code: G)*

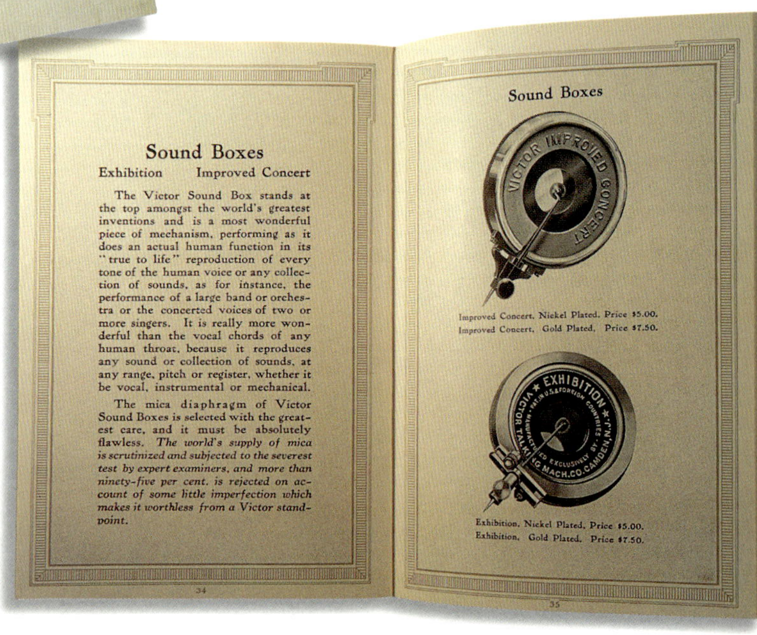

3-20. A catalog page depicting the two Victor soundboxes available in 1912. The "Improved Concert" (reviving the designation from a decade earlier) has been something of a puzzlement to latter day collectors. It was listed only in 1912, and therefore lacked the documentation enjoyed by, for instance, the long-available "Exhibition." The "Improved Concert" was available in nickel or gold finish, and was suitable for any instrument in Victor's 1912 line. *Courtesy of Ernest Carl Allen. (Value code: F, full catalog)*

3-21. The Penn Phonograph Company, Incorporated of Philadelphia has left a legacy of imaginative advertising. Shown are the November and December 1911 portions of a calendar featuring photographs of operatic recording artists Geraldine Farrar and J. Gerville-Reache. Each measures 5" x 7 1/4". *Courtesy of Ernest Carl Allen. (Value code: G)*

3-22. This stylized phonograph was the background for a 1912 calendar. 17 1/2" wide x 11" high. *Courtesy of Tom and Sandi McCarthy. (Value code: F)*

3-23. Judging by the line, "Pin your faith to….," this 4 1/8" diameter metal tray must have been intended to hold pins for a housewife, seamstress or tailor. The significant-sounding "Edison Phonograph Hall" was a retail merchant as well as a distributor (jobber). *Courtesy of Tom and Sandi McCarthy. (Value code: E)*

3-24. These three items represent examples of advertising novelties employed by phonograph and record retailers. The orange ticket for a "Victrola Concert" is dated February 29, 1912. At bottom is a glass paperweight from a Boston retailer. To the left is a hand mirror from Lamb's Music Store of Greenfield, Massachusetts. It's interesting to note that for 20 years people had been attending talking machine exhibitions, and the novelty could still be exploited in 1912. *Courtesy of Tom and Sandi McCarthy. (Value code: H, E, F)*

3-25. From a Victor catalog of 1912, the hand-decorated "Vernis Martin" gold-leaf cabinet finish, which elevated the price of this Victrola "XVI" to a towering $400.00! *Courtesy of Ron Seavers. (Value code: E, full catalog)*

3-26. In 1912, Victor offered records of the United States Presidential candidates. This red, white, and blue poster was sure to draw attention to a store window. Dissention within the Republican Party led Roosevelt to run on his own Progressive (or "Bull Moose") Party ticket. Wilson won the 1912 election, while Roosevelt garnered more votes than the incumbent Taft. Thus, all three candidates eventually served as President – with two terms for both Roosevelt (1901-1909) and Wilson (1913- 1921). Poster measures 28" x 43 1/2". *Courtesy of the Johnson Victrola Museum/ Delaware State Museums. (Value code: VR)*

3-27. A brass dealer plaque, "His Master's Voice," British, circa 1912. Measures 11" x 15 1/2". *Courtesy of the Charles Hummel collections. (Value code: D)*

3-28. A web of "Gramophone-related" companies existed in Europe, kissing cousins of Victor, the foremost of which were in Great Britain, France, and Germany. Here is evidence of other, more peripheral outlets for Victor/Gramophone merchandise: desktop glass advertising for the Czech branch of the Gramophone Company, Limited. A paperweight, 3 7/8" x 5 1/8". *Courtesy of Wilfried Sator. (Value code: F)*

3-29. Great Britain's Gramophone Company, Limited was closely associated with Victor. The British firm imported vast numbers of Victor parts (soundboxes, horns, brackets, arms, turntables and motors, among others) to be used in assembling its proprietary line of Gramophones. From a catalog, circa 1909, two instruments that bear the "Monarch" name, first applied to a Victor machine in 1901. On the left, the "Monarch Senior" retained the elaborate, oak cabinet style of the Victor "Monarch Special" (introduced 1902) and the early Style "D" (1903). Four years after this heavily ornamented look faded from the American catalog, Britons and Europeans were still endorsing it. On the right, a "Monarch," the cabinet of which was a unique creation of the Gramophone Company, Limited. *Courtesy of Ernest Carl Allen. (Value code: F, full catalog)*

3-30. To emphasize the merit of its technology, the Pathé company of France presented the purchasers of certain instruments circa 1912 with this "certificate of identity" stating that this particular Pathé reproducer (soundbox) had been tested at the factory and found to be of the utmost quality. To leave no doubt of the seriousness of this process, both the document and the envelope in which it was placed were marked with the serial number of the reproducer ("Concert" No. 211805). *Courtesy of Scott and Diana Colgrove.*

3-31. The *Victor Book of the Opera* was first issued in 1912 and continued to be published in revised editions for the next 60 years. It was ostensibly a guide for the opera-lover, more specifically, the opera record-lover, but the book's true value was to underscore Victor's (and its successors') supremacy in the field of recorded serious music. *(Value code: H, most editions)*

3-32. An embossed hanger to help dealers advertise the *Victor Book of the Opera*. Measures 11" x 15". *Courtesy of the Johnson Victrola Museum/Delaware State Museums. (Value code: G)*

3-33. In the late nineteenth/early twentieth centuries, cheap cigars could be supplied to retailers in boxes with any of a number of decorative labels attached. A customer might swear that the "Record" cigar he bought on Main Street was superior to another brand sold over on Central Avenue – but both might be the very same cigar! This sample cigar box label featured the visage of an aging Edison, and measures 4 3/4" x 7". *Courtesy of Tom and Sandi McCarthy. (Value code: F)*

3-34. An advertising postcard for Harthorn's music store, Long Beach, California. The period 1911-1913 enjoyed a Queen Anne craze in the United States – talking machines on high, shapely legs briefly dominated the market. Seen here are Grafonolas ("Baby Regent" at the left, "Princess" at the right) and a Victrola (the Victrola "X" between the other two), which exemplified the style. (The fact that the Victrola "X" appears to be missing its usual low shelf could be due to retouching by a photographer unfamiliar with the subject matter.) *Courtesy of the Charles Hummel collections. (Value code: G)*

3-35. The name "Victrola" became ubiquitous, and began to be applied to any type of talking machine. Although the evolution of its proprietary name must have worried Victor, the universality of the term continually emphasized Victor's success in the industry. 1914 sheet music, 10 1/8" x 13 3/4". *(Value code: F)*

3-36. The image of the rank hayseed thrilled by some new-fangled contraption (in this instance, a coin-operated cylinder talking machine), has recurred in American humor for centuries. A joke recently circulated on the Internet relied on the cyber-audience's willingness to believe that there was still a bumpkin dumb enough to mistake the purpose of an elevator. Shown here, 1913 sheet music, measuring 10 5/8" x 14". *(Value code: F)*

3-37. Self-aggrandizement came easily to Columbia, and for good reason – the Columbia Phonograph Company *had* created the talking machine industry, a role to which the firm frequently laid claim. In the early years of the 1890s, when Thomas Edison was tripping over his legal and business entanglements, Emile Berliner was trying to find a venue for his formative Gramophone and Eldridge Johnson was still polishing his bike, Columbia, propelled by the sheer will of Edward D. Easton, was churning out cylinder records. This effort kept Columbia alive until it could begin to make dramatic innovations in talking machine design, bringing the instruments into the hands of the public by introducing ever less expensive models. The "Talking Machine of the People" was achieved in 1899, with the release of the Type "Q" Graphophone, a high-quality, reliable machine for only $5.00. From 1913, a publicity brochure depicting a Columbia Grafonola "Grand" ($500.00), 6 5/8" x 9". *Courtesy of Ernest Carl Allen. (Value code: F)*

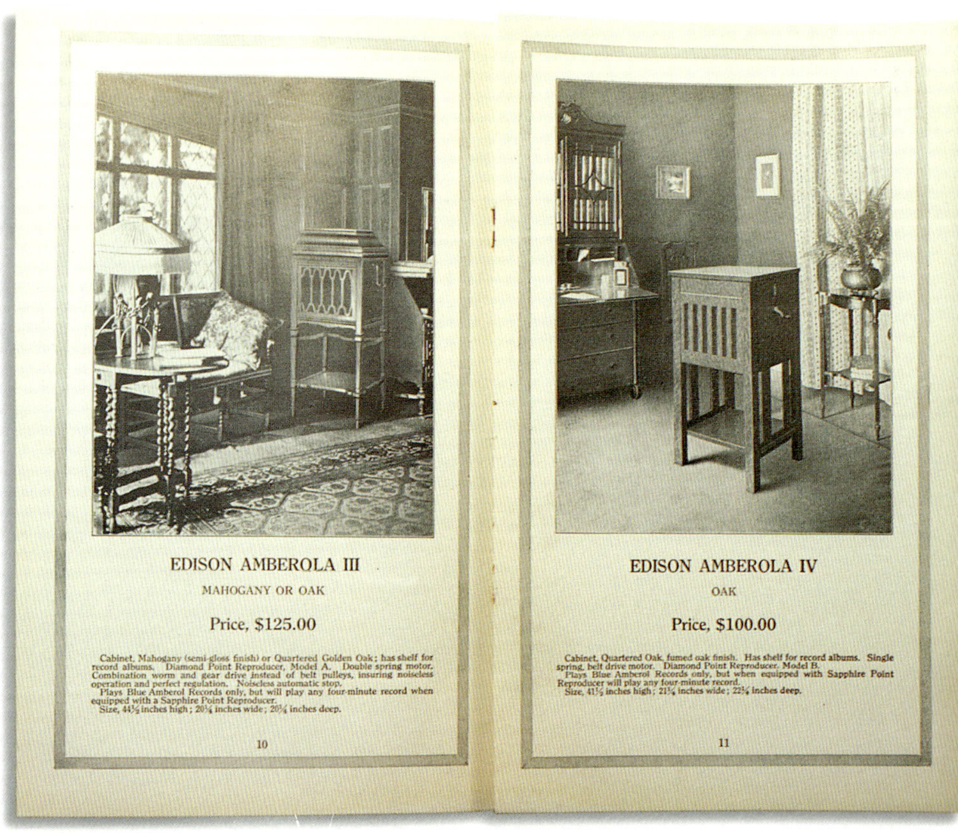

3-38. An Edison Phonograph catalog from 1913, with pages showing the Amberolas "III" and "IV." A greater disparity between two respective mechanisms could not be found. The "III" employed the technically sophisticated works of the state-of-the-art Edison "Opera" Phonograph. The "IV" awkwardly mated the upper mechanism of an obsolescent Edison "Standard" Phonograph with the motor of an equally out-of-date "Home" Phonograph. As proof that the public is not entirely without sense, the "III" sold well and the "IV" hardly at all. *Courtesy of the Tom and Joanna Broome. (Value code: F)*

3-39. The Duplex Phonograph affords us proof of life after death. The company was sued in patent court and lost its case to Victor, but a remnant of the firm managed to "slip through the cracks" and return to life. As a functioning manufacturer, Duplex was kaput following a 1909 decision in United States Circuit Court. However, the Duplex parts and inventory ended up in the hands of Mr. C.Q. DeFrance, who continued to do business "repairing" and reconstituting the unusual instruments, with their formidable double horns. Shown here, from 1913, a "close out" sale of whatever was left in stock. (Note, though, that this was not *quite* the end of the Duplex Phonograph Company. See page 87, *Discovering Antique Phonographs* by Fabrizio & Paul, Schiffer, 2000.) *Courtesy Ernest Carl Allen. (Value code: G)*

3-40. Duplex entered business in 1906 and made quite a go of its uniquely-designed machine in the direct-mail market. The oft-repeated claim that factory-to-consumer could save you a bundle rang true to Duplex buyers. It was no doubt the firm's obvious success that brought the bloodhounds out of Victor's legal department, hot on the trail of an infringer of Victor privilege (most notably, the so-called "Berliner Patent," No. 534,543, which formed the cornerstone of Victor's patent battlements). *Courtesy of Gregg Cline. (Value code: E)*

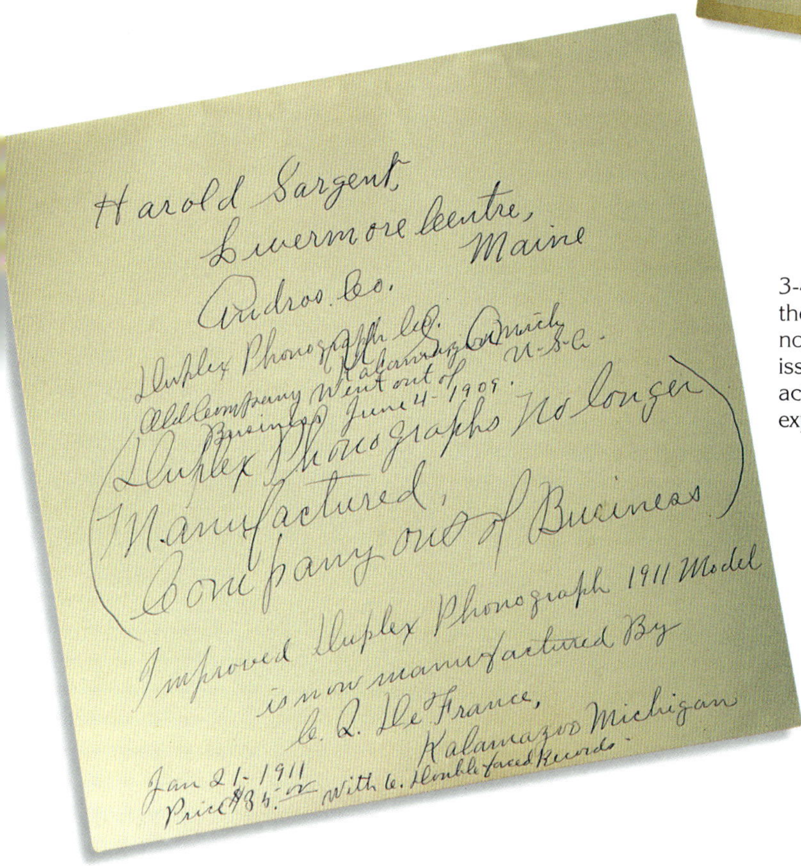

3-41. The original owner of the Duplex catalog shown in the previous illustration made remarkably accurate notes about the fate of the company. We might take issue with the suggestion that Mr. DeFrance was actually manufacturing any new parts, despite his expressed intention to do so. *Courtesy of Gregg Cline.*

3-42. A brochure of testimonials supporting the quality and superiority of the Duplex Phonograph.

> Mr. Wm. Thompson, mayor of Kalamazoo, and general manager of the Kalamazoo Stove Co, the largest stove manufacturing concern in the country, is the owner of a "Duplex." *Gentlemen: Prejudice is the greatest barrier to our happiness that I know of… I became really prejudiced against any machine that made "so-called" music… But my! how time and experience does change us: we have used the "Duplex" Phonograph long enough to consider it "one of the family," and its tone and clearness makes real melodious music…"*

Courtesy of the Michael F. Khanchalian. (Value code: F)

3-43. The Victor "XXV" (later *Victrola* "XXV") was aggressively advertised for use exclusively in schools. The machine evolved through eight variations, each looking very much like its predecessors. This August 1914 brochure shows the second variation. Measures 3 7/8" x 8 7/8". *Courtesy of Ernest Carl Allen. (Value code: G)*

3-44. It is sometimes overlooked that Columbia made persistent efforts to place its equipment in classrooms. This "Graded Selection of Educational Records" (5 1/8" x 9") and a flyer for the "Columbia School Outfit" (3 1/2" x 6 1/4") both date from 1920. *Courtesy of Ernest Carl Allen. (Value code: G)*

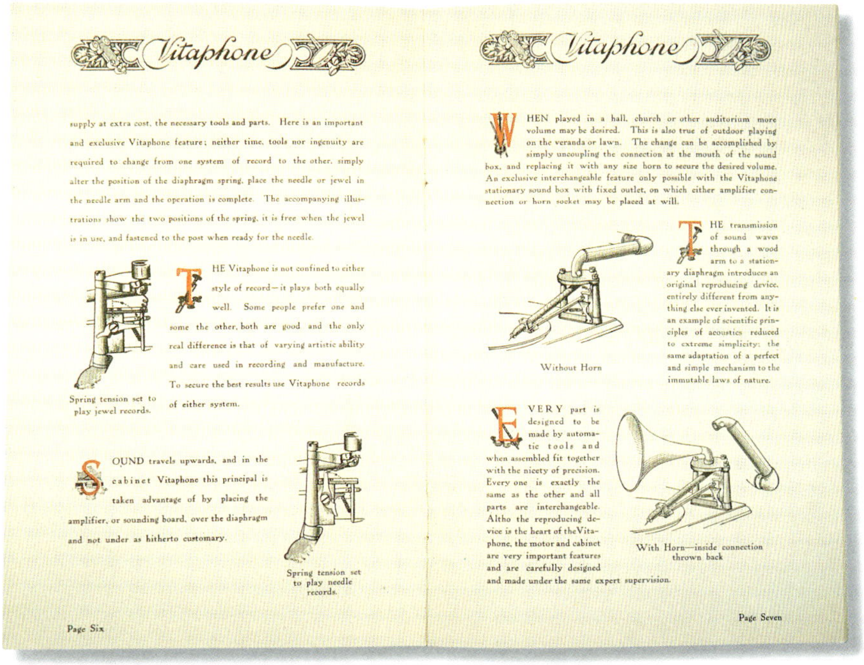

Vitaphone

supply at extra cost, the necessary tools and parts. Here is an important and exclusive Vitaphone feature; neither time, tools nor ingenuity are required to change from one system of record to the other, simply alter the position of the diaphragm spring, place the needle or jewel in the needle arm and the operation is complete. The accompanying illustrations show the two positions of the spring, it is free when the jewel is in use, and fastened to the post when ready for the needle.

THE Vitaphone is not confined to either style of record—it plays both equally well. Some people prefer one and some the other, both are good and the only real difference is that of varying artistic ability and care used in recording and manufacture. To secure the best results use Vitaphone records of either system.

Spring tension set to play jewel records.

SOUND travels upwards, and in the cabinet Vitaphone this principal is taken advantage of by placing the amplifier, or sounding board, over the diaphragm and not under as hitherto customary.

Spring tension set to play needle records.

Page Six

Vitaphone

WHEN played in a hall, church or other auditorium more volume may be desired. This is also true of outdoor playing on the veranda or lawn. The change can be accomplished by simply uncoupling the connection at the mouth of the sound box, and replacing it with any size horn to secure the desired volume. An exclusive interchangeable feature only possible with the Vitaphone stationary sound box with fixed outlet, on which either amplifier connection or horn socket may be placed at will.

THE transmission of sound waves through a wood arm to a stationary diaphragm introduces an original reproducing device, entirely different from anything else ever invented. It is an example of scientific principles of acoustics reduced to extreme simplicity; the same adaptation of a perfect and simple mechanism to the immutable laws of nature.

Without Horn

EVERY part is designed to be made by automatic tools and when assembled fit together with the nicety of precision. Every one is exactly the same as the other and all parts are interchangeable. Altho the reproducing device is the heart of the Vitaphone, the motor and cabinet are very important features and are carefully designed and made under the same expert supervision.

With Horn—inside connection thrown back

Page Seven

3-45. The Vitaphone was the conception of Clinton B. Repp. In 1899, Repp had been involved with marketing a failed disc machine known as the Vitaphone, from which he borrowed only the non-trademarked name. Repp received a patent in 1911 (No. 1,003,655) for a wooden sound-conducting arm, stationary reproducer and amplifying horn, and he founded the Vitaphone Company of Plainfield, New Jersey. This 5" x 7 1/2", 1913 catalog diagramed the basis of the Vitaphone system: vibrations from the record traveled up a solid wooden arm to a reproducing diaphragm. The sound could then be amplified by connecting an internal horn, an external horn or, in some instances, a choice of either. *Courtesy of Ernest Carl Allen. (Value code: E)*

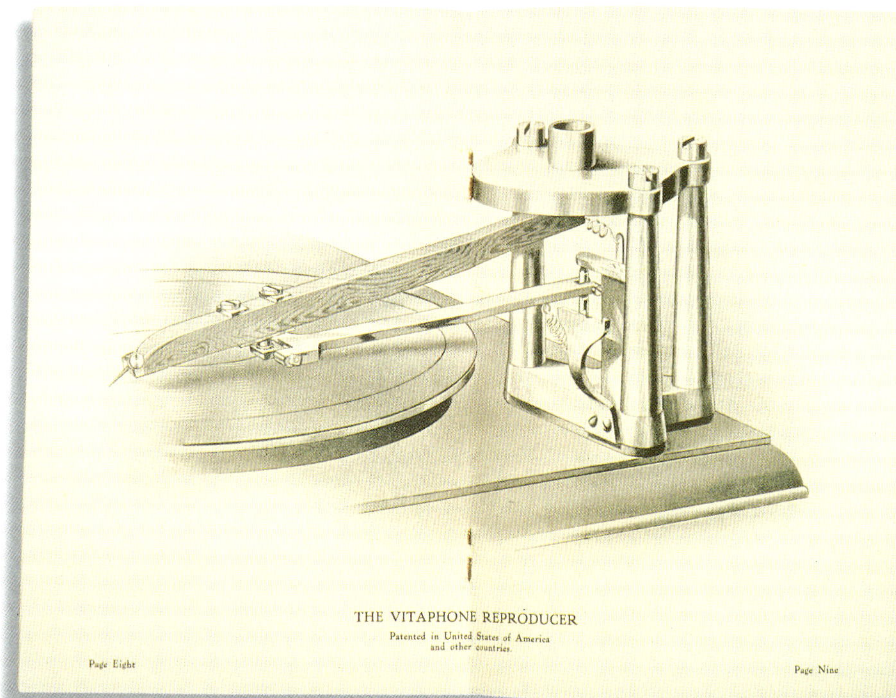

3-46. Another spread from the catalog showed the Vitaphone arm and reproducer housing in detail. Review the previous illustration and note that in this drawing, there is no rectangular weight attached to the front of the arm (which exerted downward pressure on the record, thereby increasing physical energy and volume). Both types of arm have been seen on Vitaphones. The firm's earliest model line consisted of one low-cost external-horn model, followed by four internal-horn machines of ascending price. Over the next couple of years, the styles would vary. *Courtesy of Ernest Carl Allen.*

3-47. A photograph, 5" x 9 7/8", of the Vitaphone factory in New Jersey. The company's instruments were well-made and competitively priced, but under-capitalization seems likely. A suit brought against Vitaphone in May 1914 by Fred Stern, president of the Newburgh, New York chamber of commerce sought to recover $500.00 paid for Vitaphone stock based on the understanding that the firm would locate a talking machine cabinet factory in Newburgh. The stock offering had failed, and Vitaphone bailed out of the project. Mr. Stern was seeking water from a dry well. Soon enough, the Vitaphone Company fizzled and was gone. *Courtesy of Ernest Carl Allen. (Value code: G)*

3-48. Victor dealers could order a variety of sepia-toned photogravures of famous Victor artists to display in their stores. Arguably the most popular subject was Enrico Caruso, shown here as "Dick Johnson" from *The Girl of the Golden West*. Measures 19 1/2" x 25 1/2" without frame. *Courtesy of Tom and Sandi McCarthy. (Value code: VR)*

3-49. Victor sold this framed photograph for dealer display. In this scene, Enrico Caruso has taken charge of the Victrola "XVI," but the expression on his face seems to say, "I know Frieda, but who are these other people?" *Courtesy of Norm and Janyne Smith. (Value code: F)*

3-50. The artifacts at left and bottom are Victor advertising mirrors, one unusually adorned with a portrait of Caruso. The pinback at top right promoted a distributor or jobber of Victor goods in Dallas, Texas. *Courtesy of Tom and Sandi McCarthy. (Value code: F)*

3-51. Letter openers were understated, classy advertising for phonograph firms. The example above, from the Eastern Talking Machine Company, Boston, Massachusetts, measures 8 5/8". The lower one, from the American Talking Machine Company, Brooklyn, New York, measures 9". *Courtesy of Ernest Carl Allen. (Value code: F)*

3-52. This wooden shipping box for 12" disc records measures 15 1/4" square. *Courtesy of the Johnson Victrola Museum/Delaware State Museums. (Value code: G)*

3-53. A very early example of the "Ragtime Rastus" dancing toy for phonographs manufactured by the National Toy Company of Boston, Massachusetts. The base and spindle connector differ noticeably from the common production version. The figure, too, is constructed with wooden, rather than the usual fiber, arms. The box illustrates the toy in use on a Victrola "IV." *Courtesy of Kenneth Woodbury. (Value code: E)*

3-54. The January 1915 issue of Thomas A. Edison, Incorporated's house organ. The previous month, the Phonograph Works had suffered a devastating fire, so the image of the company patriarch must have rallied morale. *Courtesy of Ernest Carl Allen. (Value code: F)*

3-55. This pamphlet from November 1915 measures 5" x 7" closed. When opened as shown here, the suspense must have seemed palpable to mom and dad... *Courtesy of Ernest Carl Allen. (Value code: F)*

3-56. Opening the tri-fold reveals an idyllic Christmas morning made memorable for the kiddies by a Columbia Grafonola. *Courtesy of Ernest Carl Allen.*

3-57. Tape measurers featuring general advertising are common, but seldom seen with phonograph advertising. The 1 1/4" diameter celluloid example at the left was from The Wittich Store (a mirror behind it shows the charming reverse). At the right are an aluminum matchbook cover and a Brunswick thimble. *Courtesy of Tom and Sandi McCarthy. (Value code: F)*

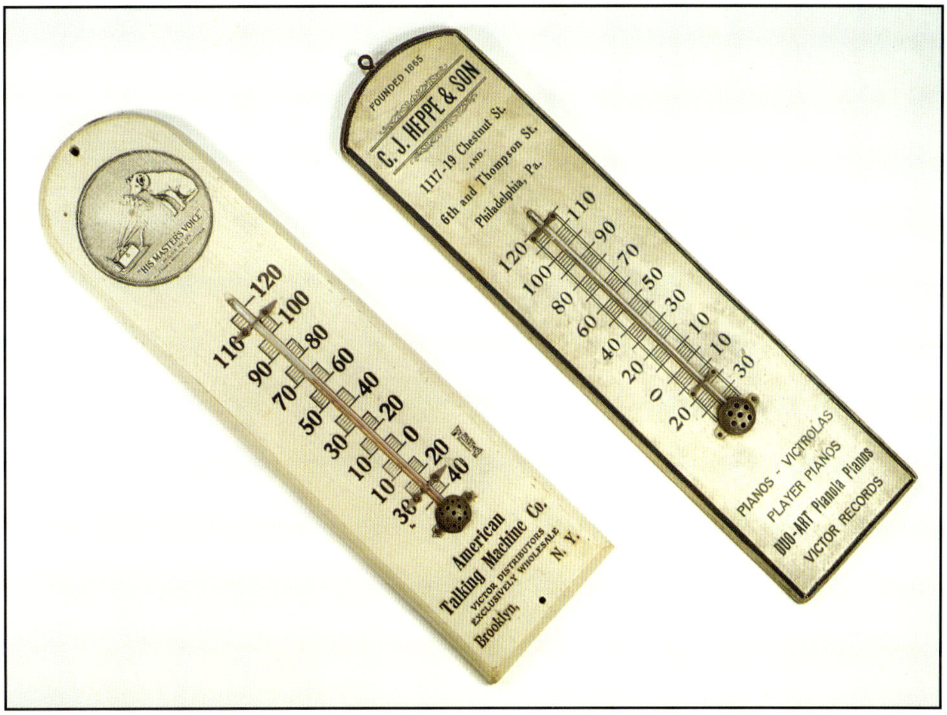

3-58. These 15" advertising thermometers seem like they would have been such good long-term investments for phonograph retailers, it's puzzling that they turn up so infrequently. The example for Heppe was made by the Met-O-Lite Pencil Company of Philadelphia. *Courtesy of Tom and Sandi McCarthy. (Value code: E, F)*

3-59. It is most unusual to see a set of original dealer "price tags" which were displayed on Grafonolas during the teens. Each measures 3 1/8" diameter. *Courtesy of Tom and Sandi McCarthy. (Value code: G)*

3-60. A catalog of Little Wonder disc records, showing the front and back covers (measures 3 1/2" x 6 1/4", folded). Little Wonder discs were single-faced, 5 1/2" in diameter, manufactured by Columbia, and featured Columbia artists of the period 1914-23. Today, it is easy to mistake these diminutive discs for children's records, which they decidedly were not. The repertoire was strictly grown-up – Little Wonders simply were small and therefore inexpensive records (10 cents each). Note the advertisement for Irving Berlin sheet music on the back cover. Henry Waterson (publishing partner of Irving Berlin and Ted Snyder) was an important figure in Little Wonder's development, though he had disassociated himself from the operation by 1917. *Courtesy of Merle Sprinzen. (Value code: VR)*

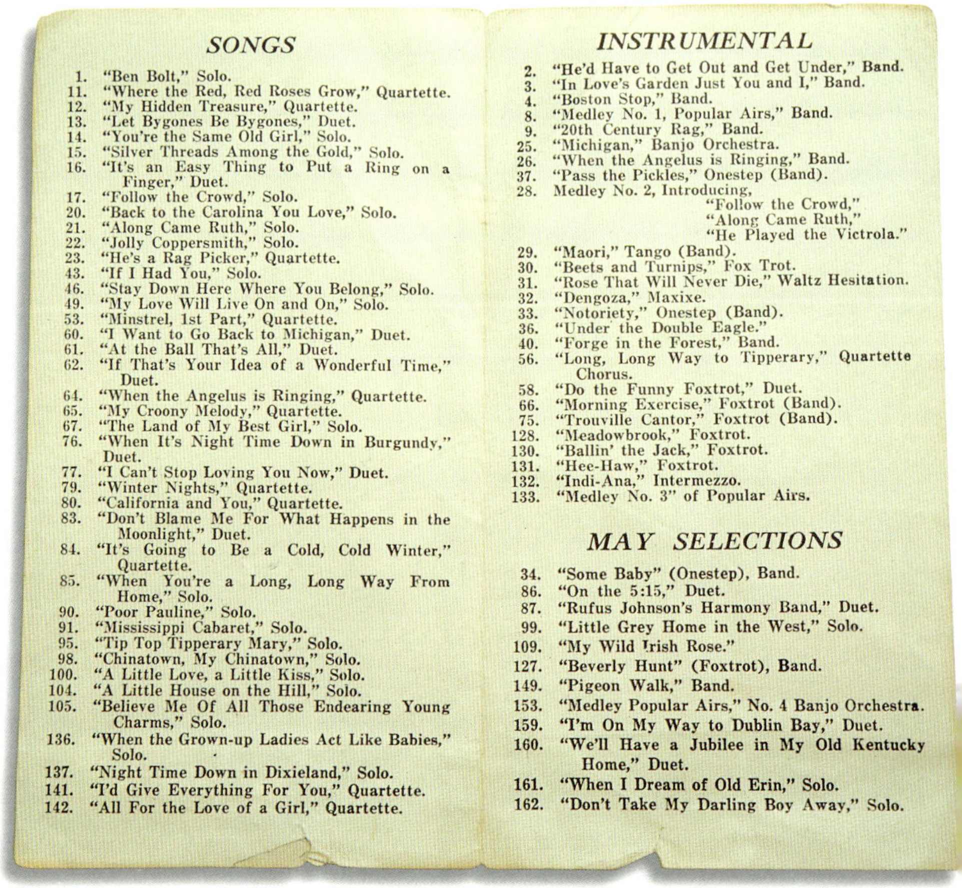

SONGS

1. "Ben Bolt," Solo.
11. "Where the Red, Red Roses Grow," Quartette.
12. "My Hidden Treasure," Quartette.
13. "Let Bygones Be Bygones," Duet.
14. "You're the Same Old Girl," Solo.
15. "Silver Threads Among the Gold," Solo.
16. "It's an Easy Thing to Put a Ring on a Finger," Duet.
17. "Follow the Crowd," Solo.
20. "Back to the Carolina You Love," Solo.
21. "Along Came Ruth," Solo.
22. "Jolly Coppersmith," Solo.
23. "He's a Rag Picker," Quartette.
43. "If I Had You," Solo.
46. "Stay Down Here Where You Belong," Solo.
49. "My Love Will Live On and On," Solo.
53. "Minstrel, 1st Part," Quartette.
60. "I Want to Go Back to Michigan," Duet.
61. "At the Ball That's All," Duet.
62. "If That's Your Idea of a Wonderful Time," Duet.
64. "When the Angelus is Ringing," Quartette.
65. "My Croony Melody," Quartette.
67. "The Land of My Best Girl," Solo.
76. "When It's Night Time Down in Burgundy," Duet.
77. "I Can't Stop Loving You Now," Duet.
79. "Winter Nights," Quartette.
80. "California and You," Quartette.
83. "Don't Blame Me For What Happens in the Moonlight," Duet.
84. "It's Going to Be a Cold, Cold Winter," Quartette.
85. "When You're a Long, Long Way From Home," Solo.
90. "Poor Pauline," Solo.
91. "Mississippi Cabaret," Solo.
95. "Tip Top Tipperary Mary," Solo.
98. "Chinatown, My Chinatown," Solo.
100. "A Little Love, a Little Kiss," Solo.
104. "A Little House on the Hill," Solo.
105. "Believe Me Of All Those Endearing Young Charms," Solo.
136. "When the Grown-up Ladies Act Like Babies," Solo.
137. "Night Time Down in Dixieland," Solo.
141. "I'd Give Everything For You," Quartette.
142. "All For the Love of a Girl," Quartette.

INSTRUMENTAL

2. "He'd Have to Get Out and Get Under," Band.
3. "In Love's Garden Just You and I," Band.
4. "Boston Stop," Band.
8. "Medley No. 1, Popular Airs," Band.
9. "20th Century Rag," Band.
25. "Michigan," Banjo Orchestra.
26. "When the Angelus is Ringing," Band.
37. "Pass the Pickles," Onestep (Band).
28. Medley No. 2, Introducing, "Follow the Crowd," "Along Came Ruth," "He Played the Victrola."
29. "Maori," Tango (Band).
30. "Beets and Turnips," Fox Trot.
31. "Rose That Will Never Die," Waltz Hesitation.
32. "Dengoza," Maxixe.
33. "Notoriety," Onestep (Band).
36. "Under the Double Eagle."
40. "Forge in the Forest," Band.
56. "Long, Long Way to Tipperary," Quartette Chorus.
58. "Do the Funny Foxtrot," Duet.
66. "Morning Exercise," Foxtrot (Band).
75. "Trouville Cantor," Foxtrot (Band).
128. "Meadowbrook," Foxtrot.
130. "Ballin' the Jack," Foxtrot.
131. "Hee-Haw," Foxtrot.
132. "Indi-Ana," Intermezzo.
133. "Medley No. 3" of Popular Airs.

MAY SELECTIONS

34. "Some Baby" (Onestep), Band.
86. "On the 5:15," Duet.
87. "Rufus Johnson's Harmony Band," Duet.
99. "Little Grey Home in the West," Solo.
109. "My Wild Irish Rose."
127. "Beverly Hunt" (Foxtrot), Band.
149. "Pigeon Walk," Band.
153. "Medley Popular Airs," No. 4 Banjo Orchestra.
159. "I'm On My Way to Dublin Bay," Duet.
160. "We'll Have a Jubilee in My Old Kentucky Home," Duet.
161. "When I Dream of Old Erin," Solo.
162. "Don't Take My Darling Boy Away," Solo.

3-61. The interior of the Little Wonder catalog, including the first-issued discs from 1914. The cheap price made Little Wonder records very popular, and they survive in great quantities today. A talking machine was sold during the teens under the name "Little Wonder," but research has shown it was not connected to the brand of discs. The talking machine was designed to play *vertical cut* records, whereas Little Wonders are *lateral cut* – and the two recording systems are not compatible. The machine was also sold under at least one other name (see *Discovering Antique Phonographs*, Fabrizio & Paul, Schiffer, 2000, figure 4-50), underscoring the coincidental use of the brand name by two distinct organizations. *Courtesy of Merle Sprinzen.*

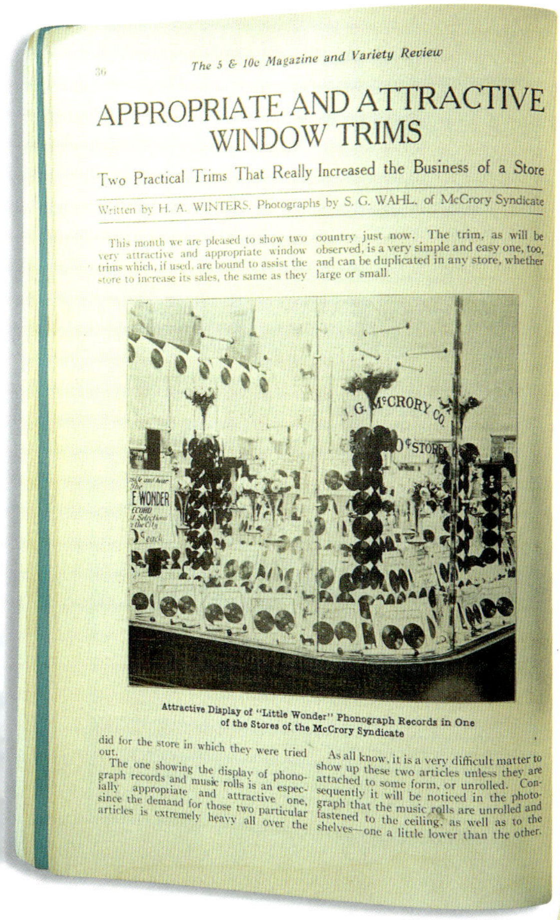

36 The 5 & 10c Magazine and Variety Review

APPROPRIATE AND ATTRACTIVE WINDOW TRIMS

Two Practical Trims That Really Increased the Business of a Store

Written by H. A. WINTERS. Photographs by S. G. WAHL, of McCrory Syndicate

This month we are pleased to show two very attractive and appropriate window trims which, if used, are bound to assist the store to increase its sales, the same as they did for the store in which they were tried out.

The one showing the display of phonograph records and music rolls is an especially appropriate and attractive one, since the demand for those two particular articles is extremely heavy all over the country just now. The trim, as will be observed, is a very simple and easy one, too, and can be duplicated in any store, whether large or small.

Attractive Display of "Little Wonder" Phonograph Records in One of the Stores of the McCrory Syndicate

As all know, it is a very difficult matter to show up these two articles unless they are attached to some form, or unrolled. Consequently it will be noticed in the photograph that the music rolls are unrolled and fastened to the ceiling, as well as to the shelves—one a little lower than the other.

3-62. A page from a dime-store trade journal from July 1915 shows the ostentatious merchandizing of Little Wonder records by the McCrory chain. Little Wonders were sold by other low-cost retailers such as Woolworth's, Kresge's, and Sears, Roebuck & Company. *Courtesy of Merle Sprinzen.*

3-63. Although Little Wonder discs are sometimes confused with children's records, the brand *was*, in fact, peripherally connected to a series of books and records for children: the "Bubble Books." Between 1918 and 1922, Columbia, in association with publisher Harper & Brothers, produced a group of 14 titles, each including three single-faced 5 1/2" diameter discs. "Bubble Books" later became associated with Victor, before returning to Columbia's stewardship. Shown here is a title that was a rare adjunct to the regular sequence of "Bubble Books." *Courtesy of Merle Sprinzen. (Value code: VR)*

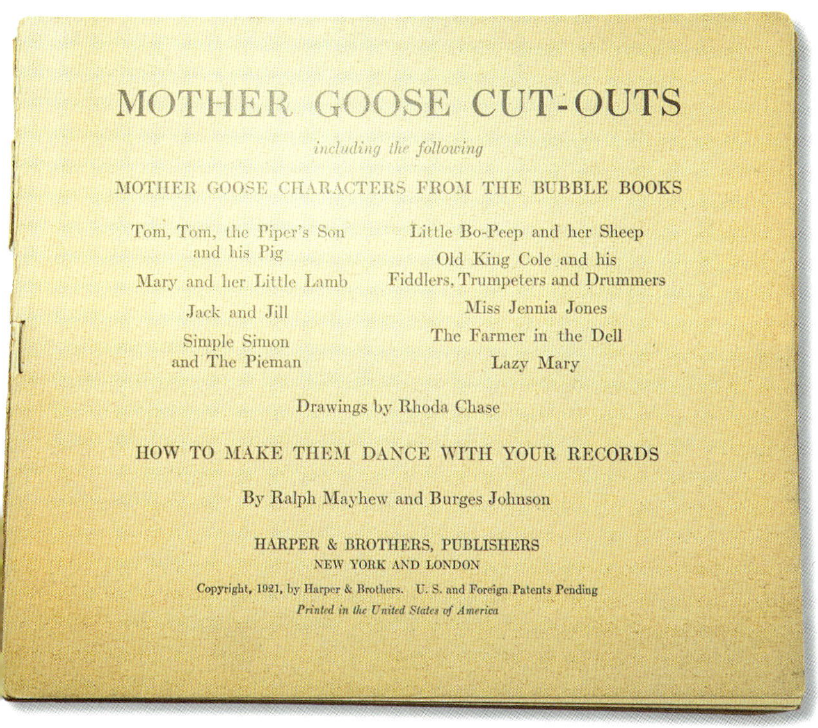

MOTHER GOOSE CUT-OUTS

including the following

MOTHER GOOSE CHARACTERS FROM THE BUBBLE BOOKS

Tom, Tom, the Piper's Son and his Pig
Mary and her Little Lamb
Jack and Jill
Simple Simon and The Pieman
Little Bo-Peep and her Sheep
Old King Cole and his Fiddlers, Trumpeters and Drummers
Miss Jennia Jones
The Farmer in the Dell
Lazy Mary

Drawings by Rhoda Chase

HOW TO MAKE THEM DANCE WITH YOUR RECORDS

By Ralph Mayhew and Burges Johnson

HARPER & BROTHERS, PUBLISHERS
NEW YORK AND LONDON

Copyright, 1921, by Harper & Brothers. U. S. and Foreign Patents Pending

Printed in the United States of America

3-64. This very special "Bubble Book" features characters from the series that were meant to be cut out and assembled into "turntable toys" for the talking machine. *Courtesy of Merle Sprinzen.*

3-65. We see here a diagram of how the "dancing" toys were meant to be mounted, as the instrument played the records that accompanied each "Bubble Book" in the series. *Courtesy of Merle Sprinzen.*

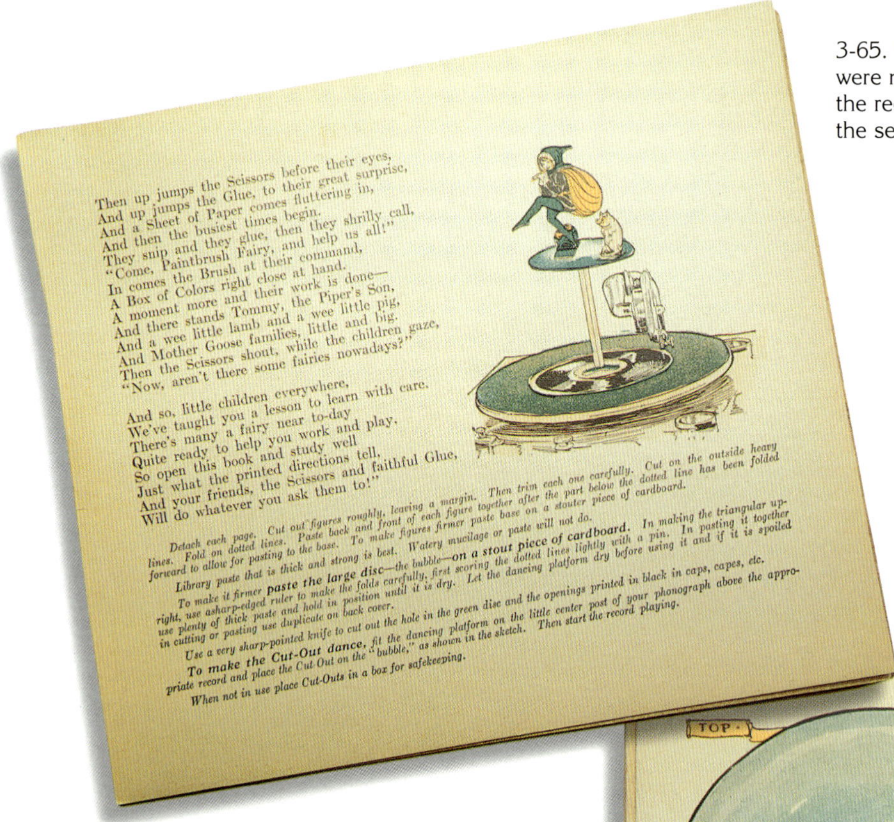

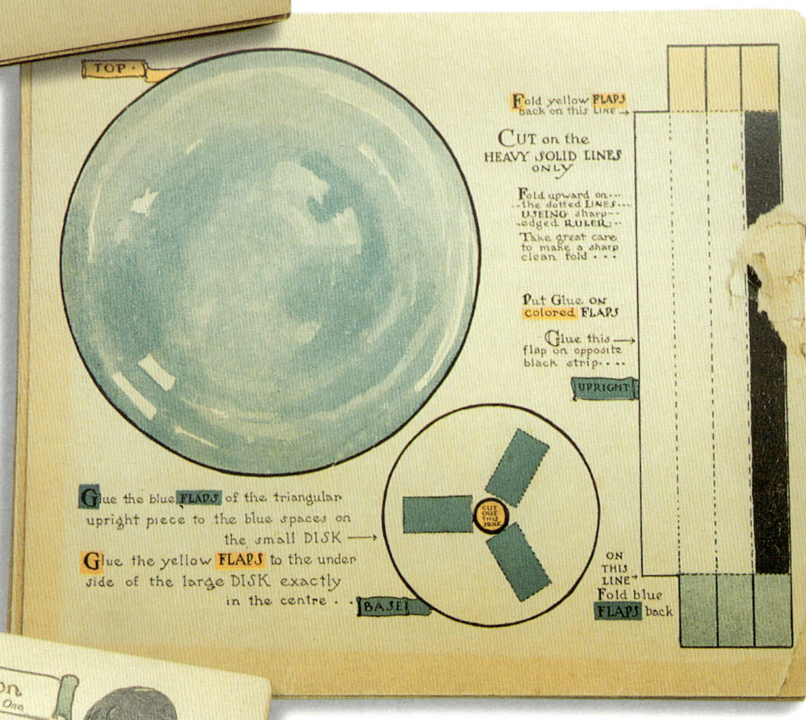

3-66. The "mechanism" was constructed entirely of paper and cardboard. One may infer the significance of the "bubble" from which the series took its name. The "bubble" was a daydream, a platform for flights of imagination, a lens through which the child could glimpse a world of fantasy. *Courtesy of Merle Sprinzen.*

3-67. An example of the familiar fairytale characters meant to be cut out and assembled. *Courtesy of Merle Sprinzen.*

3-68. Miss Jennia Jones. *Courtesy of Merle Sprinzen.*

You Can Make Them Dance, Too!

CHILDREN—just think—you can make the little people in the Bubble Books dance to the music of their own songs, for they are now in the wonderful new *Mother Goose Cut-Out Books.*

Wouldn't it be fun to cut out pictures of Tom, the Piper's Son, Little Bo-Peep, and Old King Cole, and paste them together so that they look like real people no matter from which side you look at them? You can have a whole family of little friends not just printed in a book—but on little cardboard stands so they will stand up all by themselves.

But best of all you can make up a tiny platform for them. You can place it in the center of your Bubble Book records so that while they're playing, your new, little, Bubble Book friends go round and round and round—just as though they were dancing.

Then you can work out little plays and have the Bubble Book people do the things that the records sing about.

Remember, too, that these little folks are dressed in bright colors and the directions for putting them together are clear and easy to understand.

Show this page to aunt, or uncle, or any nice grown-up and you will be sure to get a *Mother Goose Cut-Out,* especially if you are *good.*

Mother Goose Cut-Out Books can be had wherever Bubble Books are sold, and that's almost everywhere. Toy, Book, Phonograph and Department stores all have them. 60c *each.*

HARPER & BROTHERS Est. 1817 Franklin Square, New York

3-69. "Bubble Book" No. 13, 1922, contained this advertisement for the book of turntable toys. The ad copy prefigured TV's insidious cry, "Be the first kid on your block to collect the whole set!" Children were admonished to, "Show this page to aunt, or uncle, or any nice grown-up and you will be sure to get a *Mother Goose Cut-Out*, especially if you are *good*." *Courtesy of Merle Sprinzen. (Value code: G, average Bubble Book)*

3-70. Kimball, like many firms that produced cabinet talking machines in the teens, was a piano manufacturer. The firm's logo featured a little girl listening to a Kimball, accompanied by the slogan "I want to see the lady COME OUT!" Who's to say whether a bouncy tot is any coyer than a fox terrier? *Courtesy of Ernest Carl Allen. (Value code: G)*

3-71. Aeolian Vocalion cabinet talking machines and records began as an offshoot of the Aeolian Company, a well-known maker of pianos. Piano and furniture manufacturers, with considerable woodworking facilities already in place, found it convenient to enter the talking machine trade during the teens. 3 1/4" x 6 1/4". *Courtesy of Ernest Carl Allen. (Value code: G)*

3-72. "Mr. Edison's Sublime Gift to Man" was ordinarily published in a smaller size and employed as a giveaway in 1917 to encourage sales of Diamond Disc Phonographs. This deluxe slip-cased version measures a whopping 12 3/4" x 16 1/4" and features a tissue paper dust jacket. Probably intended for use in the dealer's own facility, it is shown with a typical, lesser catalog (3 1/2" x 6 1/4") to illustrate its comparative size. *Courtesy of Tom and Sandi McCarthy. (Value code: VR)*

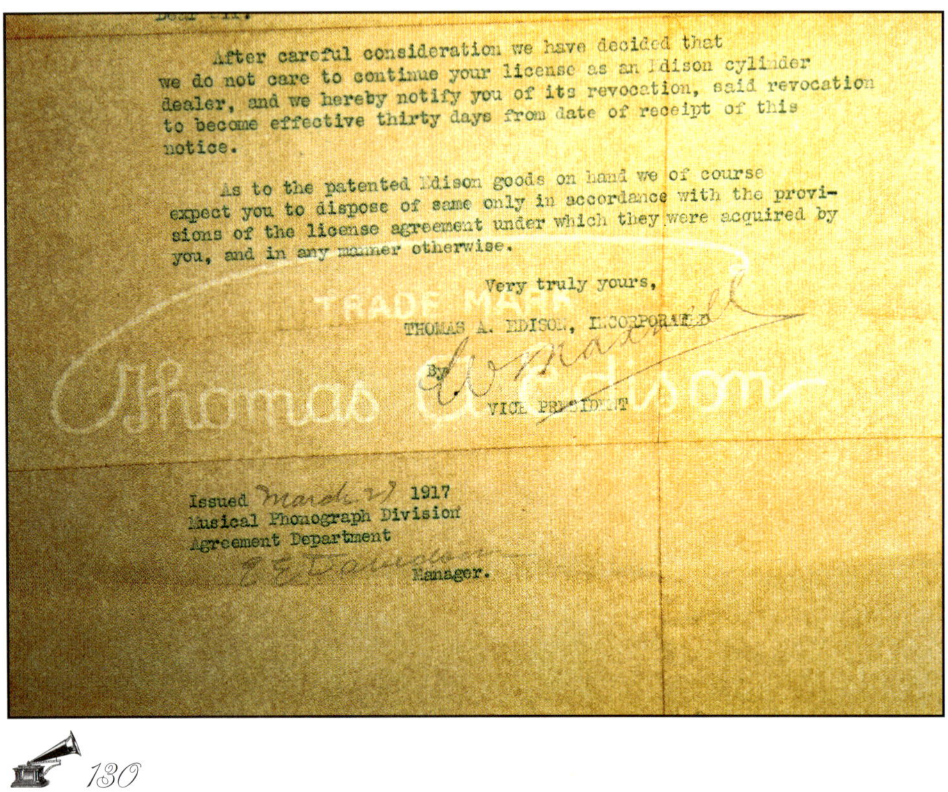

After careful consideration we have decided that we do not care to continue your license as an Edison cylinder dealer, and we hereby notify you of its revocation, said revocation to become effective thirty days from date of receipt of this notice.

As to the patented Edison goods on hand we of course expect you to dispose of same only in accordance with the provisions of the license agreement under which they were acquired by you, and in any manner otherwise.

Very truly yours,

TRADE MARK

THOMAS A. EDISON, INCORPORATED

Thomas A. Edison

By

VICE PRESIDENT

Issued March 27 1917
Musical Phonograph Division
Agreement Department
Manager.

3-73. The watermark from a 1917 Edison document. *Courtesy of Tom and Joanna Broome.*

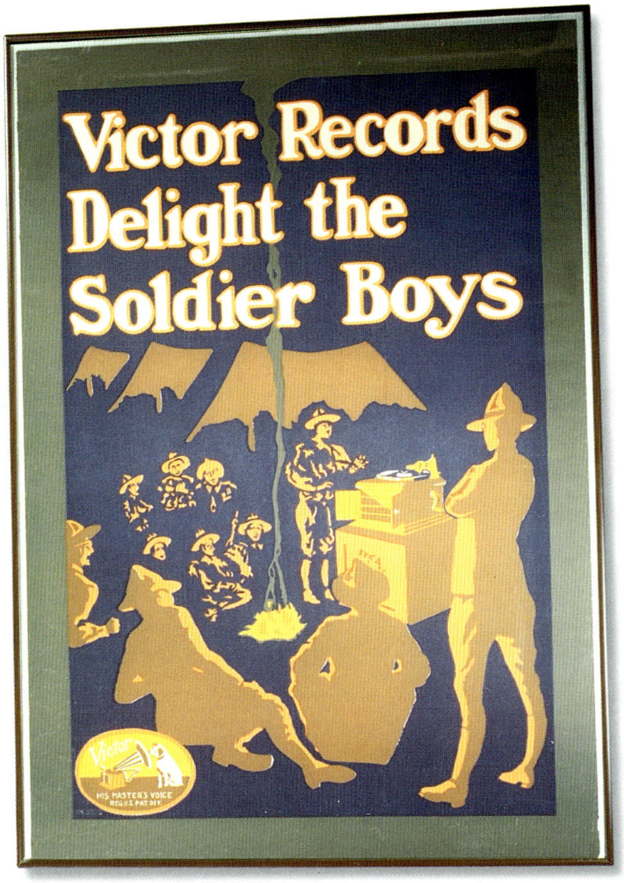

3-74. A World War I vintage poster pressed the point that the Victor was performing a valuable national service. Measures 21 3/4" x 30 1/4". *Courtesy of the Johnson Victrola Museum/Delaware State Museums. (Value code: VR)*

3-75. What better way to express your patriotism during the Great War, or solidarity for the Allies, than to dress up the talking machine with the "Liberty" turntable flags? Get the No. "2" Outfit churning and you'll think you've walked into one of Childe Hassam's Impressionist paintings of flags flying on Fifth Avenue in New York City. A flyer, 3 1/2" x 6 1/2". *Courtesy of Harvey P. Kravitz. (Value code: G)*

3-76. In the political climate of the early twenty-first century, with its emphasis on visible manifestations of patriotic spirit, the advertising copy of the "Liberty" flyer, complete with reference to the Liberty Bell, rings true. *Courtesy of Harvey P. Kravitz.*

נייע
וויקטאר רעקארדס
אין
העברעאיש אונד יודיש
(New Victor Hebrew and Yiddish Records)

(July, 1917) 1917 יולי

Der Romanen Laiser (B. Kowner) (Humorous Monologue) (Yiddish) — דער ראמאנען לעזער — Albert Harris
Oif an Ocshion Sale (B. Kowner) (Humorous Monologue) (Yiddish) — אויף אין אקשען סייל — Albert Harris — 69406 10 $0.75
Menshheit erwach (H. Meisel-A. Perlmutter) (Tenor with Orchestra) (Yiddish) — מענשהייט ערוואך — William Schwartz
Sheloi Osany Yishe (H. Meisel-Perlmutter-Wohl) (Tenor with Orchestra) (Yiddish) — שלא עשני אשה — William Schwartz — 69407 10 .75

העברעאישע אונד יודישע רעקארדס דיא לעצטע צייט ארויס געגעבען.
(Hebrew and Yiddish Records recently issued)

Es is gewehn a meidel (Soprano with Orchestra) — עס איז געווען א מיידעל (פון "דיא פוילישע חתונה") — Mme. Regina Zuckerberg
Shir hashirim Gavote (Soprano with Orchestra) — שיר השירים גאוואטע (פון "שיר השירים") — Mme. Regina Zuckerberg — 67920 10 $0.75
Kadisch Nuch Der Mamen (Duet with Orchestra) — קדיש נאך דער מאמען (פון "דיא ערשטע כלה") — Fanny Lubritsky-David Lubritsky
Kerachem ov al Bonim (Soprano with Orchestra) — כרחם אב על בנים — Fanny Lubritsky — 67921 10 .75
Nur Eins in Leben (J. Silbert) (Soprano with Orchestra) — נור איינס אין לעבען — Mme. Zvibel
Rundo Voltz (Bomberg) (Soprano with Orchestra) — רונדא וואלץ — Mme. Zvibel — 67934 10 .75
Tateneu Kim a Heim Tzurik (Yiddish) (Solomon Small) (Shmulov) (Tenor with Orchestra) — טאטעניו קום א היים צוריק — Nahen Rosenshak
Milchomo Korbones (War Victims) (Anshel Schorr-Yudele Belzer) (Tenor with Orchestra) (Yiddish) — מלחמה קרבנות — Nahen Rosenshak — 69134 10 .75
Leiba (Love) (From "The Victims") (Brody-Friedsell) (Boy Soprano with Orchestra) (Yiddish) — ליעבע — David Lubritsky
Galt iz roond (Money is Round) (From "The False Step") (Brody) (Soprano with Orchestra) (Yiddish) — געלד איז רונד — Fanny Lubritsky — 69161 10 .75
The Sweat Shop (Morris Rosenfeld) (Dramatic Recitation in Hebrew) — סוועט שאפ — Albert Harris
Gechapt (I Caught) (David Opptaiker) (Comic Recitation in Hebrew) — געכאפט — Albert Harris — 69165 10 .75

3-77. At the time that this Victor record supplement was issued, Hebrew was an ancient, "dead" language, used primarily in sacred ritual. In addition to the Roman alphabet, Hebrew characters could be used to phonetically express the vital, European language Yiddish. Well-educated Jews might understand Hebrew; the common man spoke Yiddish. So, we see here represented both Yiddish and a couple of Hebrew selections, expressed in both Roman and Hebrew characters. *Courtesy of Ernest Carl Allen. (Value code: G)*

3-78. Many phonograph dealers made their own display materials. Some surviving examples are quite good, showing imagination and intelligence. Others are more…well… "human." These Rishell display cards are artistically rendered, but spelling evidently was a challenge for the maker. These cards average 6" x 13 1/2". *Courtesy of Tom and Sandi McCarthy. (Value code: G, each)*

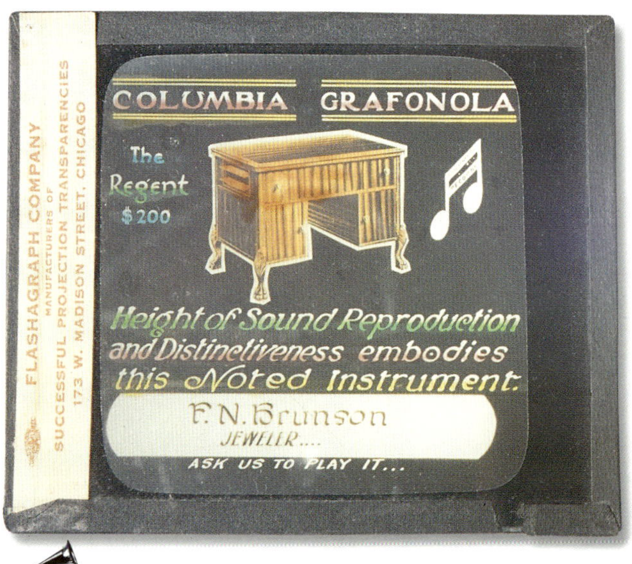

3-79. Prior to the features in moving picture shows, advertising slides would be projected for the audience's "edification." This circa 1917 glass slide illustrated the Columbia "Regent," a talking machine in the shape of a desk. *(Value code: F)*

3-80. These packing slips record a dealer's sales of various model Victrolas in 1916 and 1917. Each measures 6" x 11 1/4". *(Value code: H)*

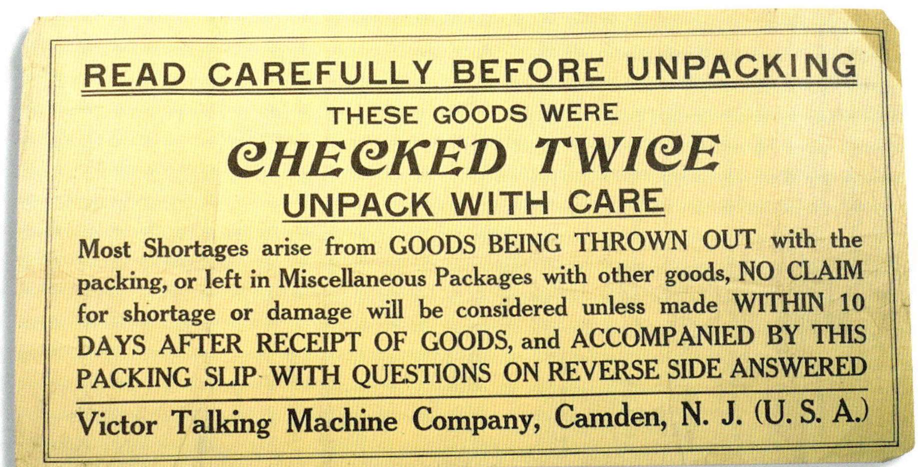

READ CAREFULLY BEFORE UNPACKING

THESE GOODS WERE

CHECKED TWICE

UNPACK WITH CARE

Most Shortages arise from GOODS BEING THROWN OUT with the packing, or left in Miscellaneous Packages with other goods, NO CLAIM for shortage or damage will be considered unless made WITHIN 10 DAYS AFTER RECEIPT OF GOODS, and ACCOMPANIED BY THIS PACKING SLIP WITH QUESTIONS ON REVERSE SIDE ANSWERED

Victor Talking Machine Company, Camden, N. J. (U. S. A.)

3-81. Each packing slip carries this message on its reverse.

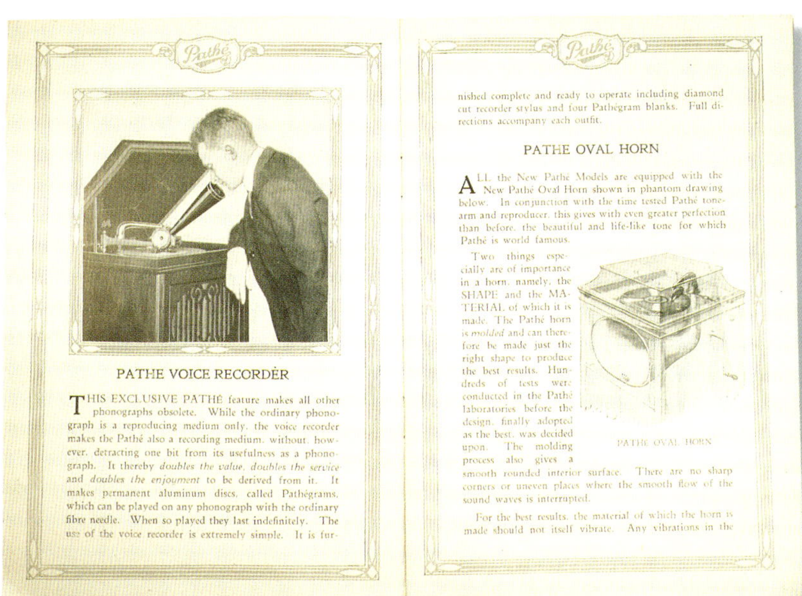

PATHE VOICE RECORDER

THIS EXCLUSIVE PATHÉ feature makes all other phonographs obsolete. While the ordinary phonograph is a reproducing medium only, the voice recorder makes the Pathé also a recording medium, without, however, detracting one bit from its usefulness as a phonograph. It thereby *doubles the value, doubles the service* and *doubles the enjoyment* to be derived from it. It makes permanent aluminum discs, called Pathégrams, which can be played on any phonograph with the ordinary fibre needle. When so played they last indefinitely. The use of the voice recorder is extremely simple. It is furnished complete and ready to operate including diamond cut recorder stylus and four Pathégram blanks. Full directions accompany each outfit.

PATHE OVAL HORN

ALL the New Pathé Models are equipped with the New Pathé Oval Horn shown in phantom drawing below. In conjunction with the time tested Pathé tone-arm and reproducer, this gives with even greater perfection than before, the beautiful and life-like tone for which Pathé is world famous.

Two things especially are of importance in a horn, namely, the SHAPE and the MATERIAL of which it is made. The Pathé horn is *molded* and can therefore be made just the right shape to produce the best results. Hundreds of tests were conducted in the Pathé laboratories before the design, finally adopted as the best, was decided upon. The molding process also gives a smooth rounded interior surface. There are no sharp corners or uneven places where the smooth flow of the sound waves is interrupted.

PATHE OVAL HORN

For the best results, the material of which the horn is made should not itself vibrate. Any vibrations in the

3-82. Pathé and Brunswick were allied for a short period. The Brunswick, Balke, Collender Company, maker of commercial interiors and sports equipment, began manufacturing a line of cabinet talking machines under its own name (having done "job work" for Edison and others) in 1916. Lacking record production facilities, Brunswick agreed to a short-term contract (until its own record making could be initiated) with Pathé to distribute that firm's vertical discs. This leads us to this interesting spread from a Pathé catalog of the presently-discussed period. Although a Pathé phonograph was being described, a *Brunswick* is most unmistakably pictured, even including the "signature" Brunswick molded white internal horn. *Courtesy of Ernest Carl Allen. (Value code: G)*

3-83. In the late teens, Brunswick, and all the other independents, were faced with the same obstacles. Lateral recording was being jealously guarded by Victor and Columbia, vertical recording had become the default of the competition, and then Edison had muddied the field with his Diamond Disc system. Many talking machine firms designed their instruments to be adaptable between lateral and vertical reproduction. A few, such as Vitaphone, promoted the notion of playing Edison Diamond Discs, with the proper stylus. Brunswick, which was manufacturing excellent products, tried to give its customers a full range of choices. There was a "Large Reproducer" and a "Small Reproducer," and three different styli, not to mention ordinary steel needles. Brunswick soon simplified matters, eliminating the two interchangeable reproducers and introducing the "Ultona," a masterpiece of engineering able to play all types of records. *Courtesy of Ken Wood. (Value code: F)*

3-84. The September 1917 Columbia record supplement featured an understandably patriotic cover. *(Value code: H)*

ORIGINAL

Deferred Payment Contract

The undersigned H. B. Lanigan of No. 5348 Race Ave Street, City of Austin, Ill hereby acknowledges to have this day received from S. K. Turek a licensed dealer or distributor of Victor patented goods of the Victor Talking Machine Co. of Camden, N. J., one Victor Talking Machine, Style VV-XI No. 3407096 subject to all the conditions of the special license for said machine printed upon the label affixed thereto; and will pay for the conditional ownership of, and patent license to use the same, the full cash royalty of $110 00/100 stated on said license label.

In consideration of the present acceptance of a partial payment of $60 00/100, and of the remainder of said fixed cash royalty in monthly installments of $10 00/100 with interest at the rate of 6 per cent. per annum, the undersigned hereby agrees and declares that, until payment is made of the full amount above stated his right of possession of said machine is and shall be that of lessee or bailee only; that if any of said deferred installments, as specified above, remain unpaid after it becomes due, as above provided, the other party hereto may at his or their option, with or without legal process, enter into or upon any premises where said Victor Talking Machine may be, and take and carry away the same as its or their property; and all damages for such entry, taking and carrying away are hereby waived.

The undersigned further agrees to pay for all loss or damage to said Victor Talking Machine by reason of fire, or from any cause whatsoever, not to exceed the aggregate of the deferred and unpaid part of the royalty sum above specified; that he will not allow said Victor Talking Machine to be removed from the place at which it is delivered to him as specified above, without the consent of the other party hereto; that if by reason of default in payment of any installment, the other party hereto shall retake possession of the machine by voluntary surrender thereof by the undersigned or otherwise, then the partial payments theretofore made shall be considered as having been made solely for the use and enjoyment of the machine during the preceding period; that if the undersigned shall make default in any partial payment he will return the machine to the other party hereto within 30 days after such default; but if and when all of the deferred payments are made then the right of the undersigned to retain possession of said machine shall be subject only to the conditions, limitations and restrictions imposed by the Victor Talking Machine Co. as set forth in the label license affixed to the machine.

In witness whereof, the undersigned has set his hand and seal to this agreement, in duplicate, the 19th day of February 1918.

Signed, Sealed and Delivered in the presence of

H B Lanigan [SEAL]

Endorsement

In consideration of the above agreement, I hereby guarantee that .... will either pay the amount specified in the above contract or will return the said Victor Talking Machine to .... within 30 days after default or violation of any of the above mentioned conditions.

Date .... 191 .... Endorser's Signature [SEAL]

Accepted .... 191 Address .... Dealer's Signature

3-85. On February 19, 1918, Mr. H.B. Lanigan of Austin, Illinois purchased a Victrola "XI" from dealer S.K. Turek, for $60.00 down and $10.00 per month to total $110.00. This contract recorded the transaction. *(Value code: H)*

3-86. A Pied Piper leads children on a hopefully less sinister journey than the fairy tale describes – to delight in the joys of a Pathéphone. An American cousin of the French Pathé company had been operating in the United States since 1913, selling vertical cut records and a full range of instruments. 3 3/4" x 7 1/2". *Courtesy of Ernest Carl Allen. (Value code: G)*

3-87. A Pathé catalog showing typical designs of the World War I period. *Courtesy of Ernest Carl Allen. (Value code: G)*

3-88. "The Sapphire Ball." Explaining to prospective customers the difference and merit of the Pathé vertical cut system, which employed a permanent sapphire stylus (as opposed to frequently-changed steel needles), was a major hurdle for the company. 3 1/2" x 6 1/4". *Courtesy of Ernest Carl Allen. (Value code: G)*

3-89. This immense 10-foot x 3-foot canvas banner hung outside a dealer's store in the late teens. *Courtesy of Tom and Sandi McCarthy. (Value code: VR)*

3-90. This 1918 Cheney catalog cover is unusually artistic for the genre. The Cheney Talking Machine Company was begun in 1914 by former classical violinist Forest Cheney. The firm meticulously designed an internal sound passage that would not infringe Victor patents. Victor sued anyway, though Cheney eventually prevailed in court. Cheney focussed its efforts on substantial (in some cases ponderous) floor-models. Measures 8 1/8" x 10 1/8". *Courtesy of Ernest Carl Allen. (Value code: G)*

3-91. This 1918 pamphlet illustrated "Puritan Phonographs," and the side shown here featured its two best models: the "Merrimac" and the "Baroque." "Puritan" began, like a number of other talking machine brands, with a furniture manufacturer. The Wisconsin Chair Company of Port Washington started making cabinets for phonograph companies in the mid-teens. This was a period of hell-for-leather expansion in the talking machine market. Late in 1916, the United Phonographs Corporation, a subsidiary of Wisconsin Chair, began operating out of a factory in Sheboygan. The trade name "Puritan" was used, although the firm also produced machines that were sold under various "client" brands. The "Puritan's" principal design conceit was an elongated internal sound chamber that channeled sound to the lower portion of the cabinet. The result was peculiar in appearance – and paunchy! Pamphlet measures 3 3/8" x 6 1/4". *Courtesy of Ernest Carl Allen. (Value code: G)*

3-92. A calendar illustrating the Sonora "Baby Grand" included poetry along with the Pennsburg Furniture Company's address. Measures 8 1/4" x 14". *Courtesy of Tom and Sandi McCarthy. (Value code: F)*

3-93. In June 1918, Brunswick showcased its range of models in a brochure titled "Captive Music." Shown in this spread are the luxurious "Style 1500" and the "Style 350." *(Value code: E)*

3-94. Here, the oriental decorations of "Style 275" makes the similar "Style 225" (the grille seems to be the only structural difference) look rather plain in comparison.

3-95. "Style 200" and "Style 175" show successively decreasing amounts of cabinet decoration.

3-96. "Style 150" and "Style 125" were less elaborate and less expensive, but Brunswick clients seem to have preferred the better models, based on surviving evidence. The penciled notations differ from the prices noted at the back of the brochure, and may represent either the dealer's cost with freight or a "deal" offered to a family member or a friend.

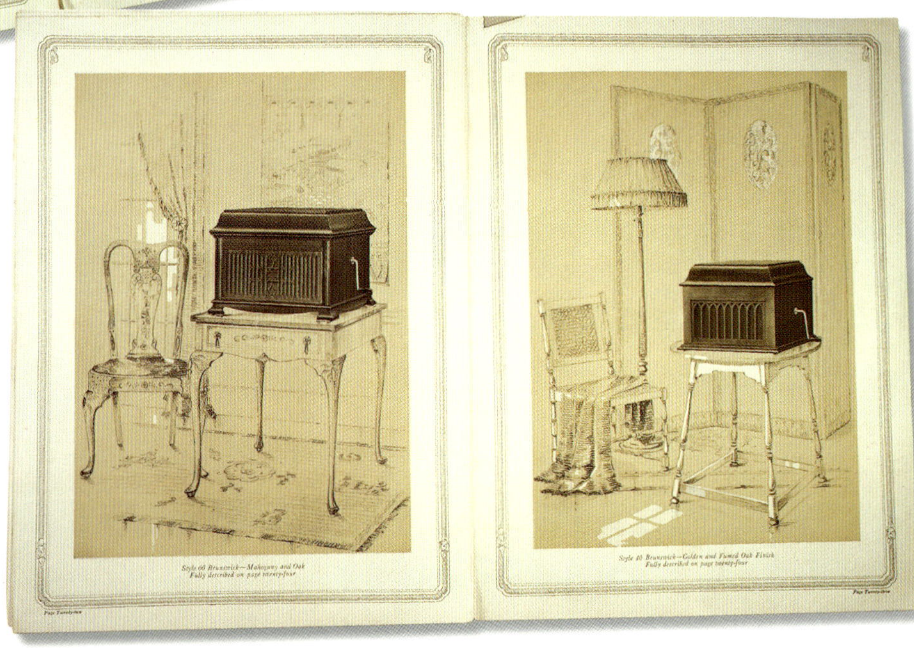

3-97. "Style 60" and "Style 40," contrary to the formal settings, were meant for more humble surroundings.

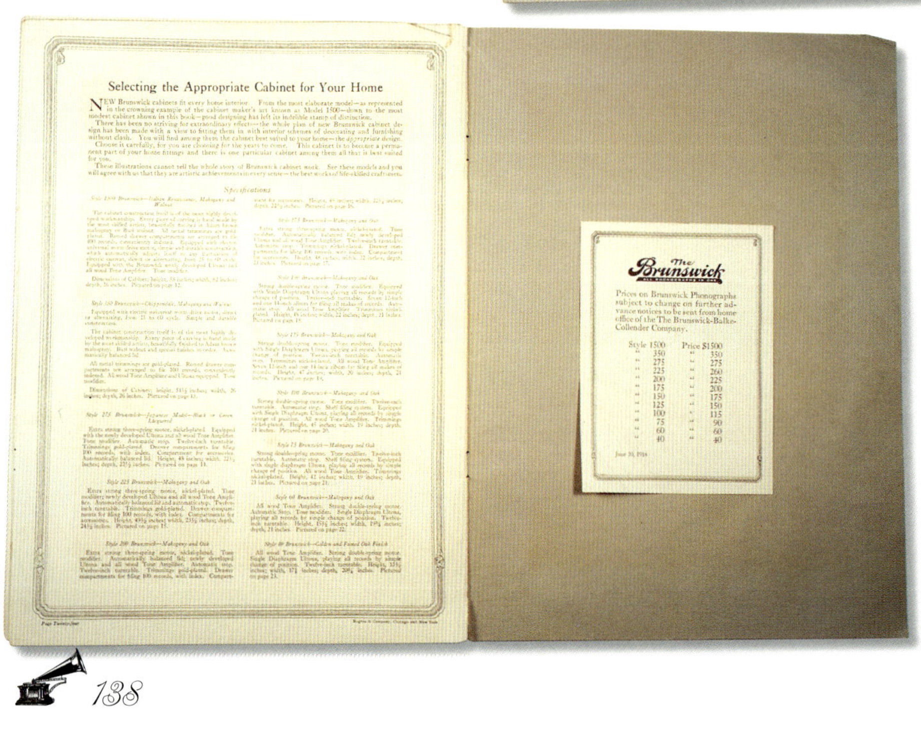

3-98. The Brunswick, Balke, Collender Company offered helpful suggestions on "Selecting the Appropriate Cabinet for Your Home," but the principal determinant more likely was found at the right.

3-99. Modern collectors are occasionally confronted with cabinets whose decorations were not factory-applied, but undeniably the product of a professional hand. Shown are 4 3/4" x 8" cards illustrating the designs available from one such freelance decorator, M. Takacs, of 145 E. 42nd Street, New York City. *(Value code: G, each)*

3-100. This trolley card dates from the late teens, and illustrates the Victrola "XVII." 10 1/8" x 20 1/2" without frame. *Courtesy of Tom and Sandi McCarthy. (Value code: E)*

3-101. This 14 1/4" fan advertising the Columbia Grafonola appears to be made of woven a palm frond, but these details are merely printed on card stock. *Courtesy of Tom and Sandi McCarthy. (Value code: G)*

3-102. The 10 1/2" fan on the right dates from the late teens, and illustrates the Edison "C-250" Diamond Disc Phonograph. The account book from the Metropolitan Phonograph Company offers an alternate spelling for "Grafonola." Columbia suffered through years of its products' names being misspelled – sometimes on its own box labels! *Courtesy of Tom and Sandi McCarthy. (Value code: G)*

3-103. Three advertising postcards featuring Edison Diamond Disc artists accompanying their own records as played by Model "C-250." The Edison company sponsored a series of "Tone Tests," beginning in 1915, in cities and towns across the United States. Edison artists would switch between live and recorded performances with clever lighting effects to slightly disorient the audience. Theatrical distractions aside, many testimonials exist of people utterly unable to distinguish between *real* and *canned. Courtesy of Ernest Carl Allen. (Value code: G)*

3-104. Following the First World War, Thomas Edison recorded a message to the American people, asking them never to forget the sacrifice of United States troops or the support of the Allies in the struggle against Germany and Austria. "Let Us Not Forget" was issued on both Diamond Disc and Blue Amberol cylinder records. In this flyer promoting the disc, an artist's impression shows Edison examining a disproportionately small record. *Courtesy of the Charles Hummel collections. (Value code: VR)*

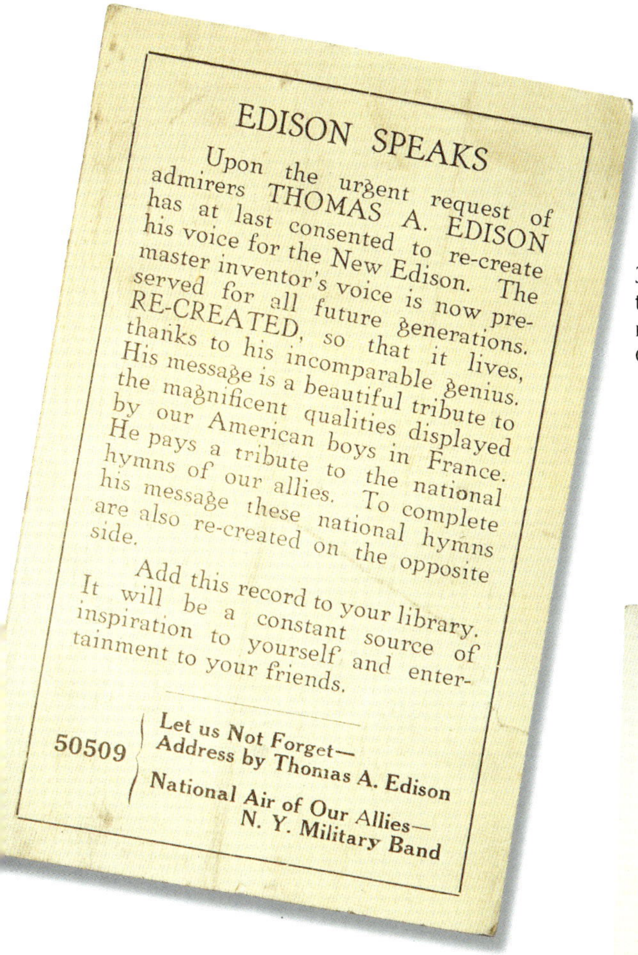

EDISON SPEAKS

Upon the urgent request of admirers THOMAS A. EDISON has at last consented to re-create his voice for the New Edison. The master inventor's voice is now preserved for all future generations. RE-CREATED, so that it lives, thanks to his incomparable genius. His message is a beautiful tribute to the magnificent qualities displayed by our American boys in France. He pays a tribute to the national hymns of our allies. To complete his message these national hymns are also re-created on the opposite side.

Add this record to your library. It will be a constant source of inspiration to yourself and entertainment to your friends.

50509 { Let us Not Forget—Address by Thomas A. Edison
National Air of Our Allies—N. Y. Military Band

3-105. On the reverse, the historical significance of the record is touted. In fact, it would be the only recording of Edison's voice issued to the public. *Courtesy of the Charles Hummel collections.*

Scientific Construction of Tone Arm

THE tone arm, or sound conductor, is scientifically constructed by means of a series of graduating sections, perfectly joined and smoothed as if one piece, to permit the even and perfect flow of sound waves which gradually increase in volume after leaving the reproducer until the sound chamber is entered. The designers of the "Mandel" have succeeded in blending the tone arm with the sound chamber so perfectly that the softest note, or the whispered conversation, is clearly carried to the listener in its recorded tone volume.

Why the "Mandel" is Supreme

THE sound chamber is the heart of a phonograph and in the "Mandel" it is truly everything it should be. It has been demonstrated that the reproduction of sound depends upon the elasticity of the medium through which it is transmitted, and wood is the best sound reproducer known to science. The "Mandel" sound chamber is as scientifically made as the master's violin, and is designed and combined with the "Mandel" graduating tone-arm to reproduce in true tone, every voice modulation of the singer, the rhythm and beauty of the dance music, the incomparable productions of great orchestras, or the strong melody of Sousa's band.

The speed-regulator of the "Mandel" is conveniently located and permits the playing of any record at any desired speed while the brake gives positive action and, when applied, instantly stops the record. The turn table is cast from steel, is well balanced and runs true, with a firm grip on the record.

The Adaptability of the "Mandel"

ONE of the many excellent features of the "Mandel" is the operation of the tone graduator which permits the playing of any instrumental or vocal selection with loud or soft tone, to suit the wishes of the listener or the acoustics of the room. This feature makes the "Mandel" readily adaptable for use in public entertainments, churches, schools, or the home.

Model No. 10

Equipped with Spring Motor, $250
Equipped with Electric Motor, 275

Dimensions:
Height—51 (with castors)
Depth—25¼
Width—23¾

Finishes:
Mahogany
Golden Oak
Fumed Oak

Specifications:

All exposed metal parts heavily gold plated.
Twelve-inch velvet covered turn-table.
Double spring worm drive motor.
Plays seven ten-inch records with one winding.
Tone Modifier.
Hinged motor board.
Three metal removable needle cups in motor board. Genuine mahogany horn.
Velvet padded indexed record compartment for thirty-eight records.
Two record albums. Two drawers for storing needles.
One drawer at bottom of cabinet for storing records, polishing cloth and miscellaneous. Beautifully hand carved.

3-106. This spread from a Mandel Phonograph catalog shows the top-of-the-line Model "10" along with an explanation of its merits. Measures 4 3/4" x 7 3/4" closed. *Courtesy of Tom and Sandi McCarthy. (Value code: G)*

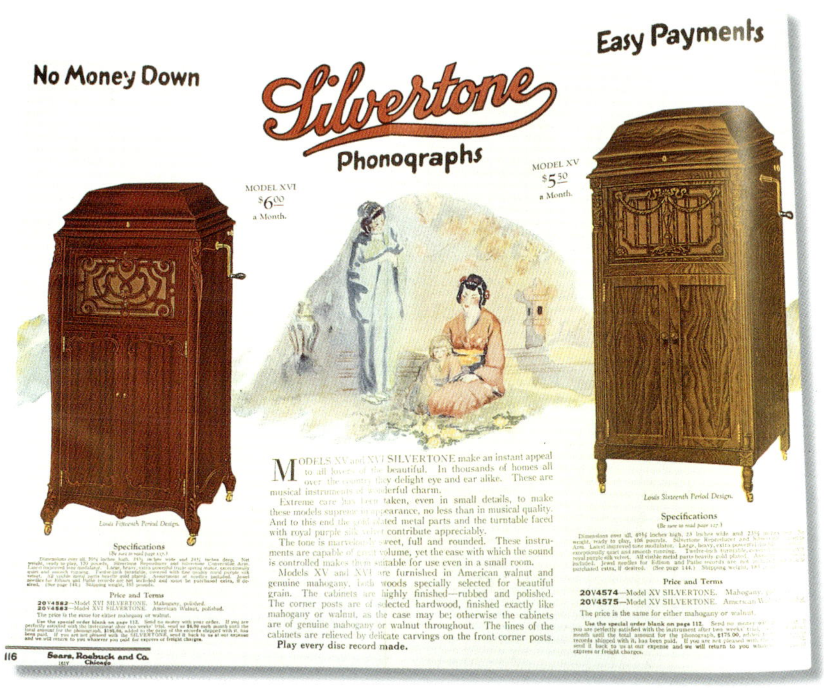

No Money Down

Easy Payments

Silvertone Phonographs

MODEL XVI $6.00 a Month.

MODEL XV $5.50 a Month.

MODELS XV and XVI SILVERTONE make an instant appeal to all lovers of the beautiful. In thousands of homes all over the country they delight eye and ear alike. These are musical instruments of wonderful charm.

Extreme care has been taken, even in small details, to make these models supreme in appearance, no less than in musical quality. And to this end the gold plated metal parts and the turntable faced with royal purple silk velvet contribute appreciably.

The tone is marvelously sweet, full and rounded. These instruments are capable of great volume, yet the ease with which the sound is controlled makes them suitable for use even in a small room.

Models XV and XVI are furnished in American walnut and genuine mahogany, both woods specially selected for beautiful grain. The cabinets are highly finished—rubbed and polished. The corner posts are of selected hardwood, finished exactly like mahogany or walnut, as the case may be; otherwise the cabinets are of genuine mahogany or walnut throughout. The lines of the cabinets are relieved by delicate carvings on the front corner posts.

Play every disc record made.

Specifications

Price and Terms

116 Sears, Roebuck and Co.

3-107. Sears, Roebuck and Company, the premier mail-order merchandiser in the United States called its talking machines of the teens and 1920s by the "Silvertone" brand name (it also sold Silvertone disc records). On a page from a 1920 catalog, we see the finest of the Silvertone models, heavily influenced by the design of Victor's Victrolas. *Courtesy of Merle Spritzen. (Value code: H, full catalog)*

3-108. Victor could supply its dealers with this hanger of "Samples showing various Finishes of Victrola Cabinets." Our moderns eyes may be shocked to see how fresh and bright these finishes were when new. Box measures 10 1/4" x 20 1/4", circa 1919. *(Value code: VR)*

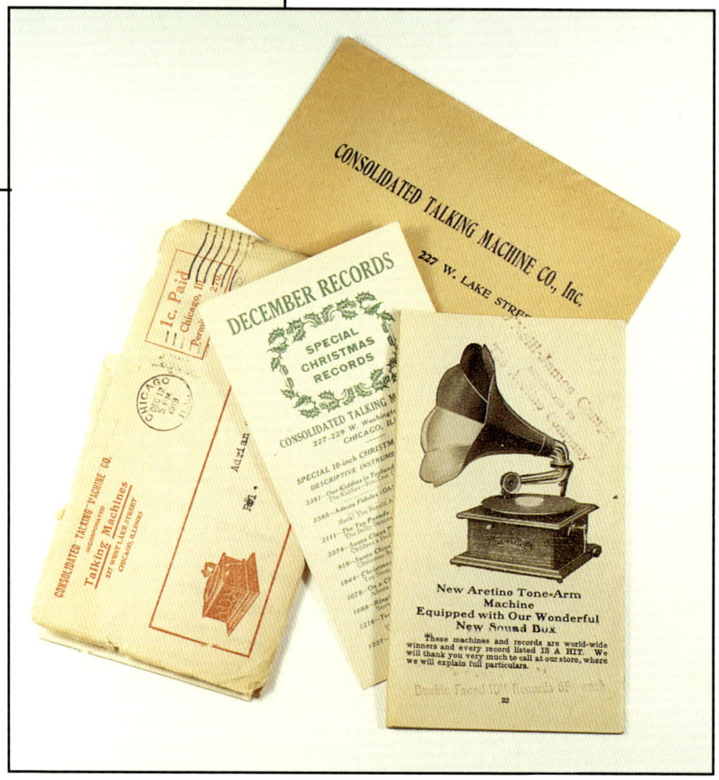

3-109. The "Chicago brands" of disc talking machine, those which engaged in a largely successful experiment with turntable spindle sizes during the period 1906-1915, had a significant impact on the phonograph industry in the United States. Although a number of firms competed against one another (Standard with its 1/2" spindle, Harmony with 3/4", United with 1 1/2", Aretino with 3", and O'Neill-James/"Busy Bee" with effectively *two* spindles), all these enterprises ended up spending eternity together when they were absorbed by the Consolidated Talking Machine Company of Chicago. This envelope of promotional material, mailed December 12, 1919, included a flyer for the Aretino machine, ten years out of date but lingering, several years after the death of the man who created it: Arthur J. O'Neill. *Courtesy of Harvey P. Kravitz. (Value code: G)*

3-110. The Cornish Company of Washington, New Jersey produced upright talking machines in the typical style of the late teens. Included in these documents is an explanation of the firm's credit terms. The admonition "Don't Be Misled By 'No Interest' Claims" might be heard today in almost any television commercial for a furniture store. *Courtesy of Ernest Carl Allen. (Value code: G)*

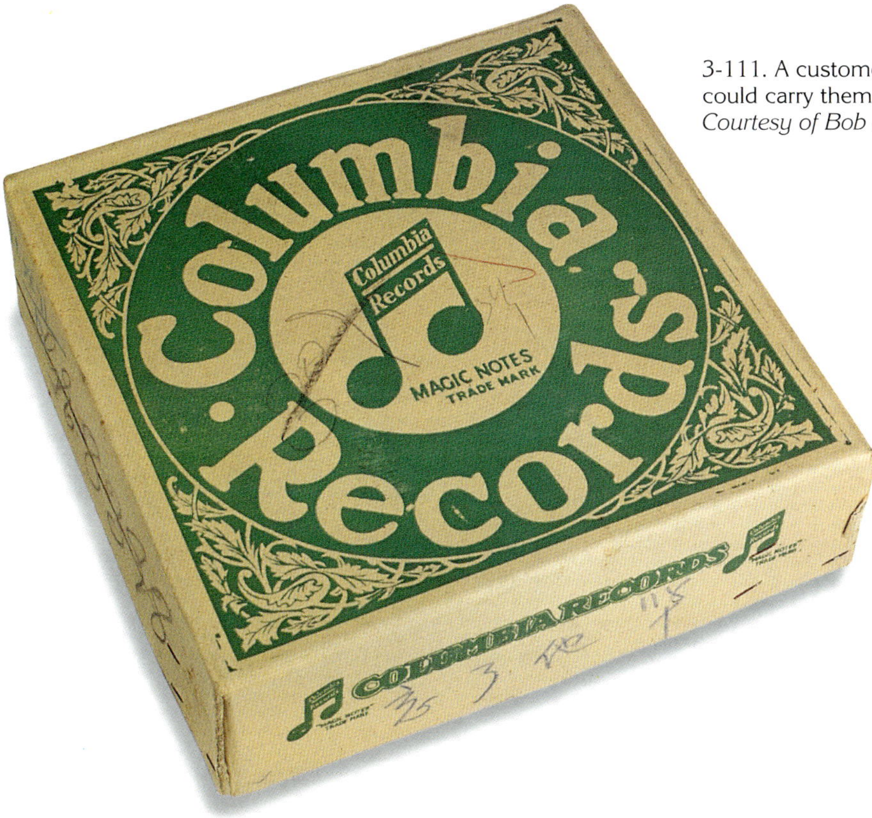

3-111. A customer buying a number of Columbia records could carry them home in this attractive cardboard box. *Courtesy of Bob and Karyn Sitter. (Value code: G)*

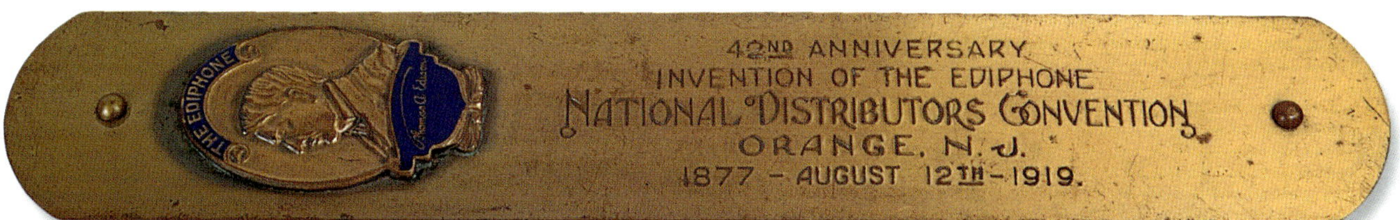

3-112. The office, or commercial, application of the Edison Phonograph was handled through the Ediphone Division of Thomas A. Edison, Incorporated. Although always a "poor relation" of the larger entertainment side of the business, it was the Ediphone that would survive through the 1920s and well beyond, as the last vestige of the once-great Edison Phonograph. Attendees at the 1919 Ediphone National Distributors Convention received this blotter (paper missing). 8 1/2" long. *(Value code: G)*

3-113. This 9 1/2" x 12 1/2" calendar from 1920 featured a sentimental illustration of "Baby's first Music Lesson on the Columbia Grafonola." *Courtesy of Tom and Sandi McCarthy. (Value code: F)*

3-114. This 18 3/4" x 42" wooden sign was mounted on heavy stanchions and equipped with fasteners, which suggests that it may have been on a delivery truck. *Courtesy of Tom and Sandi McCarthy. (Value code: E)*

3-115. This 18" x 72" (six feet wide!) wooden sign was once framed and hung outside Henrys Music Store. *Courtesy of Tom and Sandi McCarthy. (Value code: F)*

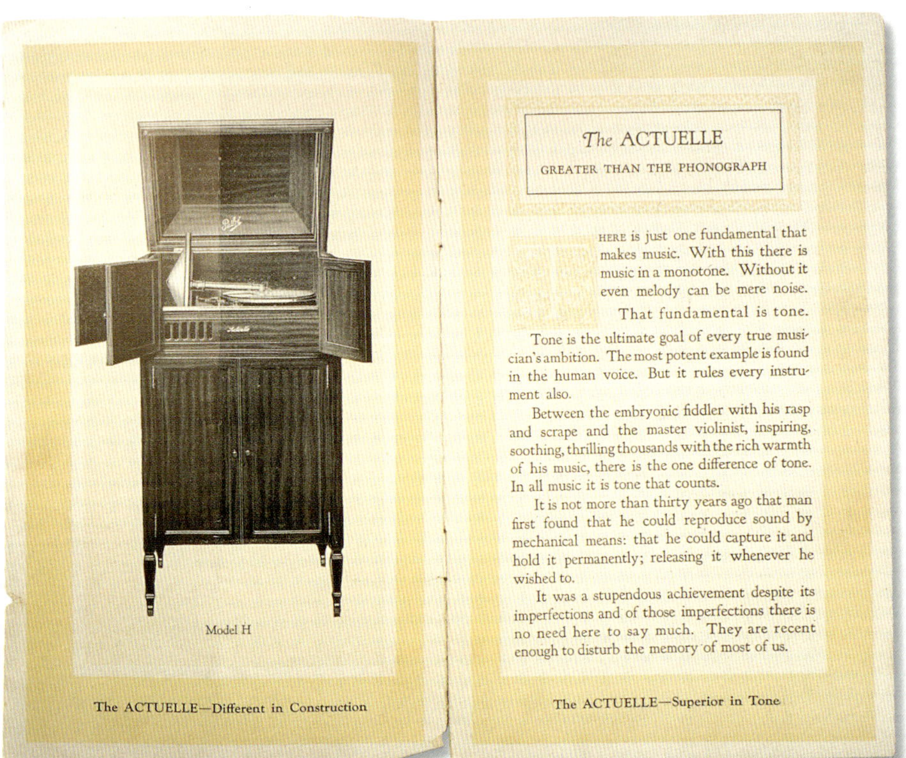

Model H

The ACTUELLE—Different in Construction

The ACTUELLE

GREATER THAN THE PHONOGRAPH

THERE is just one fundamental that makes music. With this there is music in a monotone. Without it even melody can be mere noise.

That fundamental is tone.

Tone is the ultimate goal of every true musician's ambition. The most potent example is found in the human voice. But it rules every instrument also.

Between the embryonic fiddler with his rasp and scrape and the master violinist, inspiring, soothing, thrilling thousands with the rich warmth of his music, there is the one difference of tone. In all music it is tone that counts.

It is not more than thirty years ago that man first found that he could reproduce sound by mechanical means: that he could capture it and hold it permanently; releasing it whenever he wished to.

It was a stupendous achievement despite its imperfections and of those imperfections there is no need here to say much. They are recent enough to disturb the memory of most of us.

The ACTUELLE—Superior in Tone

3-116. The vertical cut disc record boom in the United States ended right after World War I. The American branch of the Pathé company had navigated the teens, selling only vertical cut records and Pathéphones. However, when Victor experienced a rare 1919 reversal in patent court, those companies in the United States that had supported the vertical system rapidly faded or defected to the lateral side of the fence. Included in this rush was the formerly stalwart Pathé Frères Phonograph Company of Brooklyn, New York. It introduced the "Actuelle" (up-to-date) lateral cut discs, and instruments that employed a paper cone to diffuse the sound. 5" x 8". *Courtesy of Ernest Carl Allen. (Value code: G)*

3-117. After Theodore Roosevelt's death in late 1919, prominent Chicago Edison retailer Fred K. Babson issued this flyer advertising Edison Blue Amberol cylinder recordings of three speeches made by the "Happy Warrior." These records were reissues of 1912-catalogued offerings on earlier 4-minute wax Amberols. *Courtesy of Ernest Carl Allen. (Value code: F)*

3-118. The reverse of the Babson flyer featured a large photo of Roosevelt, as well as a pitch for regular Blue Amberol records. Measures 8 1/2" x 11". *Courtesy of Ernest Carl Allen.*

3-119. Columbia took a back seat to nobody in the realm of holiday advertisements. At the left is a flyer listing records for Christmas 1920 (4" x 7"). At the right is a machine brochure from 1916 (4 5/8" x 7"). *Courtesy of Ernest Carl Allen. (Value code: G)*

Chapter Four:

1921-1930

4-1. The vast facilities of the Victor Talking Machine Company, Camden, New Jersey, at the height of its power. 17" x 38 1/4" including frame. *Courtesy of Ernest Carl Allen. (Value code: D)*

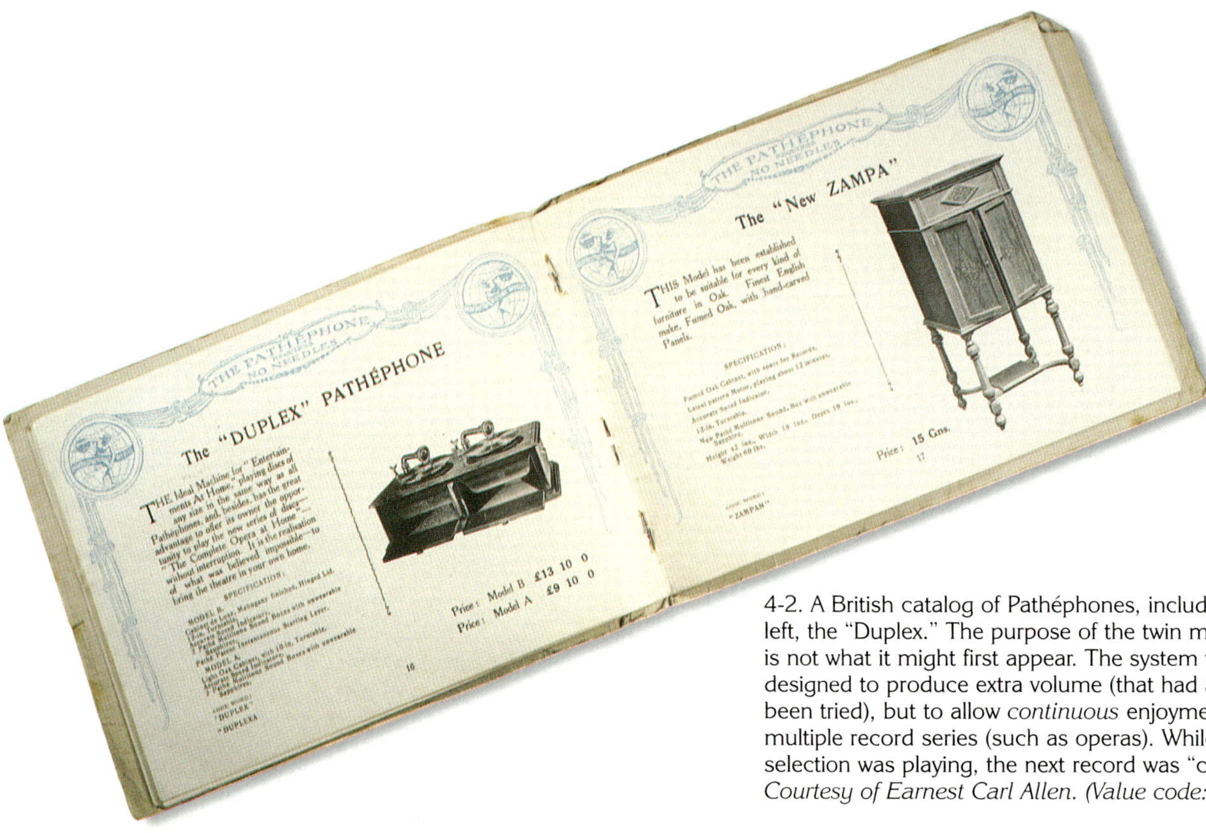

4-2. A British catalog of Pathéphones, including, at the left, the "Duplex." The purpose of the twin mechanisms is not what it might first appear. The system was not designed to produce extra volume (that had already been tried), but to allow *continuous* enjoyment of multiple record series (such as operas). While a selection was playing, the next record was "cued up." *Courtesy of Earnest Carl Allen. (Value code: G)*

4-3. Jobbers and dealers of Edison Diamond Disc Phonographs attended annual conventions, which in 1920 and 1921, for instance, were known as "Caravans." This issue of *Edison Diamond Points*, the dealer magazine for Diamond Disc Phonograph merchants, gave a report of the activities. This 3" diameter paperweight was a souvenir. *Courtesy of the late Dave Abfalter. (Value code: F)*

4-4. A close-up of a similar paperweight from Chicago. The Model "C-250" was the featured Diamond Disc illustration during this period. *(Value code: F)*

4-5. This pinback with ribbon was worn at the 1920 Caravan Convention by a dealer from Hamburg, New York. 4 1/2" long. *(Value code: F)*

4-6. Also from the 1920 Edison Caravan Convention was this menu, and painted chalk camel. *Courtesy of the late Dave Abfalter. (Value code: VR)*

4-7. When the Philadelphia Victor Jobbers hosted a convention of Victor dealers in March 1922, this program was supplied – in the shape of a Victrola packing crate! *Courtesy of the Johnson Victrola Museum/Delaware State Museums. (Value code: F)*

4-8. The idea of dressing in evening clothes in order to sit before the Victrola may seem utterly astonishing to us today, but this imagery recurs often in talking machine advertising. 9 3/8" x 11 7/8", 1920. *Courtesy of Ernest Allen. (Value code: F)*

4-9. A large, silk-screened Columbia banner, measuring 23 1/2" x 59". *Courtesy of the Charles Hummel collections. (Value code: VR)*

4-10. A courtly old duffer busses his babe – typical of the high-falutin imagery that pervaded Columbia and Victor advertising circa 1921. 10" x 13 1/8". *Courtesy of Ernest Allen. (Value code: F)*

4-11. The art of building one's own phonograph cabinet and equipping it with readily available "generic" parts could rightly be called a "cottage industry" during the 1919-1922 period. Firms such as the Modern Phonograph & Supply Company ("MAKAPHONE") and the Choraleon Phonograph Company (as shown here) assured would-be carpenters that a "handsome machine [can be] built out of an ordinary dry goods box" by sending for a set of plans and purchasing the motor and parts from the company. Most surviving examples of such home-built cabinets fall short of the handiwork displayed in Mr. Charles Leaderer's workshop. *(Value code: H)*

4-12. These letter openers all date from the late teens and early 1920s. Mechanical advertising pencils such as the example at bottom are not often seen. *Courtesy of Tom and Sandi McCarthy. (Value code: F)*

4-13. Not an advertising piece, but an instance where the phonograph brought a gift to life. Manufactured by the Talking Book Corporation, New York City. 7 1/2" long. *Courtesy of Walter and Carol Myers. (Value code: F)*

4-14. This 8 1/2" high figurine makes a pretty appealing pitch for Columbia records. We would like to have seen this robust motif reflected in more of Columbia's advertising... *Courtesy of Tom and Sandi McCarthy. (Value code: VR)*

4-15. This substantial Columbia counter display plaque dates from the late teens or early 1920s, and was manufactured by the Goodman Company of New York City. Measures 8 1/4" x 24 1/4". *Courtesy of the Charles Hummel collections. (Value code: B)*

4-17. A lithographed bottle of Columbia Grafonola Oil, from the early 1920s. Needless to say, Grafonola Oil did not differ significantly from any comparable oil intended for light machinery – but selling it under the "house brand" reinforced recognition of all Columbia products. *Courtesy of the Domenic DiBernardo collection. (Value code: G)*

4-16. A store advertising placard, 13 1/4" x 21 1/2", promoting Columbia records, from 1921. *Courtesy of the Domenic DiBernardo collection. (Value code: E)*

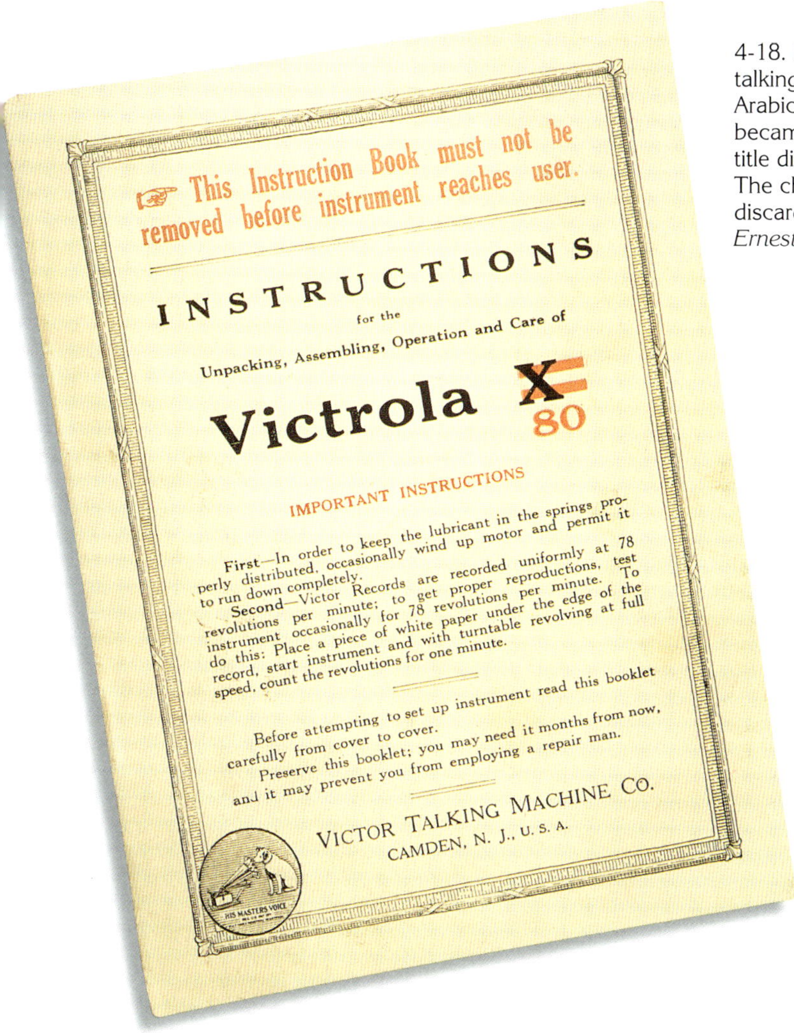

4-18. In 1921, Victor slightly redesigned its line of talking machines, and switched from Roman numeral to Arabic number designations. The popular Victrola "X" became the "80" (unlike the policy of some firms, this title did not reflect the retail price, which was $100.). The changes in the machine were insufficient to warrant discarding the existing instruction booklets. *Courtesy of Ernest Allen. (Value code: H)*

4-19. A very unusual circular flyer (5" diameter) from the early 1920s illustrating the Edison Diamond Disc line inside. Edison publicists carefully nurtured the "favorite invention" sentiment throughout the life of the Edison Phonograph. *Courtesy of Tom and Sandi McCarthy. (Value code: G)*

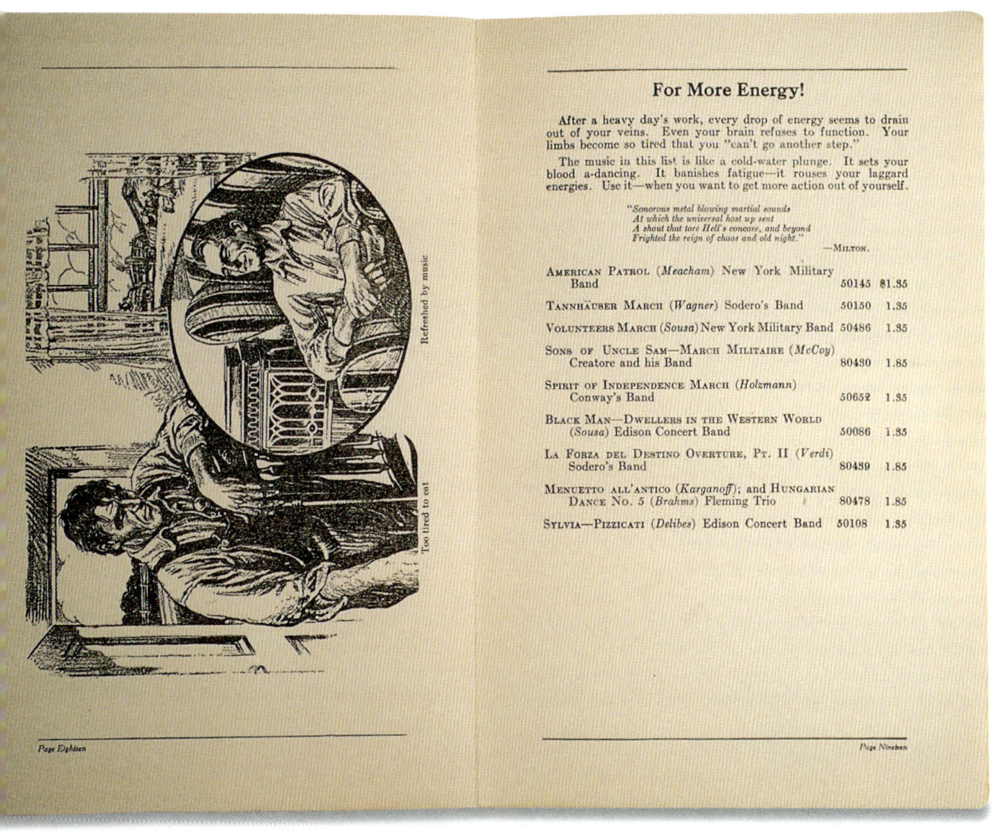

For More Energy!

After a heavy day's work, every drop of energy seems to drain out of your veins. Even your brain refuses to function. Your limbs become so tired that you "can't go another step."

The music in this list is like a cold-water plunge. It sets your blood a-dancing. It banishes fatigue—it rouses your laggard energies. Use it—when you want to get more action out of yourself.

"Sonorous metal blowing martial sounds
At which the universal host up sent
A shout that tore Hell's concave, and beyond
Frighted the reign of chaos and old night."
—MILTON.

AMERICAN PATROL (*Meacham*) New York Military Band	50145	$1.35
TANNHÄUSER MARCH (*Wagner*) Sodero's Band	50150	1.35
VOLUNTEERS MARCH (*Sousa*) New York Military Band	50486	1.35
SONS OF UNCLE SAM—MARCH MILITAIRE (*McCoy*) Creatore and his Band	80430	1.85
SPIRIT OF INDEPENDENCE MARCH (*Holzmann*) Conway's Band	50652	1.35
BLACK MAN—DWELLERS IN THE WESTERN WORLD (*Sousa*) Edison Concert Band	50086	1.35
LA FORZA DEL DESTINO OVERTURE, PT. II (*Verdi*) Sodero's Band	80439	1.85
MENUETTO ALL'ANTICO (*Karganoff*); and HUNGARIAN DANCE NO. 5 (*Brahms*) Fleming Trio	80478	1.85
SYLVIA—PIZZICATI (*Delibes*) Edison Concert Band	50108	1.35

Page Nineteen

4-20. Forget herbal potions and vitamin cocktails – energize the enervated spirit with Edison Diamond Disc records. From an Edison brochure entitled "Mood Music," 1921. *Courtesy of Martin F. Bryan. (Value code: F)*

4-21. "Imitation" disc records, made to scale, were produced for a number of record companies in the 1920s. They were manufactured from a durable substance, such as lithographed steel, as display pieces for stores, and were intended to occupy positions of exposure that would quickly ruin a conventional record. The heat of store windows, ultraviolet light and accumulated dust could do little harm to imitation discs. The example shown here was meant to represent a 10" Edison Diamond Disc record, but is less a doppelganger than a suggestion, in this case fashioned of thick pasteboard. The sales slogan is somewhat ironic. It's true that the Edison vertical cut disc system, introduced in 1912 and manufactured until the middle of 1929, was technically advanced, but the preponderance of stodgy and out-of-date repertoire in the catalog must have seemed "real" only to the Old Oaken Bucket crowd of middle age and beyond. *Courtesy of the Charles Hummel collections. (Value code: F)*

4-22. These three Edison electrotypes from the early 1920s were used for newspaper advertising. The image has been reversed here so that the text may be easily read. *Courtesy of the Gfell family collection. (Value code: F)*

4-23. This cut for Columbia children's records featured an area in the center where the local dealer could insert the latest selections, and have the entire ad printed in the hometown newspaper. The image has been reversed so that it may be easily read. Measures 6 1/8" x 8 3/8". *Courtesy of Norm and Janyne Smith. (Value code: F)*

4-24. Two extremely stylish catalogs. Left, Starr phonographs, manufactured by the Starr Piano Company of Richmond, Indiana, 5 1/8" x 7 1/8". Below, a light and fluid drawing completes the visual connection between the billowing outline of the woman in motion and the sinuous silhouette of the Sonora. 3 1/2" x 6". *Courtesy of Ernest Allen. (Value code: G)*

4-25. The March 1, 1922 issue of the *Phonograph and Talking Machine Weekly* featured this advertisement for "Da-Lite" electric displays. The use of removable gel display panels would keep the latest Victor records in the public eye, and cost the dealer $6.00 a month. *Courtesy of the Johnson Victrola Museum/Delaware State Museums.*

4-26. The actual display illuminated. Gel measures 11" x 16 3/4." *Courtesy of the Johnson Victrola Museum/Delaware State Museums. (Value code: VR)*

4-27. Another "Da-Lite" gel advertising one of Victor's most popular records. *Courtesy of the Johnson Victrola Museum/Delaware State Museums. (Value code: VR)*

4-28. A similar lighted display used a reverse-painted glass panel. Measures 7" x 13". *Courtesy of Tom and Sandi McCarthy. (Value code: VR)*

4-29. This cigarette dispenser represents a stylized table model Victrola with Nipper seated on top. Base measures 3 1/4" x 4". *Courtesy of Tom and Sandi McCarthy. (Value code: VR)*

4-30. This 5 1/2" high silver-plated teapot and tray are engraved with the Victor trademark – almost certainly a presentation piece. *Courtesy of Tom and Sandi McCarthy. (Value code: VR)*

4-31. Though "Valuphone" may have been a line of unpretentious cabinet talking machines during the early 1920s, the model shown to the left was equipped with a first rate "Motor of Quality" mechanism, a surprise to those who might have expected rubbish. *Courtesy of Ernest Carl Allen. (Value code: G)*

4-32. This stock certificate was issued in April 1922 by a little-known company: Canadian Edison Phonographs, Limited. *Courtesy of Tom and Sandi McCarthy. (Value code: F)*

4-33. What talking machine enthusiast wouldn't be happy to find this 4" x 9" Victor Gift Certificate in his or her stocking? *Courtesy of Fred Guenther. (Value code: H)*

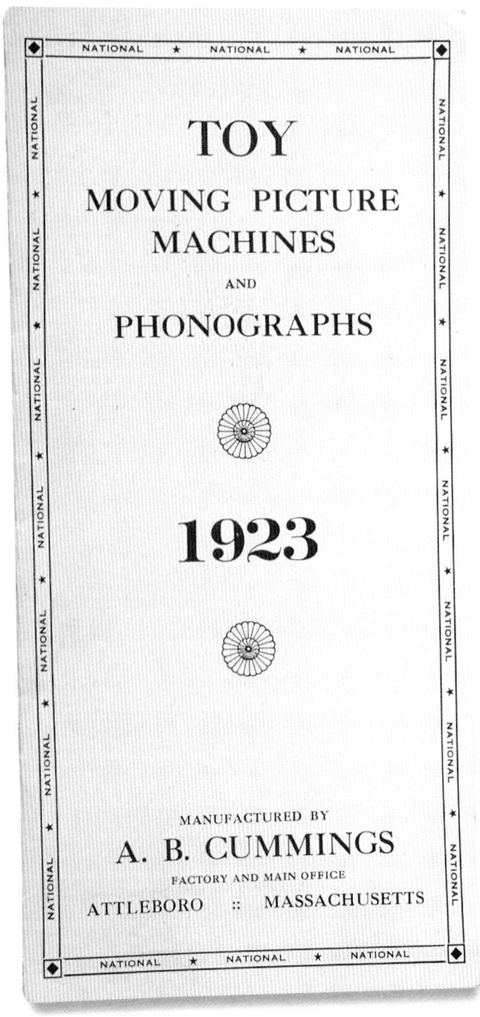

4-34. A. B. Cummings of Attleboro, Massachusetts, issued this catalog of toy film projectors and phonographs in 1923. Measures 4 7/8" x 8 3/4". *Courtesy of Ernest Carl Allen. (Value code: G)*

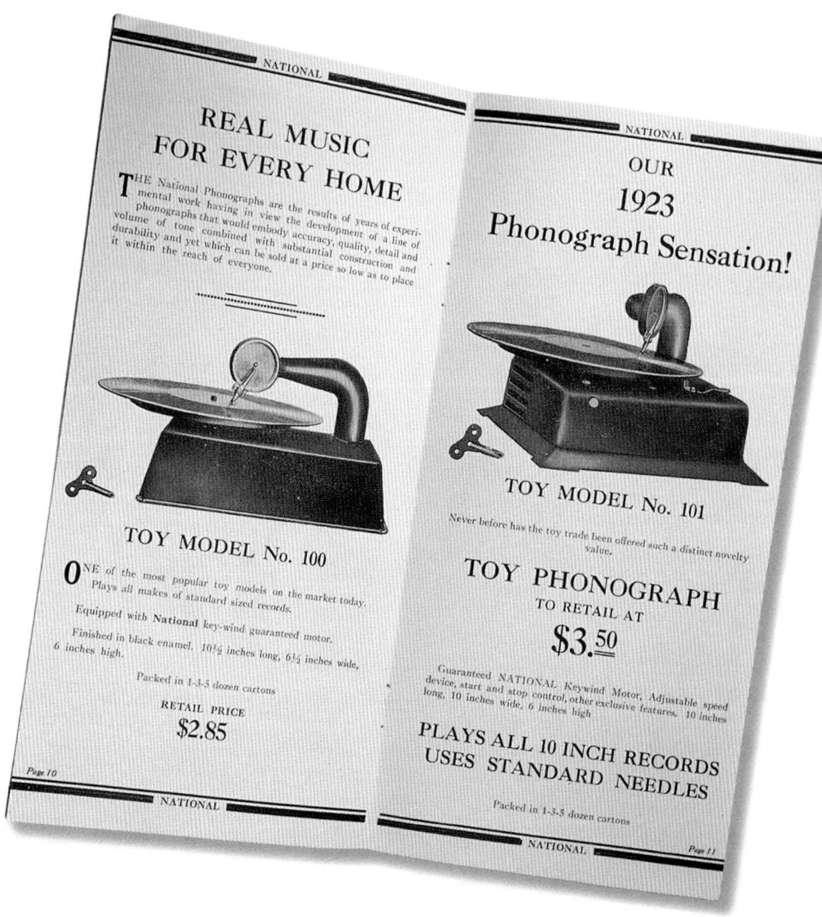

4-35. Included are models intended to retail for as little as $2.85! *Courtesy of Ernest Carl Allen.*

NATIONAL

TOY MODEL No. 103

Equipped with Crank Wind Motor

Special Reproducer

PLAYS TEN AND TWELVE INCH RECORDS

SPECIFICATIONS

SIZE—
10 inches long, 10 inches wide, and 6 inches high.

MOTOR—
National silent motor, side wind, with adjustable speed regulator.

FINISH—
Grained enamel. Nickeled fittings.

TONE ARM—
Standard size tapered arm with high grade reproducer.

RETAIL PRICE
$5.50

Page 12

NATIONAL

AT LAST! A LOW PRICED PORTABLE.

MODEL No. 104

THIS popular-priced portable phonograph satisfies the demand for a sport model so small, light and efficient that it may be carried at any time, anywhere without inconvenience.

JUST the thing for camp or bungalow. A first class portable within the price reach of everyone. An ideal Xmas gift for the kiddies.

CLOSED

SPECIFICATIONS

DIMENSIONS — when closed 10½ inches long 10½ inches wide 4 inches high.

MOTOR — carefully constructed of finest materials. Guaranteed.

FINISH—Imitation leather, nickeled fittings throughout.

RETAIL PRICE
$10.00

Page 13

4-36. For the big spender, the $10.00 Model "104" was described as, "An ideal Xmas gift for the kiddies." *Courtesy of Ernest Carl Allen.*

4-37. This cork blotter added a touch of elegance to the desk of a music dealer who handled products of the Victor Talking Machine Company. 2 3/4" x 5 7/8". *Courtesy of the Johnson Victrola Museum/ Delaware State Museums. (Value code: F)*

4-38. An attractive metal dealer's display, this could accommodate several copies of six "Latest Victor Hits" on 10" diameter discs. *Courtesy of the Johnson Victrola Museum/ Delaware State Museums. (Value code: VR)*

4-39. A May 1923 brochure promoted "summer music" during a traditionally slow time for the business. Note the Victrola "50" being carried by the jaunty lad preparing to board a yacht. *Courtesy of Tom and Sandi McCarthy. (Value code: F)*

4-40. Reverse of previous item illustrated other Victrolas and described the joys of a "Victrola vacation." *Courtesy of Tom and Sandi McCarthy.*

4-41. Schwegler Brothers of Buffalo, New York offered playing cards to its patrons in order to keep the firm's name close at hand. *Courtesy of Tom and Sandi McCarthy. (Value code: F)*

4-42. These 3 1/2" record dusters date from the mid 1920s. The yellow "Mayberry's" duster illustrates the 1922 model of a Victrola "260," one of the often derided "humpback" series. Victor resisted the "flat top" cabinet configuration for years – to the exasperation of its dealers who saw the more conventional console models of other brands selling while Victors remained in the store. *Courtesy of Tom and Sandi McCarthy. (Value code: F, G)*

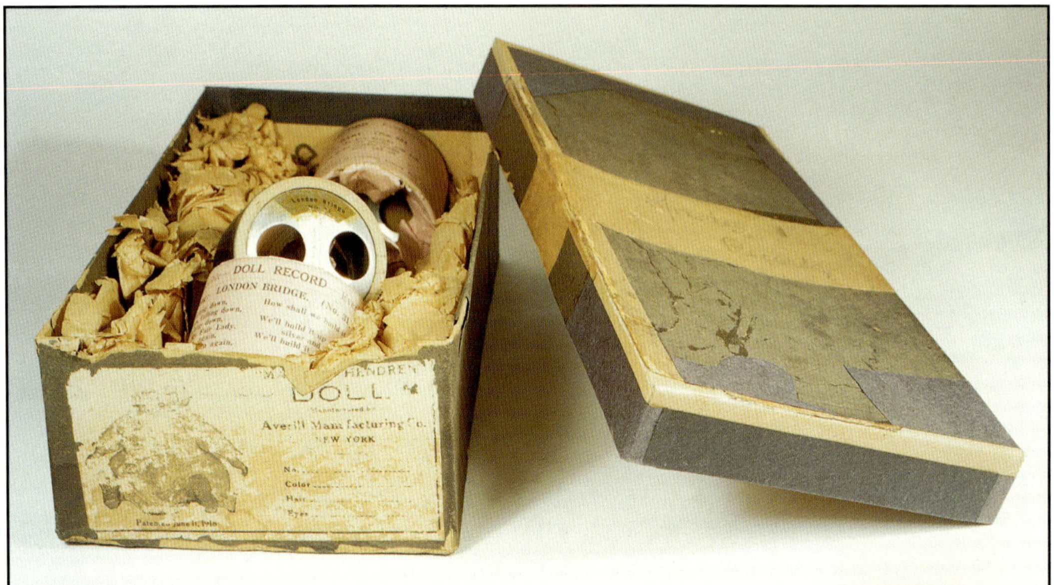

4-43. Madame Hendren talking doll cylinder records, from the Averill Manufacturing Company, New York City, 1923. The records were celluloid over cardboard, and each had a metal top with two punched holes to facilitate handling. The composition and cloth dolls enclosed a miniature phonograph. *Courtesy of Jean-Paul Agnard. (Value code: G, each early style cylinder in printed container)*

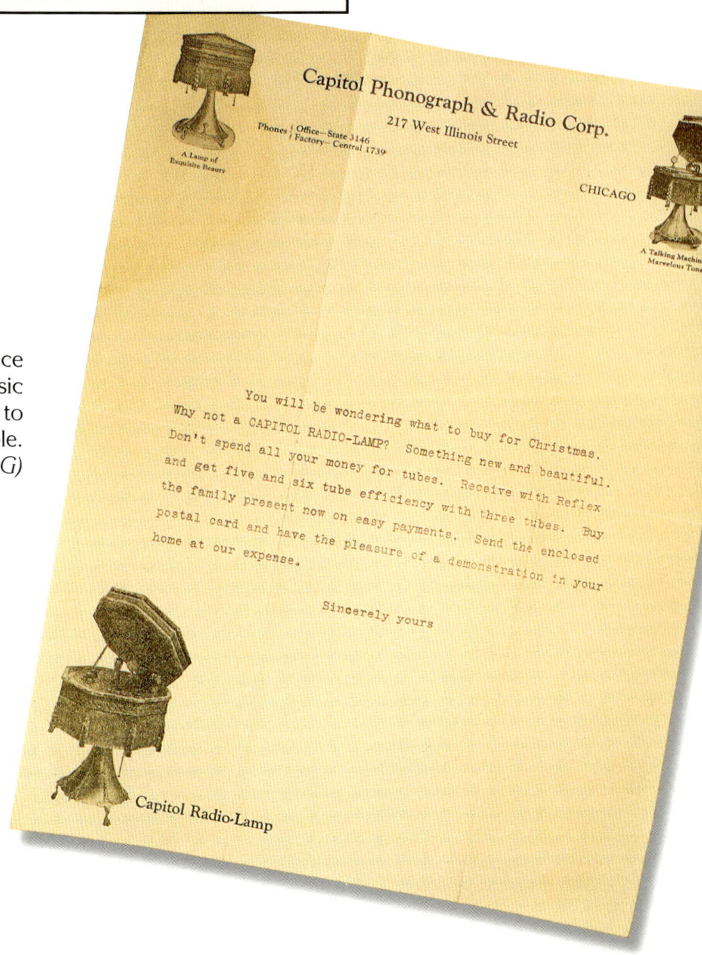

Capitol Phonograph & Radio Corp.

217 West Illinois Street

Phones | Office—State 3146 | Factory—Central 1739

CHICAGO

A Lamp of Exquisite Beauty

A Talking Machine Marvelous Tone

You will be wondering what to buy for Christmas. Why not a CAPITOL RADIO-LAMP? Something new and beautiful. Don't spend all your money for tubes. Receive with Reflex and get five and six tube efficiency with three tubes. Buy the family present now on easy payments. Send the enclosed postal card and have the pleasure of a demonstration in your home at our expense.

Sincerely yours

Capitol Radio-Lamp

4-44. Capitol Lamp Phonographs had been around since 1919, but why not extend the idea of recorded music emanating from a functional element of décor to *broadcast* music? Easy payment terms available. *Courtesy of Ernest Allen. (Value code: G)*

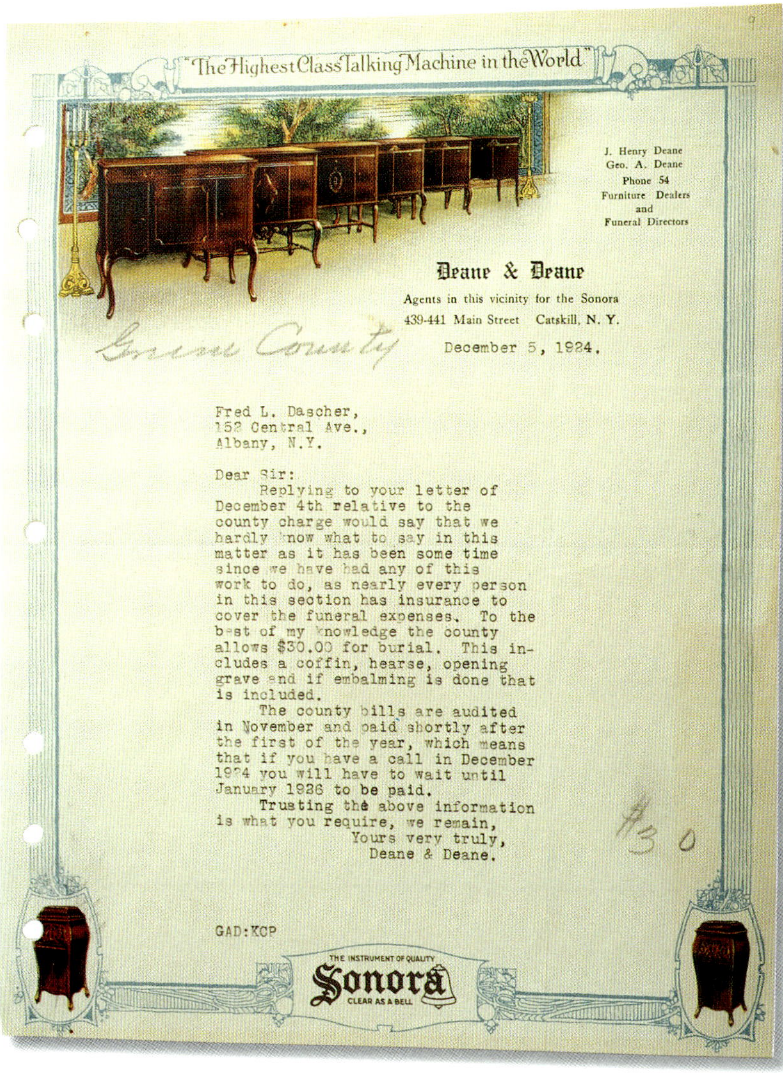

"The Highest Class Talking Machine in the World"

J. Henry Deane
Geo. A. Deane
Phone 54
Furniture Dealers
and
Funeral Directors

Deane & Deane

Agents in this vicinity for the Sonora
439-441 Main Street Catskill, N. Y.

Greene County

December 5, 1924.

Fred L. Dascher,
152 Central Ave.,
Albany, N.Y.

Dear Sir:

Replying to your letter of December 4th relative to the county charge would say that we hardly know what to say in this matter as it has been some time since we have had any of this work to do, as nearly every person in this section has insurance to cover the funeral expenses. To the best of my knowledge the county allows $30.00 for burial. This includes a coffin, hearse, opening grave and if embalming is done that is included.

The county bills are audited in November and paid shortly after the first of the year, which means that if you have a call in December 1924 you will have to wait until January 1926 to be paid.

Trusting the above information is what you require, we remain,

Yours very truly,
Deane & Deane.

#30

GAD:KCP

THE INSTRUMENT OF QUALITY
Sonora
CLEAR AS A BELL

4-45. Certain professions became allied with selling talking machines. Bicycle shops, clock stores, merchants who sold sewing machines and other domestic appliances – but *funeral homes*? In fact, undertakers frequently sold furniture, and after the talking machine began to inhabit floor-standing cabinets, phonographs became a logical extension of the mortician's fare. One supposes that the bereaved families could be subtly wooed into a buying mood by soothing demonstrations of recorded art such as "Beautiful Isle of Somewhere." Note, in the text, that a complete burial, including coffin, hearse and embalming (if wanted), could be accomplished for $30.00. *Courtesy of Tom Sideleau. (Value code: H)*

4-46. This 14" Victor advertising fan featured a winsome young lass whose parasol virtually begs to be twirled. *Courtesy of the Johnson Victrola Museum/Delaware State Museums. (Value code: F)*

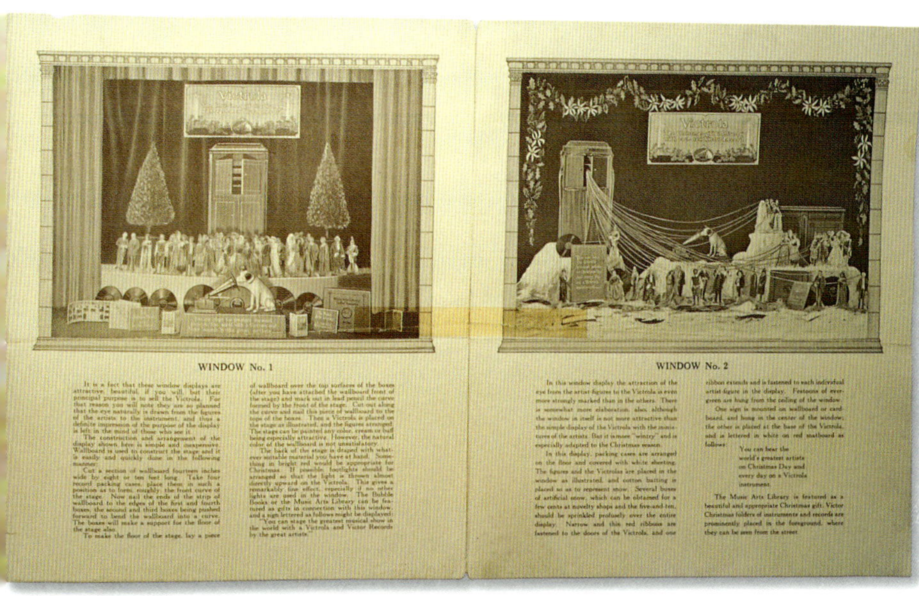

4-47. This 9 1/2" x 16 1/8" publication, showing Victor's suggestions for Christmas window displays, dates from 1924. Note the use of die-cut "artists" in both windows. These cardboard figures were lithographed representations of Victor's roster of recording stars. *Courtesy of Tom and Sandi McCarthy. (Value code: G)*

4-48. Oddly, this Nipper figurine on a walnut base was a complimentary item from an electroplating corporation. Keulman's Jewelry Store may have sold Victor merchandise, but the connection is uncertain. *(Value code: E)*

4-49. From October 1924, the colorful cover of a Victor brochure. *Courtesy of Ernest Allen. (Value code: G)*

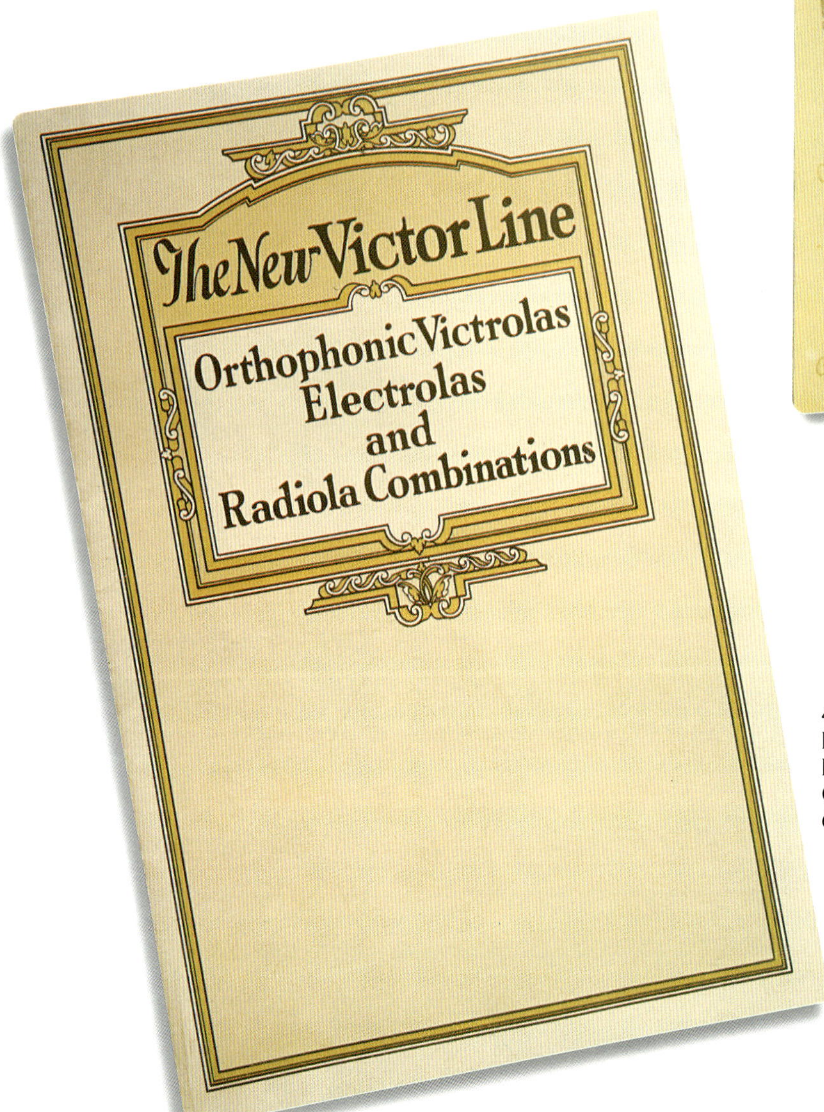

4-50. A catalog announcing the "New Victor Line" of late 1925. The most noteworthy advance was the logarithmically-designed internal horn found in the Orthophonic acoustic Victrolas. Measures 6" x 9". *Courtesy of Ernest Carl Allen. (Value code: E)*

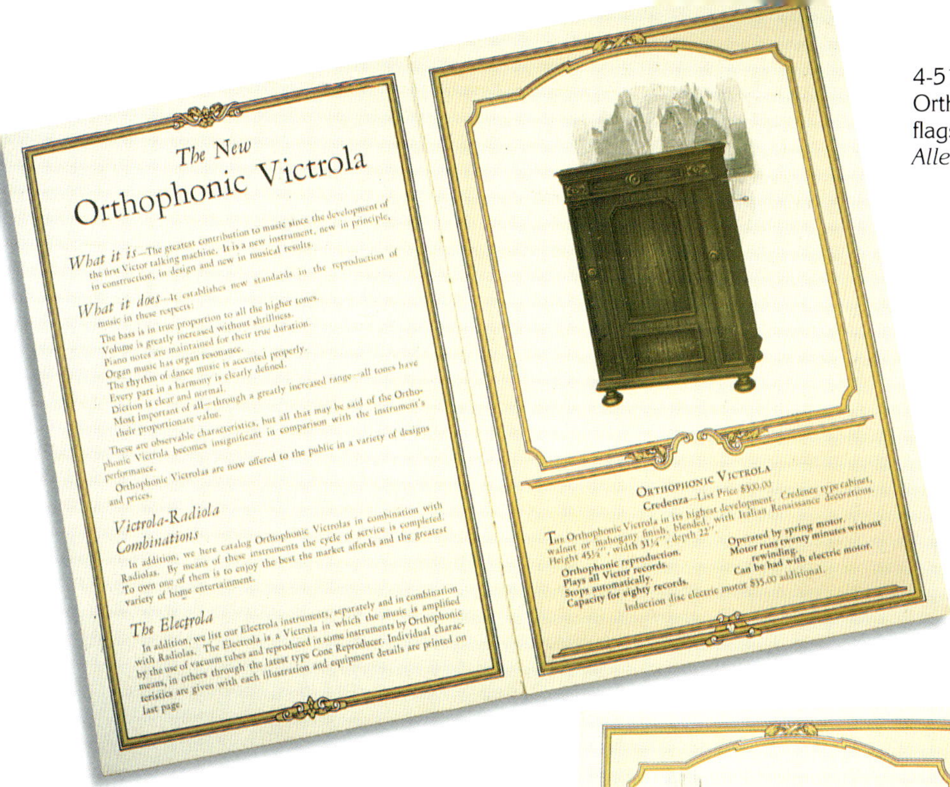

4-51. The opening pages described the new Orthophonic models, while showcasing the flagship "Credenza." *Courtesy of Ernest Carl Allen.*

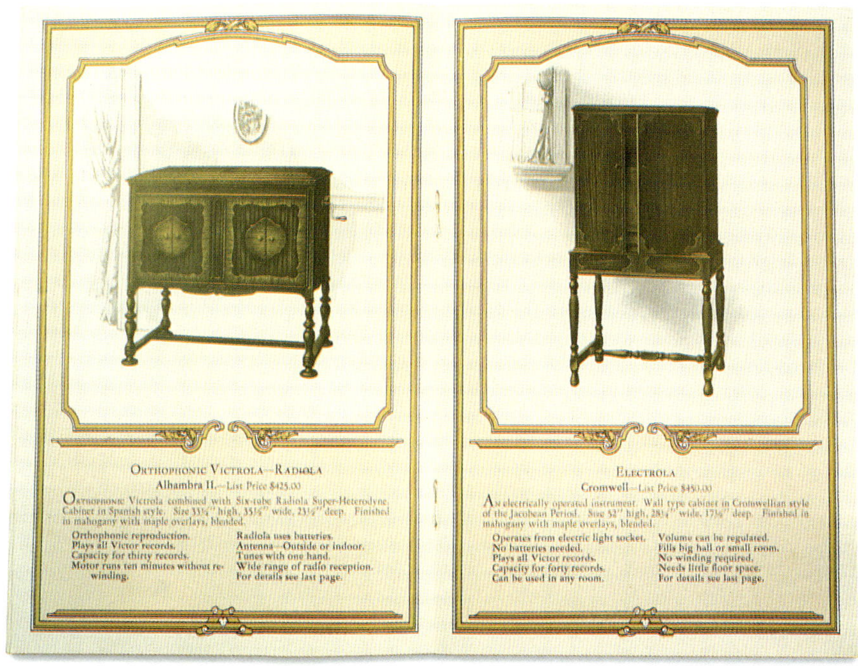

4-52. Victrolas from this period could be had with spring motors, mains current electric motors, battery-powered radios, and, of course, the highly touted Orthophonic reproduction from the mathematically-designed internal horn and ball-bearing soundbox. *Courtesy of Ernest Carl Allen.*

4-53. For those with some money to spend, Victor could oblige with expensive models. *Courtesy of Ernest Carl Allen.*

Orthophonic Victrola—Orthophonic Electrola—Radiola

Borgia II.—List Price $1000.00

Orthophonic Victrola and Orthophonic Electrola in their highest development, combined with Eight-tube Radiola Super-Heterodyne. Exceptional Radio reception through Orthophonic system. Cabinet in Italian Renaissance style finished in walnut, blended. Size 45½″ high, 44″ wide, 22¼″ deep.

Operates from electric light socket.
Plays all Victor records on Orthophonic Victrola or gives electrical reproduction of records on Orthophonic Electrola.
No batteries needed.
Capacity for eighty records.
Radiola tunes with one hand.
Locates stations easily.
Built-in loop antenna in cabinet.
Can be used in any room.
Volume can be regulated.
Fills big hall or small room.
Reproduces all that can be had on records or from the air.
Economical—uses little current.
Cabinet contains all equipment.
For details see last page.

Specifications

ORTHOPHONIC VICTROLA—RADIOLA

Alhambra I.—List Price $350.00

Five-tube Radiola tuned radio frequency. Operates on dry-cell batteries. Indoor or outside antenna with ground. Victor inspected radio tubes provided, Four UX-199, One UX-120. Equipped with concealed control valve permitting Orthophonic reproduction of records or radio reception. Radio reception may also be had through separate loud speaker—jack provided. Spring motor. Non-set automatic eccentric groove brake. Speed indicator and regulator. Twelve-inch turntable. Complete with record albums. Two-piece lid, one-hand lid supports.

ORTHOPHONIC VICTROLA—RADIOLA

Alhambra II.—List Price $425.00

Six-tube Radiola Super-Heterodyne. Operates on dry-cell batteries. Indoor or outside antenna with ground; or with minor alterations indoor loop antenna may be used. Victor inspected radio tubes furnished, Five UX-199, One UX-120. Equipped with concealed control valve permitting Orthophonic reproduction of records or radio reception. Radio reception may also be had through separate loud speaker—jack provided. Spring motor. Non-set automatic, eccentric groove brake. Speed indicator and regulator. Twelve-inch turntable. Complete with record albums. Two-piece lid, one-hand lid supports.

ELECTROLA

Cromwell—List Price $450.00

Operates on alternating current, 100 to 120 volts, 60 cycles. Volume control. Electric pick-up with latest type ten-inch cone reproducer. Instrument is a splendid loud speaker and amplifier for an independent radio set—jack provided. Victor inspected radio tubes furnished, One UX-199, One UX-210, One UX-216-B. Induction disc electric motor drives turntable. Non-set automatic eccentric groove switch. Combination compartment and pilot lamp. Speed indicator and regulator. Twelve-inch turntable. Complete with record albums. Fixed top, drop panel over operating compartment. Double doors opening in front.

ELECTROLA

Tuscany—List Price $550.00

Operates on alternating current, 100 to 120 volts, 60 cycles. Volume control. Electric pick-up with latest type six-inch cone reproducer. Instrument is a most remarkable loud speaker and amplifier for an independent radio set—jack provided. Victor inspected radio tubes furnished, One UX-199, One UX-210, Two UX-216-B. One UV-876 ballast tube. Induction disc electric motor drives turntable. Non-set automatic eccentric groove switch. Combination compartment and pilot lamp. Speed indicator and regulator. Twelve-inch turntable. Complete with record albums. Fixed top, drop panel over operating compartment. Double doors opening in front.

ORTHOPHONIC VICTROLA—RADIOLA

Florenza—List Price $550.00

Six-tube Radiola Super-Heterodyne. Operates on dry-cell batteries. Concealed directional door loop antenna. Victor inspected radio tubes provided, Five UX-199, One UX-120. Equipped with concealed control valve permitting Orthophonic reproduction of records or radio reception. Radio reception may also be had through separate loud speaker—jack provided. Spring motor. Non-set automatic eccentric groove brake. Speed indicator and regulator. Twelve-inch turntable. Complete with record albums. Two-piece lid with spring-balanced lid supports. Double doors opening in front.

ORTHOPHONIC VICTROLA—RADIOLA

Borgia I.—List Price $675.00

Eight-tube Radiola Super-Heterodyne. Concealed directional built-in loop antenna. Victor inspected radio tubes provided, Seven UX-199, One UX-120. Operates on dry-cell batteries. Equipped with concealed control valve permitting Orthophonic reproduction of records or radio reception. Jack also pro vided for separate speaker. Spring motor. Nor set automatic eccentric groove brake. Spe indicator and regulator. Twelve-inch turntab Complete with record albums. Two-piece with air lid supports.

4-54. For those with A LOT of money to spend, only the "Borgia II" would do. *Courtesy of Ernest Carl Allen.*

4-55. Trolley cards were a common form of advertising during the first half of the twentieth century. This sign from Christmastime 1926 touts one of the "New Orthophonic" Victor records. 10 1/8" x 20 1/2". *Courtesy of the Johnson Victrola Museum/Delaware State Museums. (Value code: E)*

4-56. This homey trolley card promoted Victor's records of "Little Mother," as well as the 1928 Fox film, "Four Sons." 10 1/8" x 20 1/2". *Courtesy of the Johnson Victrola Museum/Delaware State Museums. (Value code: E)*

4-57. The "modern" look of this trolley card is quite a contrast to the previous example! 10 1/8" x 20 1/2". Courtesy *of the Johnson Victrola Museum/Delaware State Museums. (Value code: E)*

4-58. It's interesting to consider that the first talking machines designed for home use were housed in wooden boxes modeled after sewing machine cabinets. The horn had to be appended to the instrument. After a dozen years had passed, the public wanted a phonograph that didn't *look* like a phonograph, and the Victrola, with its hidden horn, was introduced. Several companies, notably Columbia, marketed phonographs that masqueraded as something else: pianos, desks, tables, lamps, and secretaries. As time passed, the "Victrola's" name became generic, and its domed lid, upright configuration, and curved corner moldings became icons. By the mid-to-late 1920s, the phonograph had become so familiar that more utilitarian appliances, such as furnaces, emulated its cabinet. Shown is a 14" advertising fan for Washington Furnaces. The model on the left is accompanied by the text: "No – This Is Not A Victrola." The "Martha Washington" furnace in the center was a close copy of Victor's popular Orthophonic "Credenza." *Courtesy of the Johnson Victrola Museum/Delaware State Museums. (Value code: F)*

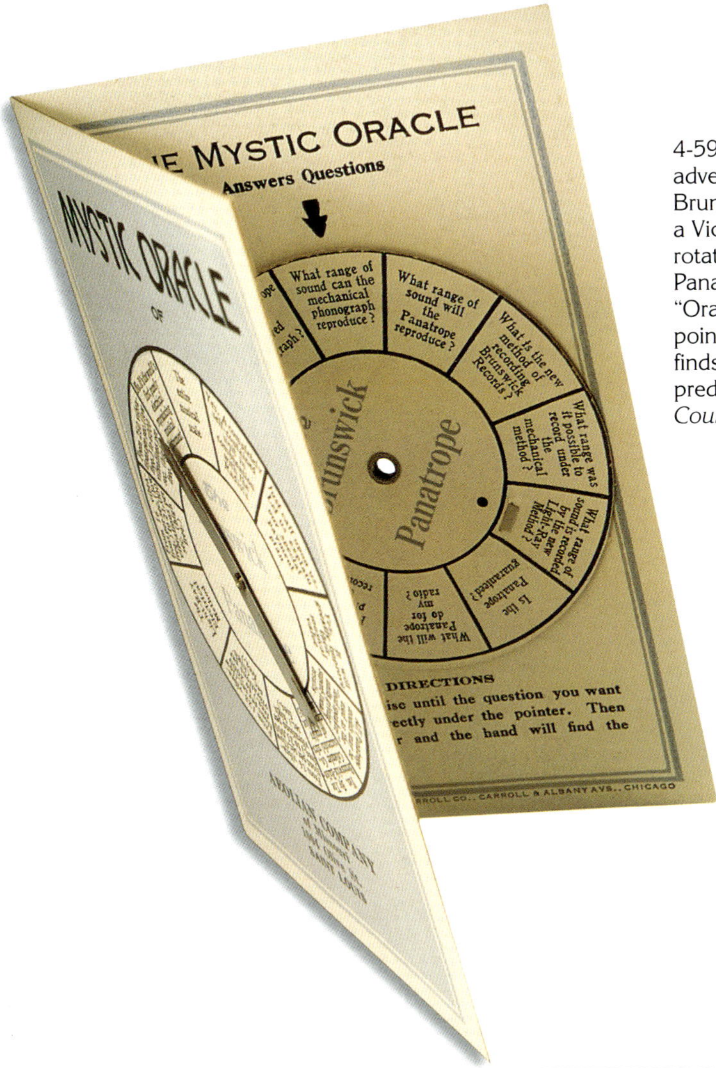

4-59. The "Mystic Oracle" was an advertising novelty, here promoting the Brunswick "Panatrope" but also seen in a Victor version. The interior disc is rotated to position a question ("Is the Panatrope guaranteed?") at the top, the "Oracle" is closed, and a magnetic pointer on the front face "magically" finds the correct (and mechanically predetermined) "answer." 3 1/2" x 5 1/2". *Courtesy of Ernest Allen. (Value code: F)*

4-60. A full set of documents, as accompanied a Brunswick "Panatrope" (acoustic style, Brunswick also designed an electric version) to the customer. *(Value code: H)*

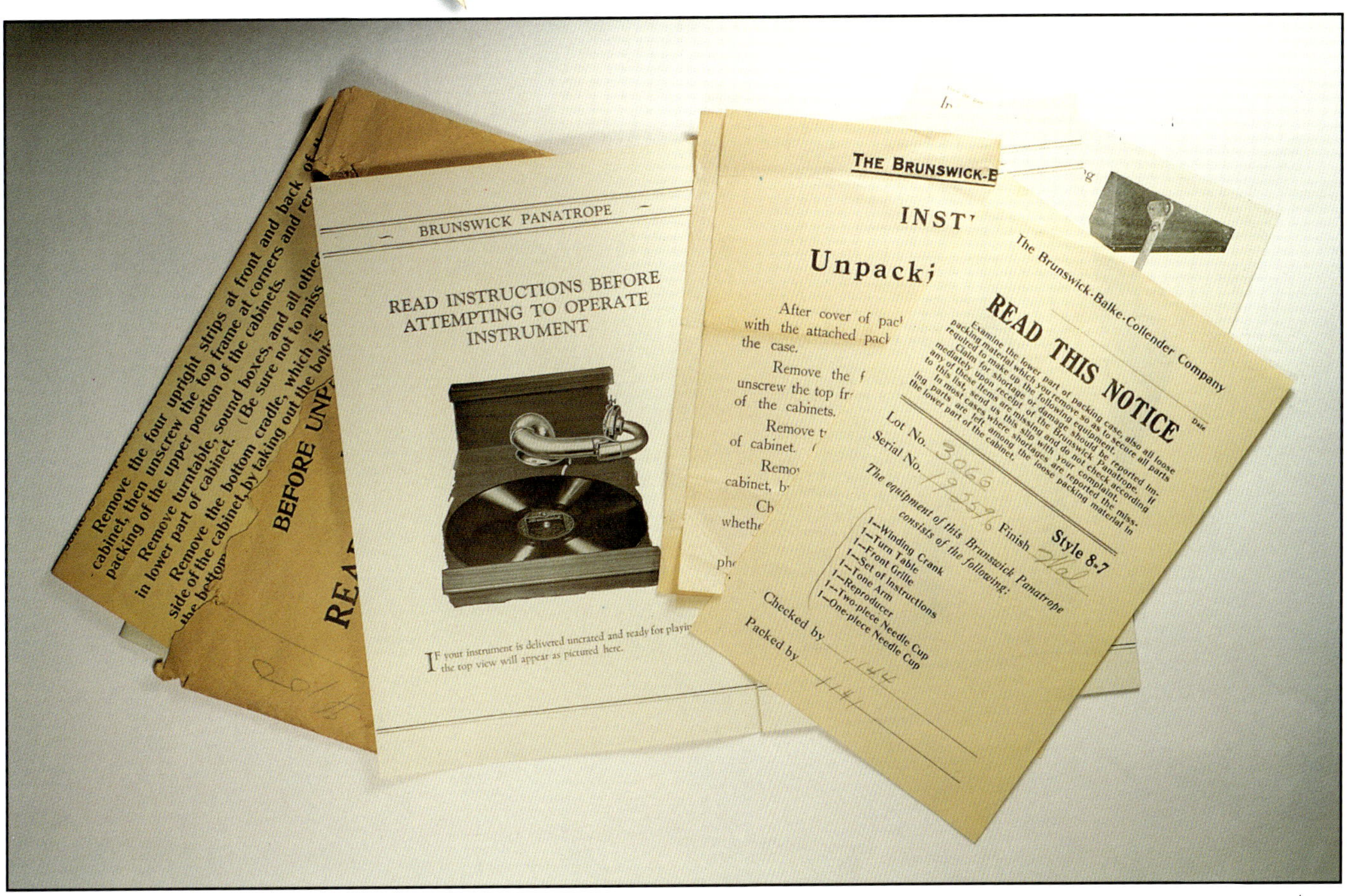

4-61. Columbia acquired the Okeh label in 1926, which revived a brand that had been in the doldrums. From this same period, a record supplement in Deco style. *Courtesy of Ernest Allen. (Value code: H)*

4-62. These small celluloid disc recordings were affixed to postcards as novel advertising for anything from carpet sweepers to Chevrolets. In the case of the lower example, the phonograph was being utilized to advertise itself. *Courtesy of Tom and Sandi McCarthy. (Value code: G)*

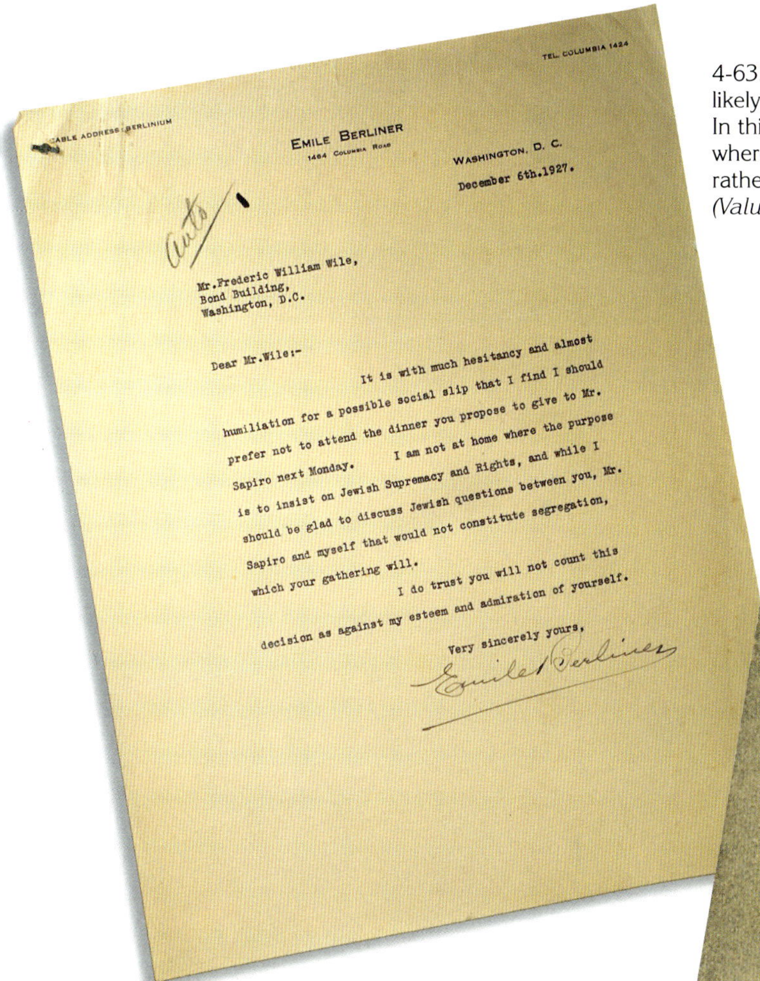

TEL. COLUMBIA 1424

CABLE ADDRESS: BERLINIUM

EMILE BERLINER
1464 COLUMBIA ROAD

WASHINGTON, D. C.

December 6th.1927.

Ans/

Mr.Frederic William Wile,
Bond Building,
Washington, D.C.

Dear Mr.Wile:-

It is with much hesitancy and almost humiliation for a possible social slip that I find I should prefer not to attend the dinner you propose to give to Mr. Sapiro next Monday. I am not at home where the purpose is to insist on Jewish Supremacy and Rights, and while I should be glad to discuss Jewish questions between you, Mr. Sapiro and myself that would not constitute segregation, which your gathering will.

I do trust you will not count this decision as against my esteem and admiration of yourself.

Very sincerely yours,

Emile Berliner

4-63. Emile Berliner was Jewish, but as a German Jew, he would likely have been punctilious about matters regarding his religion. In this letter from 1927, Mr. Berliner demurs to attend a dinner where he seems to feel he might be "put on the spot." He would rather, he wrote, discuss Jewish issues in a more private context. *(Value code: VR)*

4-64. The Ediphone was Thomas A. Edison, Incorporated's version of the office dictating machine. Here, the company namesake can be seen noodling with a speaking tube. After the Entertainment Division of the firm gave up the ghost late in 1929, the Ediphone branch continued to chug along, making brand new anachronisms: cylinder "Phonograph" mechanisms for commercial use. *Courtesy of the Tom and Sandi McCarthy. (Value code: H)*

4-65. An indoor thermometer in the shape of a radio microphone, a clever reminder of the Buffalo [New York] Talking Machine Company. 4 1/2" high. *Courtesy of Alan H. Mueller. (Value code: F)*

Greatest in volume, richest in tone
Sonora's newest Portables present an entirely new reproduction principle that far surpasses anything heretofore developed!
Now the Portable is a real musical instrument—the deep booming notes and low crooning cries are brought out with a richness and wealth of volume never before believed possible in Portables—and comparable only with the new type reproducing phonograph.
$35.00
$50.00

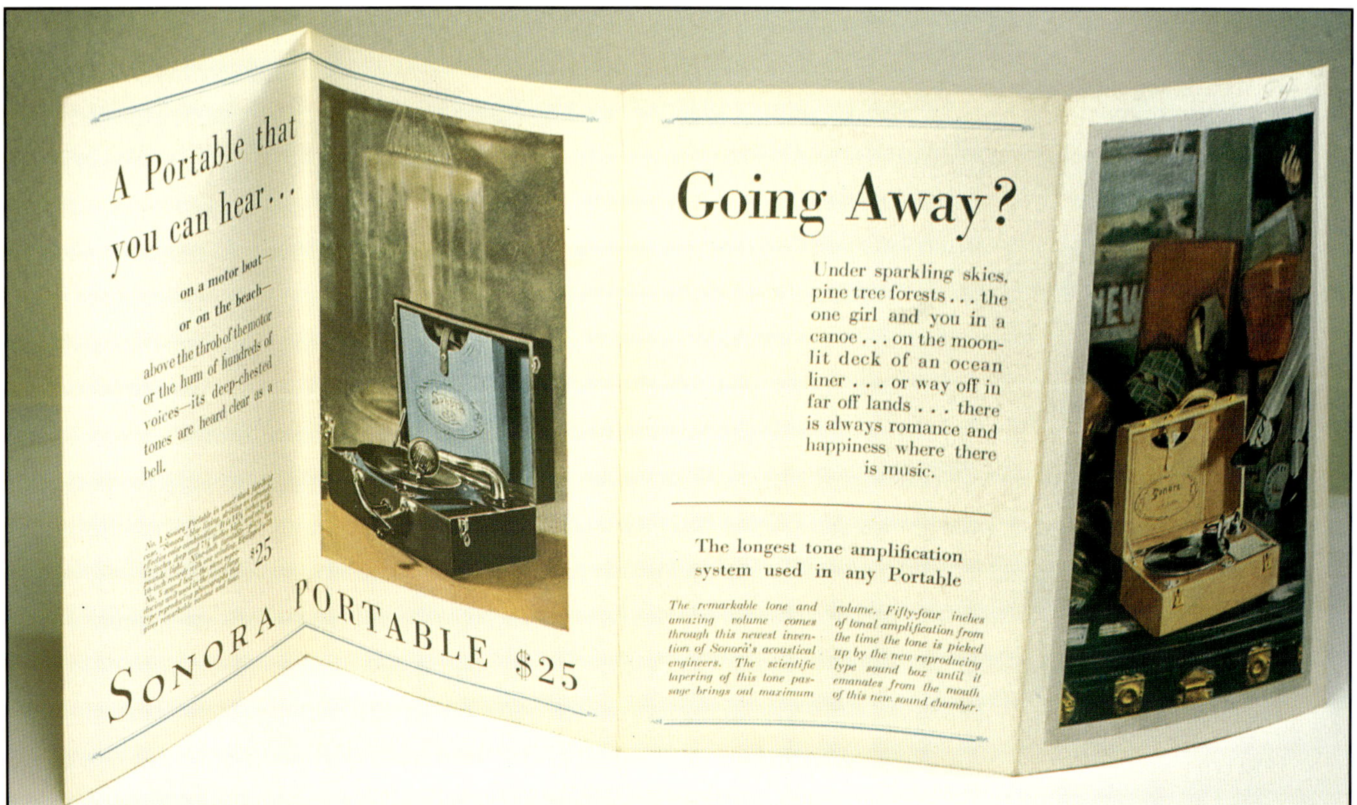
A Portable that you can hear...
on a motor boat—
or on the beach—
above the throb of the motor or the hum of hundreds of voices—its deep-chested tones are heard clear as a bell.
$25
Sonora Portable $25
Going Away?
Under sparkling skies, pine tree forests ... the one girl and you in a canoe ... on the moonlit deck of an ocean liner ... or way off in far off lands ... there is always romance and happiness where there is music.
The longest tone amplification system used in any Portable
The remarkable tone and amazing volume comes through this newest invention of Sonora's acoustical engineers. The scientific tapering of this tone passage brings out maximum volume. Fifty-four inches of tonal amplification from the time the tone is picked up by the new reproducing type sound box until it emanates from the mouth of this new sound chamber.

Opposite page, top: 4-66. The Sonora Phonograph Company was known for its attractive, colorful advertising. As the 1920s came to a close, the company applied this same merchandising to a relatively inexpensive product: a portable phonograph. Measures 3 1/2" x 6" closed. *Courtesy of Ernest Carl Allen. (Value code: G)*

Bottom: 4-67. The reverse was no less colorful. After the stock market crash of 1929, expensive presentations like this would virtually disappear for years. *Courtesy of Ernest Carl Allen.*

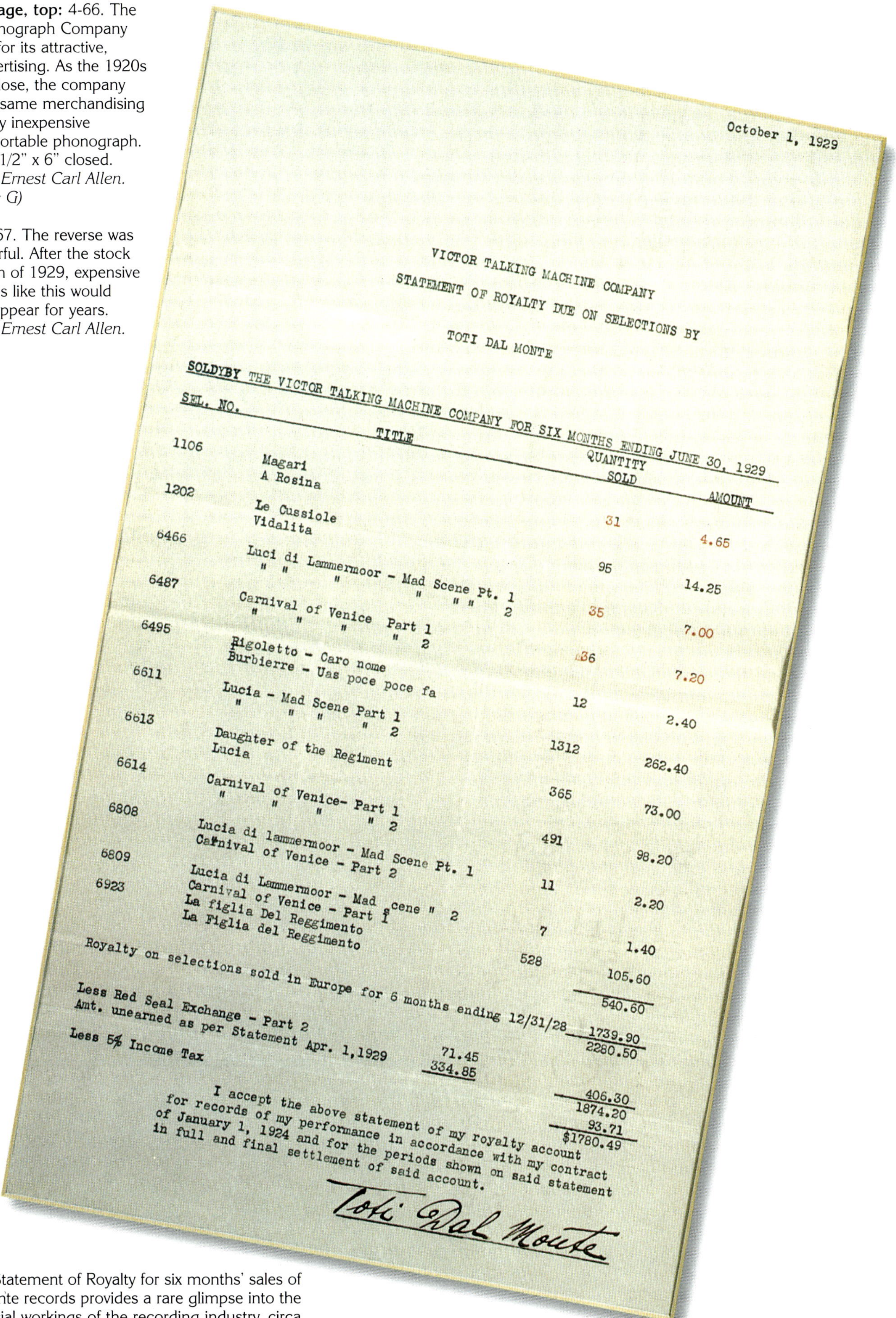

October 1, 1929

VICTOR TALKING MACHINE COMPANY
STATEMENT OF ROYALTY DUE ON SELECTIONS BY
TOTI DAL MONTE

SOLDYBY THE VICTOR TALKING MACHINE COMPANY FOR SIX MONTHS ENDING JUNE 30, 1929

SEL. NO.	TITLE	QUANTITY SOLD	AMOUNT
1106	Magari A Rosina	31	4.65
1202	Le Cussiole Vidalita	95	14.25
6466	Luci di Lammermoor - Mad Scene Pt. 1 " " " " " 2	35	7.00
6487	Carnival of Venice Part 1 " " " " 2	36	7.20
6495	Rigoletto - Caro nome Burbierre - Uas poce poce fa	12	2.40
6611	Lucia - Mad Scene Part 1 " " " " 2	1312	262.40
6613	Daughter of the Regiment Lucia	365	73.00
6614	Carnival of Venice- Part 1 " " " " 2	491	98.20
6808	Lucia di lammermoor - Mad Scene Pt. 1 Carnival of Venice - Part 2	11	2.20
6809	Lucia di Lammermoor - Mad Scene " 2 Carnival of Venice - Part 1	7	1.40
6923	La figlia Del Reggimento La Figlia del Reggimento	528	105.60

		540.60
Royalty on selections sold in Europe for 6 months ending 12/31/28		1739.90
		2280.50
Less Red Seal Exchange - Part 2	71.45	
Amt. unearned as per Statement Apr. 1,1929	334.85	
		406.30
		1874.20
Less 5% Income Tax		93.71
		$1780.49

I accept the above statement of my royalty account for records of my performance in accordance with my contract of January 1, 1924 and for the periods shown on said statement in full and final settlement of said account.

Toti Dal Monte

4-68. This Statement of Royalty for six months' sales of Toti Dal Monte records provides a rare glimpse into the financial workings of the recording industry, circa 1929. What might the statements of tenor John McCormack or famed violinist Fritz Kreisler contain? *Courtesy of David Werchen. (Value code: F)*

4-69. A postcard of the Victor plant in Camden, New Jersey in the late 1920s. Time, neglect and demolition have eliminated all but the building supporting the tower with round windows seen at the right, which has been converted to condominiums. *(Value code: H)*

4-70. A portable phonograph craze gripped the United States during the late 1920s. The economy was prospering, more automobiles were being sold, people were more mobile than ever – and the "suitcase" portable was just the ticket. The sluggish administration at Thomas A. Edison, Incorporated, finally recognized a bandwagon and jumped on it. Two lateral record playing portable models were introduced in 1929, just in time for the company to discontinue its home entertainment operations. *Courtesy of Gregg Cline. (Value code: F)*

4-71. With radio encroaching on the formerly sovereign territory of the phonograph during the late 1920s, the two entertainment media began to share the same advertising space. The firm that gave away this consequential woven fan might have still been known as a "Talking Machine Company," but its wares had begun to reflect the popularity of broadcasting. Measures 13" long. *Courtesy of the Scott and Denise Corbett collection. (Value code: F)*

Chapter Five:

Real Life Images

A note about the pricing of original photographic images. In recent years, Internet auctions have significantly affected the prices of "real life images." Photos have risen in value as quickly as other items (lithographed postcards of the "old couple" logo widely used by Edison in the 1906-1912 period, for instance) have declined. As a general rule, images on postcards (one of the most popular formats) sell in the "G" range. "Cabinet" or studio shots inhabit the "G" to "F" ranges. Larger images of store or other interiors, especially if they depict external-horn instruments, occupy the F range.

5-1. An image from the dawn of the talking machine industry, yet with a surprising amount of history already represented. This is the Edison Phonograph display from the Paris Exposition of 1889. The heart palpitates in contemplation of the precious artifacts shown here in a casual jumble. Left rear: the "Improved Phonograph," the instrument that defined the general characteristics of Edison Phonographs for the next 40 years. Near left: a "Tin Foil" Phonograph constructed by Hardy, 1878. At the center: Edison's invention, the very first Phonograph. Right rear: a Brobdingnagian "Tin Foil" Phonograph by Bergmann, 1879. Near right: an Edison Talking Doll, containing the first commercially-offered home entertainment Phonograph mechanism, cradling an Edison "Military" Phonograph (a half-sized version of the "Improved" that used miniature cylinder records, none of which have been found). *Courtesy of Ray Phillips.*

5-2. "A rose by any other name..." This dapper gentleman *could* have painted "Type K Graphophone" on his talking machine and its horn, but there was no real publicity in that. Everyone knew who invented the Phonograph. People would *pay* to hear Edison's Phonograph – so that's what this circa 1895 exhibitor gave them. Edison's Concert Phonograph, no less! *Courtesy of Julian Anton.*

5-3. Listening-tubes cascade down from a showman's Class "M" Edison electric Phonograph of the mid-1890s. The dapper exhibitor affects a casual air, his right hand lightly resting on his trusty mechanical marvel. Note the cylinder record storage box to the left, covered in the buff canvas frequently used on the earliest record carrying cases. *Courtesy of Tom and Sandi McCarthy.*

5-4. This stereo view of the Negro Building of the Atlanta (Georgia) Exposition of 1896 was meant to emphasize the dramatic sculpture in the foreground. Beyond, however, our eye is drawn to a row of Edison Kinetoscope movie machines and coin-operated Phonographs. Since the early 1890s, automatic Phonographs were found in public places. Nearly all were created out of Edison Class "M" electrically-driven mechanisms, modified with various coin-op devices such the "Kansas [Phonograph Company}" apparatus invented by George Tewksbury.

5-5. Mr. Connor wasn't a rich man, but he did well enough that he wanted the best Edison Phonograph money could buy. He went at it hammer and tongs – saved for over a year – but that first winter was when Mrs. Connor was taken by the whooping cough. Mr. Connor didn't leave the house for three weeks. Later on, in 1897, the store dropped the price of an Edison "Spring Motor" Phonograph to $75.00 – $25.00 off – and Mr. Connor found that he had enough money, so he bought one. I don't know that it brought him much happiness, though. I don't believe he ever got over the loss of his wife. *Courtesy of Tom and Sandi McCarthy.*

5-6. A Graphophone Type "A" (sold 1896-1900) entertains three intent listeners. Ear-tubes were being used in this case, but a black-japanned "tin" horn may be seen lying next to the gentleman on the left. *Courtesy of Loran and Genetta Hughes.*

5-7. Uncle Atwell was always looking for get-rich-quick schemes. One fall, he showed up at his brother Frank's house carrying a Type "A" Graphophone equipped with ear-tubes, and demanded that everyone give it a listen. Frank's wife thought everyone looked so funny with tubes in their ears, she insisted that the outfit be taken outside and a photograph taken. There they are – Atwell, nephew Tom, Frank, and his wife Lena – all trying not to laugh while I snapped the Kodak. *Courtesy of Tom and Sandi McCarthy.*

5-8. These exquisitely-dressed, commercially-posed Asian women convey a recurrent theme in phonograph imagery: live performance and the talking machine (in this instance a Columbia Type "AT" Graphophone, circa 1900). *Courtesy of the Charles Hummel collections.*

5-9. Some of the very rarest recordings are of obscure ethnic music, which might otherwise have been lost and forgotten. Pictured on this postcard, mailed from Saigon, Vietnam, March 25, 1900, a Laotian singer records a (standard-sized) cylinder on a Pathé No. 4 machine. Note the small horn that is being used to concentrate the sound. *Courtesy of Philippe Le Ray.*

5-10. "Thisby, the flowers of odours savours sweet… But hark, a voice!" The hand cupped to his ear to convey the visual impression of listening is the equivalent of a hand extended out from the brow to suggest "looking." This neatly turned out fellow auditions a typical Graphophone setup, circa 1900: Type "AT" cylinder machine, tripod floor stand, long brass horn, and boxes of records. *Courtesy of Tom and Sandi McCarthy.*

5-11. Was it coincidental that the parents of this quaintly-attired French baby paired him with the exact style of Berliner "Trade Mark" Gramophone that had been depicted in the famous "His Master's Voice" logo? Note the highly-polished brass bell of the horn – which should instruct the modern collectors' debate over whether bells should be polished or left with a patina of age. *Courtesy of Philippe Le Ray.*

5-12. This child of the twentieth century will live to see even greater marvels than the telephone that she grasps, and the Victor "Royal" Talking Machine nearby. This particular "Royal," circa 1902, has been equipped with an extended set of arms to support a horn of uncommonly large dimensions. *Courtesy of Dennis Peterson.*

5-13. In a scene as beautifully composed and shaded as a Pre-Raphaelite painting, a bicycle shop owner stands next to his sideline: a Columbia Type "AA" Graphophone, circa 1902. *Courtesy of Gilbert "Bert" Pasley.*

5-14. What every conservative patriarch feared most: wild women. The loosening of their corset stays became a veritable erosion of the underpinnings of Western civilization – or not. This card is from a European series of shady ladies driven to sheer abandon by the music of a "Trade Mark" Gramophone.

5-15. The old and the new. An aged gentleman – possibly an octogenarian, Civil War veteran, or born during the term of John Quincy Adams – sits beside a marvel of the twentieth century. *Courtesy of Mr. R. Chase.*

5-16. A miscellaneous crowd of music lovers gathered in Durban, South Africa at the turn of the twentieth century. Edison "Standard" Phonographs with multiple listening-tubes are in use. *Courtesy of Philippe Le Ray.*

5-17. A gentleman demonstrates a Columbia Type "BS" coin-operated Graphophone to a rather sedate family. It is unusual to see a coin-in-slot machine being operated in a private home (instead of a restaurant, bar, train station or arcade), especially to such an unenthusiastic audience. *Courtesy of Tom and Sandi McCarthy.*

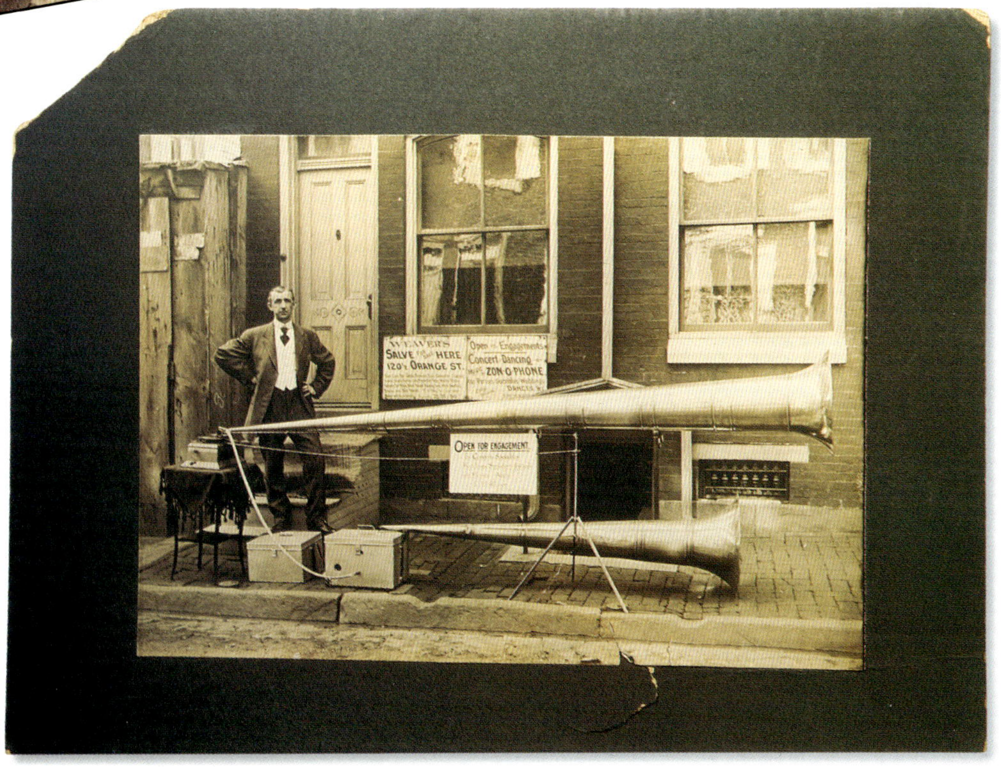

5-18. If one is to believe that size matters, this gentleman must have been the envy of his peers. Despite the posted claim for "Concert Dancing" from a Zonophone playing through one of these outsized horns, the results would not be dramatically improved over a smaller, conventional horn. What is of most interest is the mysterious rubber tube running into an equally cryptic box. Were these merely props designed to impress the locals, or was a compressed air amplification system in use? Compressed air could be employed to dramatically increase the sound energy emanating from a vibrating diaphragm. Several "loud-speaking" talking machines exploited this concept at the time. One thing seems certain: the exhibitor's records must have quickly worn out due to the inertia created by the large horns. *Courtesy of Ernest Carl Allen.*

5-19. These two girls were posing with a Victor "E" and rather incongruous ladies' watches one Sunday after church. The timepieces were probably new gifts, to be carefully put away in their hope chests. The memory of the day was recorded in this inadvertently haunting portrait of youth, quite literally shackled by time. *Courtesy of Tom and Sandi McCarthy.*

5-20. What is extraordinary about this scene is the care taken to display the proud owner's Victor "E" and records. A table and chair were arranged outside against a plain wall. The presence of the daisies suggests that a woman may have participated in staging the tableau. Three sizes of record – 7", 10", and 12" – were arranged near the machine and a record storage box. The gentleman's derby was, for some reason, placed on the table shelf. What resulted was this contrived yet sincere expression of the owner's happiness with his purchase. *Courtesy of Fred Guenther.*

5-21 . John Phillip Sousa worried that recorded entertainment – "canned music" – would discourage people from playing musical instruments. This scene might have comforted him: a group of amateur musicians warming up despite the presence of a Columbia Disc Graphophone. Sousa, however, might have cringed at the sight of the forlorn young woman standing in the rear – playing a kazoo.

5-22. May Day. Three generations of kin converged at the homestead for a day of fun and celebration. That afternoon, the tables were carried outdoors and set with white tablecloths. Play clothes were removed and the children dressed for dinner. As the shadows lengthened, the entire family was gathered on the front yard. It was a day to remember. *Courtesy of Gilbert (Bert) Pasley.*

5-23. With the majestic architecture of Paris behind her, this turn-of-the-twentieth-century street vendor has attracted a crowd along the banks of the Seine with the help of her "Puck" cylinder phonograph. She proudly displays the pendant watches she is selling, while the onlookers, who might have been sent down from Central Casting to play *citoyens*, gawk. As the visual evidence suggests, at least two disadvantages of the uncomplicated Puck were already recognizable: the difficulty of finding somewhere to rest the reproducer when not in use (it sits here atop its pasteboard storage box), and the susceptibility of the zinc horn to denting. *Courtesy of Philippe Le Ray.*

5-24. Though images are mute, they can draw us into the moment they preserve. Compare this photograph to illustration 5-6. In most instances, talking machines were not actually *playing a record* when captured by the camera. It was more likely that people would pretend to be listening to music. Here, we see a notable exception. These men, comfortable in their shirtsleeves, were most certainly enjoying the joyful renderings of a Columbia Disc Graphophone Type "AH." Can you hear the infectious refrain of "Wait Till the Sun Shines, Nellie?" *Courtesy of Kurt R. Nauck, III.*

5-25. A summer's day retreat. The sky a-shimmer at midday. *Courtesy of the Charles Hummel collections.*

5-26. In a scene dated 1905, Dutch soldiers in camp with a Gramophone sold by one of Victor's European cousins. Mess tins, a riding crop, and a Model 1895 Dutch Manlicher rifle. *Courtesy of Dennis Peterson.*

5-27. These avid readers were enjoying music from a 1905 Talkophone in this posed vignette. Despite the curtain having been pulled back, poor grandma and the parents squinted in the shadows while the young folks basked in the light of the window. *Courtesy of Tom and Sandi McCarthy.*

5-28. This card was posted from South Africa on May 25, 1905. The interested Africans were listening to a circa 1900 Berliner Gramophone, whose battered horn made it look much older than it really was.

5-29. Another postcard with a similar scene, but posted almost five years later. These Philippine children were probably less impressed by "a white man's voice" than by the novelty of recorded entertainment. Ironically, a little white dog was not far from this gramophone!

5-30. A snapshot of America. The house is newly constructed – the porch is not quite finished, in fact. It's a no-nonsense dwelling, but large enough for a family of eight. There's not a lot of money in this family. The younger boys wear hand-me-downs. The two girls are somewhat better off. Today, everyone's dressed in the best he has. The father and eldest son wear ribbons. Hard work is being acknowledged. It's a proud day. Flowers bloom on the porch. One of the family's few luxuries is a Victor. It must be included in the memory of this happy moment.

5-31. The lady of the house was proud of her dining room. Aside from being equipped with a Victor and a box of records, the room featured a telephone and a built-in china cupboard – whose doors were opened for this photograph. *Courtesy of the Scott Malawski collection.*

5-32. When worlds collide. A smartly fitted-out carriage parked at the door of a dealer in Victor Talking Machines. The lithographed metal sign, circa 1905, announces "sold here." A vestige of the old, and a harbinger of the new. *Courtesy of the Charles Hummel collections.*

5-33. The 6th of July, and the sun was blinding in the afternoon. The flags we'd hung out for Independence Day were perfectly still in the breathless blast of pure, dry incandescence. Uncomfortable in my summer dress, even in the shade of the porch, I raised my hand to my damp brow, while the baby, towheaded, gleeful in the wilting heat acted out his chaste parody of our lives – trying to grow something in this atmosphere of arid calefaction. *Courtesy of the Charles Hummel collections.*

5-34. Carnival time in the French city of Aix-en-Provence. *Le directeur* conducts an aggregation of fantasy phonographs ("the bright players"). One may surmise that a musician or singer was hidden within each display. *Courtesy of the Charles Hummel collections.*

5-35. If this vignette was not posed, the photographer was either gifted, or very lucky. The contents of the table appear to have been arranged as in a still life. The composition, height of the camera, lighting, all suggest a professional at work. Yet, the elderly man offering a treat to the woman has an immediacy that decries mere staging. Whatever the circumstances, an English Gramophone graced the scene.

5-36. College boys, lost in study (was this to prove their good intentions to the folks back home?). Nearby, a Columbia "AU" Disc Graphophone – at $12.00, it was about all any impecunious student could hope for. *Courtesy of Dennis Peterson.*

5-37. There's one in every crowd. While the seated fellows put forth a dignified face for the camera, their friend mugged, gesturing into the white flower horn of a Victor "II." He held a record in his right hand, whose indecipherable title may have had something to do with the joke.

5-38. Someone was proud of this room. When the decision was made to capture it in a photograph, a Victor was, either accidentally or by design, captured in the mirror.

5-39. Piano practice. An endless hour of scales and exercises. At least Fido was always there for company. The Edison "Standard" rested in what must have been a very common state, but one rarely caught on film. Its original 14" horn was nearby, but the large Hawthorne & Sheble flower horn was perched atop the piano when not in use. A quantity of cylinder records waited off to the right, but first – finish those scales. *Courtesy of Gilbert (Bert) Pasley.*

5-40. It was Sunday afternoon, and time for an informal family portrait to send to Grandma. One wonders if mother operated the camera, of if some long-ago tragedy had reduced this family to the four members pictured here. In any event, a talking machine occasionally entertained them. *Courtesy of Martin F. Bryan.*

5-41. Something about this scene smacks of "summer resort." In fact, the tiny community of Spooner, Minnesota was in sport fishing country, Lake of the Woods. The gents in front of the sweet shop, with a little girl coyly posing, the crisp clapboard exterior, and a canvas banner announcing that Edison cylinder Phonographs were available within. Circa 1907. *Courtesy of the Charles Hummel collections.*

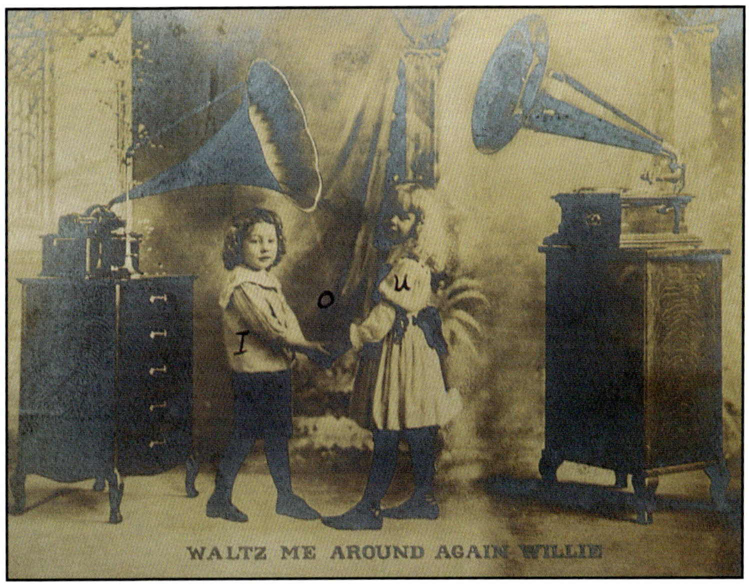

5-42. In 1906 the comic song "Waltz Me Around Again Willie" was popular on both cylinder and disc records. Can't decide which you prefer? Try both: an Edison "Home" (left) and a Victor "V" (right) both sit atop record cabinets. The "IOU" inscription is a mystery, but was probably a play on "I" and "you." Posted in August 1907. *Courtesy of Bob and Karyn Sitter.*

5-43. The firm of Buxton & Parker of Newport News, Virginia, knew the value of advertising. If you wanted an Edison record, there was no use going anywhere else. *Courtesy of Ernest Carl Allen.*

5-44. This lodge looked like thousands of others in early twentieth century America. Some remain, but these days it's doubtful that one would see a Victor "V" in the library.

5-45. Play, gypsy, play! But keep an umbrella handy in case of rain. *Courtesy of Ernest Carl Allen.*

5-46. Yaseen's Jewelry Store in Dalhart, Texas offered the services of an "Optical Parlor" as well as the charms of Victor Talking Machines. Mr. Yaseen could soothe the eyes, ears, and heart – all in one shop.

5-47. The premises of the Eastern Phonograph Company, with cylinder machines atop record storage cabinets, and records galore. *Courtesy of Tom and Sandi McCarthy.*

5-48. Why pose a baby with a talking machine? Did the child have a particular liking for it? Was a proud father snapping the two things that brought him joy? Or was some private joke being portrayed, such as "Our Two Talking Machines?" *Courtesy of Pat and Jamie Jones.*

5-49. Kids with a kid-sized talking machine, an Edison "Gem" Phonograph. The girl on the right is decidedly unhappy about having to sit still for this snapshot. *Courtesy of the Charles Hummel collections.*

5-50. Even bootleggers need some entertainment. *Courtesy of Gilbert "Bert" Pasley.*

5-51. An animated man held two cylinders aloft behind a circa 1908 Columbia "BKT" Graphophone. Like so many vintage phonograph images, the circumstances are a mystery. The chest of cylinder records appears to be purpose-made for transporting them. The lids were missing, so the records were evidently not for sale. Was this man one of the last talking machine exhibitors? *Courtesy of Mr. R. Chase.*

5-52. A Deutsche Grammophon model sitting atop a European-style chair, some rough-and-ready headgear and a Westerner in shirtsleeves, are about the only non-Stone Age elements in this scene, circa 1908. *Courtesy of the Charles Hummel collections.*

5-53. The irony of American life. Mark Twain was a rustic, but with finely-tuned cultural sensibilities. Theodore Roosevelt lived the roughneck life of the Old West, but never lost his patrician air. The unpretentiously-dressed gentleman seen here is surrounded by uplifting elements: musical instruments, a well-considered arrangement of photographs on the wall and an Edison "Home" Phonograph, circa 1908. He relaxes with a book that might just as likely be the Bible or the poems of Robert Service. Yet, he keeps his hat on, and his revolver hangs from a nail nearby. *Courtesy of the Charles Hummel collections.*

5-54. Every summer, my sister and I would travel out to the cottage to stay with our aunt for a week or two. At night we slept in the loft with the cool lake breeze wafting through the screened window. No trolleys, no trains. Not a sound but our voices, the crickets, and – during the day – the phonograph. *Courtesy of Tom and Sandi McCarthy.*

5-55. Haying time, the boys piled aboard the wagon at dawn. By noon, when Annie and the girls brought down a mountain of sandwiches, higher than the lofty hay itself, the workers were ready to interrupt their labors and relax to the sound of the phonograph playing "In the Shade of the Old Apple Tree." *Courtesy of Dennis Peterson.*

5-56. The sly looks being exchanged between these two, on March 23, 1908, may not have had much to do with the Edison "Home" Phonograph. *Courtesy of Dennis Peterson.*

5-57. In early 1908, this rakish young man could have sold you an Edison "Standard," "Home," or "Gem" Phonograph. Perhaps you would have preferred a Disc Graphophone? He would have assisted you in making a selection from the Edison cylinder records in the Syracuse wire racks along the back wall. Even if your stop were for nothing more than sheet music, he would have been glad to help you, had not the passage of a century prevented the transaction. *Courtesy of Tom and Sandi McCarthy.*

5-58. Bob's Candy Kitchen was where folks could enjoy a light lunch, coffee, sweets, and tobacco. For a penny, Bob would hand you the crank to the Standard Talking Machine Model "X2" and you could play one of his eight or nine records. Most times, Bob just handed you the crank with your coffee. *Courtesy of Tom and Sandi McCarthy.*

5-59. Silence – for the moment. The drums and bells await the eager hands of the students. The Type "BC" "Sound Amplifying" Graphophone was a good choice, it was the only thing loud enough to be heard above the din. A variety of marches and schottisches – and a teddy bear – are poised for the moment, forever. *Courtesy of Ernest Carl Allen.*

5-60. Roughing it was a rollicking good time. No electricity needed to run a cylinder phonograph. The ladies enjoyed sitting square on the ground, and the man at the front was caught in motion – rocking back in laughter, preserved in eternal joy. *Courtesy of the Charles Hummel collections.*

5-61. A cozy hunting cabin in the woods. This "BE" Graphophone sat amid bucolic surroundings, its horn suspended from the ceiling by a string. *Courtesy of Gilbert (Bert) Pasley.*

5-62. We drank coffee by the kettle-full at our old hunting camp. That might have been why we felt like sitting up most of the night around the campfire. Not that any of us talked much, but it felt good to sit with your pals and smoke a cigar; watch the smoke from the fire and the smoke from our mouths swirl together and hurry up into the darkness above. Ben tried to follow along to the Edison on his banjo, but mostly got lost. Lee Chen, the Chinese cook we had brought along, was up at dawn frying eggs and ham. Better him than me; I swear I could burn water! *Courtesy of Dennis Peterson.*

5-63. A spontaneous-looking group with fresh-caught fish proudly displayed. The Edison "Home" Phonograph was there to keep them amused. The unusually-shaped horn is a "Searchlight," made by the Searchlight horn Company of Brooklyn, New York. *Courtesy of Dennis Peterson.*

5-64. The baseball players on the left and right are obviously brothers, possibly twins. Well suited for double-plays and doubleheaders, to the accompaniment of "Take Me Out To the Ball Game" from the Edison "Home" Phonograph, circa 1908. *Courtesy of Dennis Peterson.*

5-65. These sober-looking gentlemen seem as though they're ready for a gunfight. In this case, the shoot-out appears to be a comparative judging (at a fair, circa 1909) between four prominent talking machine brands. Each of the machines is paired with a Herzog record storage cabinet. The Victor on the left is partially hidden within an unusual Herzog (Art Furniture Company) cabinet first called the "De Luxe." This formidable piece of furniture was first advertised the August 1907 *Talking Machine World* under the designation No. "834" "De Luxe Concealed Horn Cabinet" ("holds Any Size Disc machine"). One month later, Herzog discontinued the name "De Luxe," and the cabinet eventually became known simply as the "Concealed Horn Cabinet." The idea was to hide a conventional external horn disc talking machine within a gracious wooden enclosure. *Courtesy of Ernest Carl Allen.*

5-66. Small-town ostentation. The master of the house, with his wife standing dutifully behind him, sits stiffly on an uncomfortable settee and surveys the parlor, festooned with as much bric-a-brac as they can manage. The dirt on his shoes betrays his circumstances, but he has a princely $100.00 Victor "VI" in his home. It was people such as these who inspired Sinclair Lewis to write *Main Street*. *Courtesy of Ernest Carl Allen.*

5-67. This sparkling, unpretentious home featured a Standard Disc Talking Machine Style "X2" on a table with its records stacked beneath. One suspects that a child insisted that her doll and teddy bear be included in the photograph. *Courtesy of Loran and Genetta Hughes.*

5-68. I remember working up at the hotel during the summer of 1909. The city people used to come up for a few weeks – some spent the whole season. Both dining rooms had some kind of music. For a nickel, customers could play what we called the "mechanical band." The other dining room had a phonograph. One night, they had both machines fired up at the same time and the noise was something awful – but everybody was laughing. It was all in fun.

5-69. The presence of a woman is not immediately discernable in this rather cramped parlor. Regardless, the comfortable chamber had gaslights (even a gas-fed table lamp for effortless reading), fireplace, a full bookcase, upholstered furniture, and a Victor. The lady of the house was seated at the piano, her face curiously blanched to a ghostly cameo. We can't know her, beyond what her room tells us.

5-70. A general store – circa 1909. The woodsy smell of varnished pine mixed with the faint bitter scent of the coal stove. No expensive Victrolas here unless you ordered one. Interestingly, the black Victor flower horn appears to have had floral decals applied to it.

5-71. This adorable moppet was immortalized sitting beside a Royal talking machine. Royal was one of the "Chicago brands" – companies which sold their instruments through retail store promotions or mail order. The photograph was meticulously colored with pastels or enhanced with watercolor to give Junior a dash of life. *Courtesy of Norm and Janyne Smith.*

5-72. Sometimes we have no hope of distilling the essence of a scene; its meaning is lost in the secret histories of people we will never identify. Merely take in the elements of the composition as you would a work of art, the artist long dead and mute on the subject. DaVinci's "Last Supper." The anxious-looking woman working the Edison with the aid of the standing gentleman – who looks as if he could win a Harry Houdini look-alike contest. And what is it he's holding in his right hand? *Courtesy of the Charles Hummel collections.*

5-73. "How the Other Half Lives, Part I." The music room in the home of W.L. Coursen in Daytona, Florida, circa 1910. In the foreground is a baby grand piano. The room is dominated by an imposing pipe organ, whose console is to the right. Dwarfed by the organ is a piano-like Columbia "Grand" Grafonola, beneath the left window. Hidden behind the organ console is a surprise, revealed in the next photograph. *Courtesy of Tom and Sandi McCarthy.*

5-74. "How the Other Half Lives, Part II." Mr. Coursen included a Vernis-Martin Victrola "XVI" in his formidable group of musical diversions. This view also reveals the baby grand piano to be a posh "reproducing" player. It would be a hypercritical man indeed who could not spend a pleasant evening in a room with elaborately paneled architecture, art glass windows, and such interesting contents. *Courtesy of Tom and Sandi McCarthy.*

5-75. A birthday party, circa 1910. While the birthday girl (second from right) opened her presents, there sat a boy, tinkering with her parents' aging Type "B" Graphophone. Her father had told him to fix it. Considered strange by his peers and eventually ostracized by all but a sympathetic few, the boy grew into a bizarre creature. Mothers, don't let your children grow up to be collectors! *Courtesy of Tom and Sandi McCarthy.*

5-76. This no-frills showroom of furniture and housewares may have lacked style, but one couldn't complain about the variety. A Columbia Disc Graphophone lurked behind a sewing machine to the left. *Courtesy of Gilbert (Bert) Pasley.*

5-77. The Columbia Phonograph Company franchise seen in the background sponsored an impressive float, circa 1910. "Miss Columbia" herself, the firm's standard bearer since the 1890s, was accurately impersonated, with her hand resting on a Disc Graphophone Type "BD" ("Majestic"), flanked by a rampant eagle. *Courtesy of John P. Andolina, Jr. "The Early Sound Man."*

5-78. Dog names have changed with the tastes of society, just as people's names have. Around 1911, when this picture was taken, a dog was likely to be known as Bowser, Scrapper, Rex or Tippy, whom we see here. Nowadays, pet names have grown more anthropomorphic: Clancy, Jack. What *was* Tippy's favorite record as played by the Victrola "XI" – "The Whistler and His Dog," or "The Pussy Cat Rag?" *Courtesy of the Charles Hummel collections.*

5-79. Dogs were a natural prop to be placed before a phonograph, à la "Nipper," but cats occasionally were pressed into service as well. This accommodating feline was perched in the horn of a Victor "III." *Courtesy of Loran and Genetta Hughes.*

5-80. Six young men. Who they were, where they were, and what they were doing is a mystery. The boxing gloves and catcher's mitts indicate sporting activities. The building may have been a camp or boarding house. One thing is apparent: these fellows valued their music. *Courtesy of Pat and Jamie Jones.*

5-81. Camps, cabins, tents, bungalows and other rustic settings were perfectly suited to spring-driven talking machines. Here, a Victor "II" (1906) fitted with Victor's best wooden horn provides the tempo for rugged and healthful chores. *Courtesy of the Charles Hummel collections.*

5-82. Shipmates stand together, or sit as the case may be. Aboard a wooden sailing ship with a Victor "VI" – mighty fancy entertainment! *Courtesy of the Charles Hummel collections.*

5-83. It's unclear whose birthday was being celebrated, but the idea of pausing for a photograph just before cutting the cake went over like ants at a picnic. The little girl to the right had already lost control when the shutter snapped. It is doubtful that the cake lasted long. All but unnoticed was a "BZ" Disc Graphophone on a No. "809" Herzog record cabinet. *Courtesy of Fred Guenther.*

5-84. A small music shop. Or could it be a pawnshop? Note the "cage" typical of these establishments. Whatever the identity, we are suddenly reminded – of course people would pawn the family phonograph after Aunt Mildred's ring was already in hock. And it behooved the pawnbroker to keep a supply of records on hand. You might even get a cylinder storage cabinet, chock full, if you came in before the impecunious owner managed to redeem his record collection. *Courtesy of the Charles Hummel collections.*

5-85. A jug of wine, a loaf of bread and... a Pathéphone beside me in the wilderness... Two French bakers, toasting (pun intended!) their loaves to the sound of a Pathé disc record. *Courtesy of Philippe Le Ray.*

5-86. An outtake from "Fiddler on the Roof," perhaps? No, just an Eastern European holiday scene with *Di Mishpokhe*, the family: "Happy New Year… Breaking bread on Yom Kippur… After the fast, back from the synagogue, the hearts are full of faith and hope. Picture a New Year of good fortune. People sit around the table, they eat and they drink, the gramophone sings 'Hamavdil' [a closing prayer]…" *Courtesy of Robert Berkman.*

5-87. Little Wilbur, age three, posed outside his grandparents' farmhouse with a Standard Talking Machine Model "A" in 1911. *Courtesy of Tom and Sandi McCarthy.*

5-88. This postcard was copyrighted 1909, and mailed the day prior to Valentine's Day, 1911. The jest with "record" was echoed on a few other postcards over the years, but this example trumped the others by featuring a card-playing lass showing off her ankle.

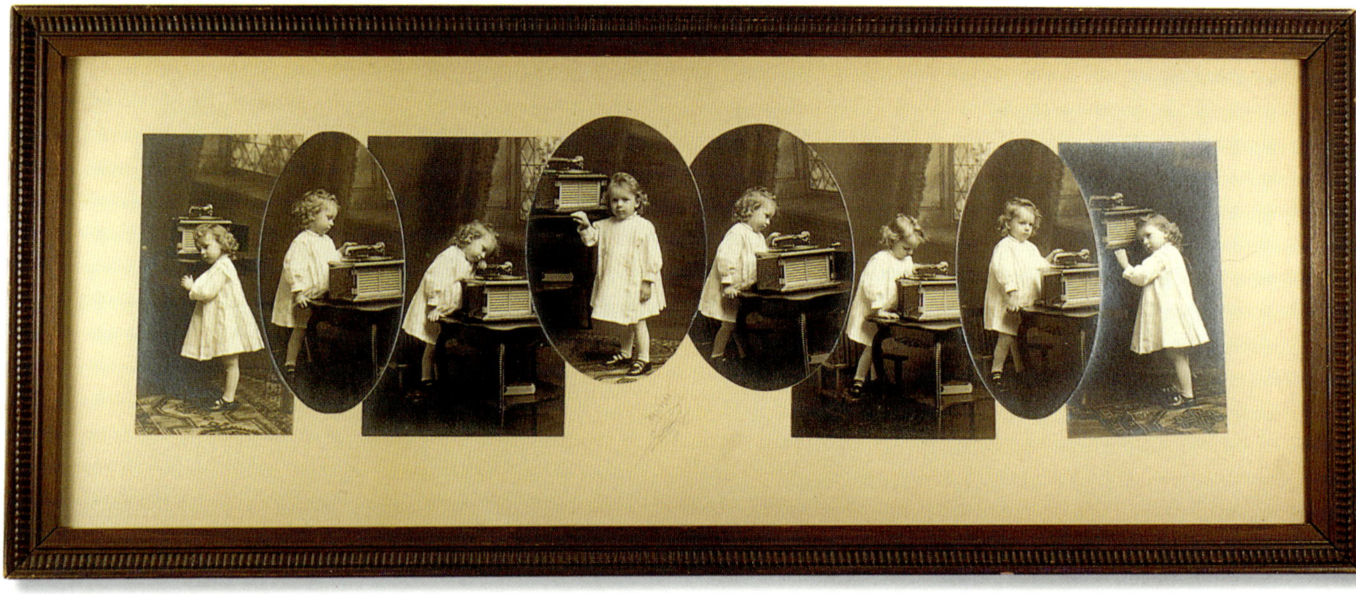

5-89. Baby Hazel was the apple of her father's eye. One day, he bundled her down to the photographer's studio, along with the Victrola "IV" that she played in her nursery. Hazel's father never wanted to forget the sight of her listening to the records. *Courtesy of Tom and Sandi McCarthy.*

5-90. Hazel and her Victrola. *Courtesy of Tom and Sandi McCarthy.*

5-91. Schmidt was a German; good with his hands. He fixed watches and clocks for most all the jewelry stores in the county. He took a room over one of them and helped out – you know, swept up, whatever needed doing when he had nothing to repair. About 1912, he took in a busted phonograph in barter. The Edison was about ten years old, but before you knew it, Schmidt had it working. He used to play marches behind the shop after hours and on Sundays. I forget the dog's name, but he was a friendly mutt. *Courtesy of Tom and Sandi McCarthy.*

5-92. "Note the pie on the table." The inhabitants of this cabin or small cottage seemed to have at least *some* of the comforts of home – including a talking machine whose horn is behind the man second from right.

5-93. A live donkey, nattily dressed people and a cardboard store display figure of Nipper, the Victor dog. This rather odd combination has already been shown in our book *Antique Phonograph Gadgets, Gizmos and Gimmicks* (Schiffer, 1999), page 218. In that previously-published image the people are not the same, and the cardboard dog is fully visible. It has "At Milwaukee" printed on it. Milwaukee must have been a riotously funny place circa 1912. *Courtesy of the Charles Hummel collections.*

5-94. The significance of certain posed snapshots may never be fully understood. These often meticulously arranged scenes have been found to include similar items – a phonograph and traditional musical instruments, for instance. Tokens of special importance to a particular family were frequently depicted – such as the photo album seen here. Or perhaps these articles were really studio props. Note the telltale backdrop of a commercial photographer. Supporting this view is the discrepancy between the age of the talking machine (Edison "Standard" Phonograph Model "B," with a flower horn manufactured by Hawthorne and Sheble, but bearing the distinctive decoration of the Babson Brothers mail-order house of Chicago, circa 1908) and the fashion of the ladies (circa 1912). *Courtesy of the Charles Hummel collections.*

5-95. Belize is a Central American country bordering the Caribbean Sea east of Guatemala. It's interesting to note that when this photo was taken, circa 1912, the country was still officially known as British Honduras (it did not receive its independence from Great Britain until 1981). The use of the name Belize here refers to Belize Town, a city. A durable canvas banner promoted Columbia products. *Courtesy of the Charles Hummel collections.*

5-96. One expects Swedes to be warmly dressed, and the "rustic lodge" backdrop seems particularly well-suited to Scandinavia. A generic German gramophone, circa 1913, provided the focus for the scene, and please note that the horn has some damage to the outer edges of the petals. Not all of the dents we see in horns today were acquired after they were put away in attics. *Courtesy of the Charles Hummel collections.*

5-97. The Edison cygnet horn by the rear wall betrays the presence of a Phonograph, and the rack of hundreds of Edison cylinder records suggests that this establishment sold them, but only as a sideline.

5-98. A Columbia retailer circa 1912, with a "Lyric" Grafonola displayed in the window – and, surprisingly, a cylinder Graphophone Type "AT." It's hard to imagine a machine introduced in 1898 could have retained any commercial value so far along in talking machine development. Perhaps the "AT" served to illustrate the arcane technology from which the up-to-date Grafonola had evolved. *Courtesy of the Charles Hummel collections.*

5-99. Somewhere in peaceful, prewar Europe, a joyful couple danced in a garden. What more can be, or needs to be said? *Courtesy of Loran and Genetta Hughes.*

5-100. McDonald's Dancing Academy reaped the benefit of the dance craze that gripped the United States about the time that the great European war began. It seemed like everybody in Pittsfield wanted to take lessons, or wanted their children to learn the art of Terpsichore. Professor McDonald himself would demonstrate the masterful steps with his graceful partner, Miss Dunn. The brand new Edison Disc Phonograph could be heard clearly throughout the spacious ballroom – it had cost $200.00! *Courtesy of the Charles Hummel collections.*

5-101. The interior of Eckman's Music Store in Jamestown, New York is shown on this card, which was posted September 25, 1913. Although at least three Victrolas are visible, older style external horn machines still predominated.

5-102. These two young women wished to be photographed with an older man – perhaps their grandfather. The light, it was thought, would be best by the window. The furniture – including a Victor – was moved out of the way, but alas, the light was too strong. Just a couple of feet to the left would have made the picture much better for all of us.

5-103. An inexpensive Victrola was among the appointments of this curtained room. The layers of rugs were meant either to cover worn areas or to prevent the same (or perhaps for purely aesthetic considerations). *Courtesy of the Scott Malawski collection.*

5-104. A man of humble means, but rich intellectual tastes. Foremost among his modest possessions are books and records. The talking machine, too, is unpretentious: a Columbia table model Grafonola of the type manufactured for clients, and most commonly sold by the United Talking Machine Company of Chicago under the brand name "Symphony." Boxes of records below leave no doubt about the owner's commitment to the life of the mind. *Courtesy of Loran Hughes.*

5-105. This image raises as many questions as it might answer. It seems to be a glimpse into the Bridgeport, Connecticut factory of the Columbia Phonograph Company, circa 1915. Workmen stand among three uncompleted Grafonolas. Yet, each instrument most notably lacks the louvers over the mouth of its horn, while other "finished" elements are already present, including mechanical parts such as the tone arm, and the company decal on the inside of the cabinet lid. So, why would the machines have finished details while they still lacked an integral part of the cabinet? Perhaps these were Grafonolas returned for reconditioning, or the louvers of which were damaged in shipping. For the sober artisans shown here, the talking machine was only a job. *Courtesy of the Charles Hummel collections.*

5-106. Late twentieth century phonograph hounds would flock to the flea market (marché aux puces) at Montreuil, environs of Paris, each weekend hoping to find talking machines, parts and records. Montreuil, like the markets at Porte de Clingancourt, was a well-known hunting ground for collectors. Those of us who have visited without success may look upon this scene from 1915 with chagrin, realizing that what we needed was a time machine. *Courtesy of Philippe Le Ray.*

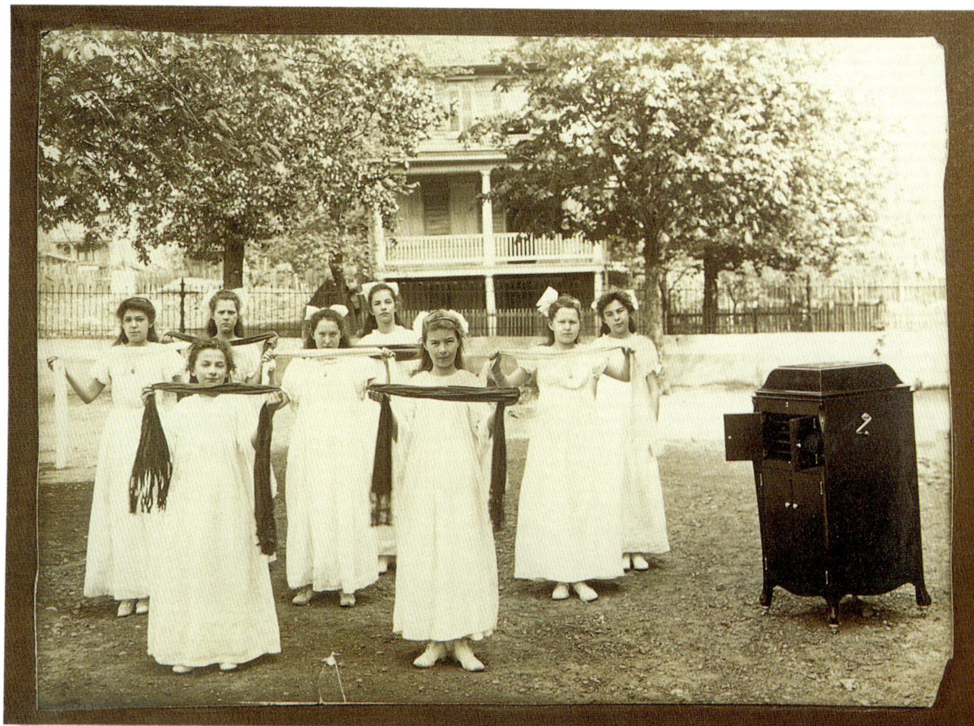

5-107. The students from Miss Natalie's School of the Dance gave a recital each summer in the park. In 1916, the musical accompaniment was handled by a Victrola "XI," which went over so well that it was included in the photograph of that year's troupe. Another Isadora Duncan in the making? *Courtesy of Tom and Sandi McCarthy.*

5-108. The White Man's Burden – or not! The intrepid European voyager — jodhpurs, pith helmet, *the works* – oversees human transport of the machinery for a night of scientific instruction: a magic lantern projector and a gramophone. Written on the reverse of this postcard image: "On our way to a 'big night' of outstation camp work along the frontier." Dated May 1917. From the context of the message, and the sparkling demeanor of the gentleman on the right, it looks like the locals were in for a multimedia missionary presentation.

5-109. It is not uncommon to find postcards of American soldiers during World War I enjoying recorded music, but the phonographs were usually more modern than their European counterparts. However, in this view an aging Victor "III" provided the music in a cavernous Soldiers' Club.

5-110. This Knight's of Columbus building sported a modern Columbia Grafonola.

5-111. A glorious summer day, far from the abominations of the trenches. Recuperating British soldiers, the music of the gramophone, and the evanescent smile of a radiant young woman who is looking right at us, across the thousands of glorious summer days that have since passed. *Courtesy Of Philippe Le Ray.*

5-112. The neat-looking establishment of D.E. Wotton, dealer in Edison Phonographs, about 1918. The "New Edison Diamond Amberolas" were actually the same three old models introduced in 1915, which the company kept in production until the late 1920s. Not only was nothing new offered in cylinder Phonographs during this period, but relentless cost reduction continued to cheapen the existing designs. Pretty much the same thing could be said of Edison's Disc Phonographs (generally known as Diamond Disc Phonographs, though the records were referred to for a time as "Re-Creations") – they were "new" in name only. *Courtesy of the Charles Hummel collections.*

5-113. It was a big day in Garden City, Kansas, when Baker's Mercantile received five wagonloads of brand new Edison Diamond Disc Phonographs. In the boom that followed the Great War it seemed like Mr. Baker couldn't order them fast enough. The horse teams, with the draymen perched above, reminded the old timers of salad days on the prairie. *Courtesy of the Charles Hummel collections.*

5-114. A not terribly enthusiastic group of young French people in Alsatian costume, with a talking machine and records, circa 1920. *Courtesy Of Philippe Le Ray.*

5-115. This rather somber-looking chamber included a Columbia Grafonola hiding in a corner. *Courtesy of the Scott Malawski collection.*

5-116. For the parade celebrating the Glorious Fourth, Yerger's Music Store decked out the delivery truck as a float. The boss insisted that the Victor trademark be enacted on the float, but by 1922, no one seemed to have an old Gramophone around anymore. No matter – a Columbia disc machine would do nicely. *Courtesy of Tom and Sandi McCarthy.*

5-117. Spengel's Furniture Store mounted quite a "Victrola Exposition" for the local fair during the summer of 1923. This photograph was taken on "Women's Club Day," and shows a carefully arranged display. Close scrutiny will reveal many interesting details. A contest was announced at the lower left: "Are you good at guessing? These 25 Victor Records in this handsome case FREE to the person guessing nearest the correct number of Victor Tungs Tone Needles in this glass jar." A bit further back on the right sat a 1901 Victor "A" (with incorrect arm, sound box, and horn!) bearing a sign: "Here's a real old timer... The granddaddy of the Victrola." A placard can be seen offering the Victrola "110" for $225.00. One of Victor's Art Models stood in the center rear. Note also the special Victrola "XVI" with glass panels in the cabinet and glass lid for dealer display. The poster behind the palm fronds at upper right dates from March 1923, thus confirming when this extraordinary summertime "Victrola Exposition" took place. *Courtesy of Robin and Joan Rolfs.*

5-118. Post cards such as this one, showing a lovey-dovey couple entranced by the music of a phonograph (in this case a "His Master's Voice" portable Gramophone) were popular in Europe during the late 1920s. This is a particularly modest scene, whereas more licentious poses, with women in various states of undress, were also available.

5-119. Oh! La la! *Courtesy of the Charles Hummel collections.*

5-120. A progression of expression, demure to coquettish, among these vacation ladies, enjoying a portable talking machine. *Courtesy of Philippe Le Ray.*

5-121. The family on a sailing spree. With portable phonograph for entertainment. *Courtesy of the Charles Hummel collections.*

5-122. Margaret always wanted a puppy. *Courtesy of Tom and Sandi McCarthy.*

5-123. As domestic accoutrements age, they travel downward among the echelons of the industry which created them. A phonograph may start out in the window of a store in the toniest street in town, but it may easily end up among the junk shops of skid row. Old window frames, used bicycles, out-of-date radios, and talking machines, all denizens of the same dreary urban landscape. In this case however, the presence of a hose, piles of debris, a shovel, and a piano on the partially wet sidewalk suggest a more sudden, dramatic end to these objects. Whether by flash flood or through the rhythmic tide of decades, the ultimate fate of most talking machines was the curbside. *Courtesy of the Charles Hummel collections.*

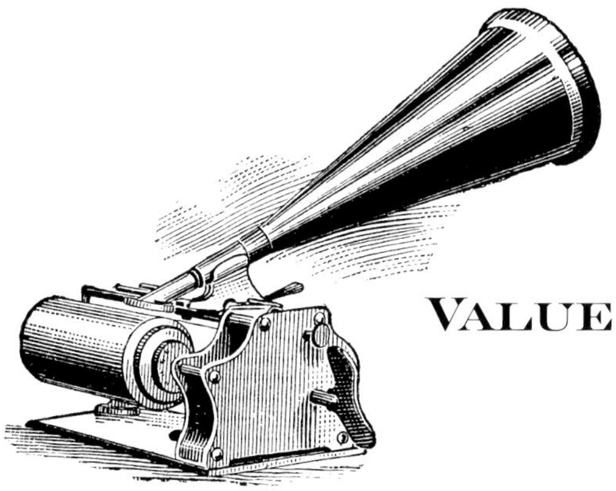

Value Code Key

Plus ça change, plus c'est le même chose.

Preparing a value system for a large number of diverse items is a task fraught with concerns. We have dealt with the job as realistically as possible. Items which are extremely uncommon will be rated "very rare" (VR). The "VR" designation is not meant to indicate a price of any kind – it is only an expression of the infrequency with which the object changes hands.

Values will be expressed *in the individual captions* by the following letter codes:

A $1500. and up
B $1000. to $1499.
C $750. to $999.
D $500. to $749.
E $250. to $499.
F $125. to $249.
G $51. to $124.
H up to $50.

This might seem to be a long list of prices. However, you will find that many of the values actually will fall within a small range of categories: E, F, G, and H. Many illustrations contain a selection of objects. If these items are similar enough, a general value code will be shown. If there is sufficient disparity in the prices, they will be individually valued: always listed from left to right and clockwise, if applicable.

By employing price ranges instead of specific dollar amounts, we leave ourselves open to various "slings and arrows" of discontent. It is our feeling that ours is the only effective approach to the subject. However, this does add an element of discretion to the use of our book.

In using the value codes, we advise *conservatism*. If you are trying to establish the value of an object in your possession, the natural instinct is to seize upon the highest price in the given range, regardless of condition. This can be a mistake. The condition of the items in the book is nearly always excellent, but the condition of *your* item is the important issue. Condition can affect the value by up to 50% — and occasionally more. The bottom end of a price range has as much validity as the top, though it seems a less attractive prospect. Different examples of the same type of object may represent the entire range of a certain price category, due to condition. The ranges are also intended to be *comparative*. Certain objects in category "B," for instance, may *never* approach $1500.00, no matter how perfectly preserved they are.

Beyond condition, many, many factors affect the price of an "antique." Exact worth is something that cannot be fixed precisely in time. We can only hint at it. Our guide is intended to help you understand the value of the objects illustrated, but we vigorously discourage the use of the phrase, "It *books* for such-and-such a price." True worth is what the market will bear. Quoting books, auction results, or what a shop is asking will probably not elevate the price higher than the general market will currently support.

Finally, in the area of phonograph advertising there are two distinct and sometimes considerably different markets: the phonograph collectors and the advertising collectors. Disputes between these two groups as to the value of certain objects never will be fully resolved — so we content ourselves by hoping to satisfy the needs of the most number of people.

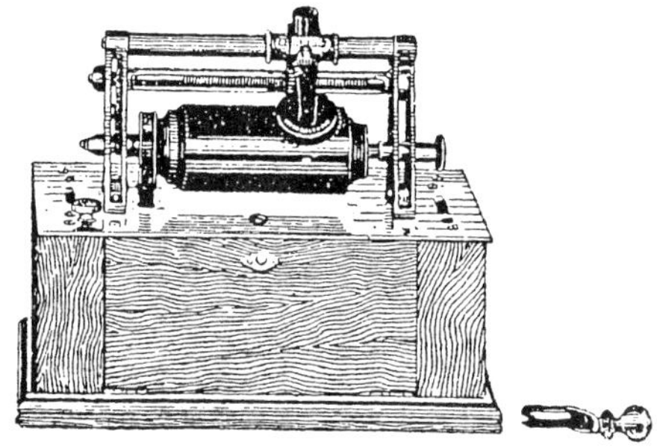

BIBLIOGRAPHY

Association for Recorded Sound Collections, various ed.s: *ARSC Journal*, various issues to present.

Barnum, Frederick O. III: *His Master's Voice in America*, Camden, General Electric Company, 1991.

Barr, Steven C.: *The Almost Complete 78 RPM Record Dating Guide (II)*, Huntington Beach, Yesterday Once Again, 1992.

Baumbach, Robert W.: *Columbia Phonograph Companion, Vol. II (Disc Graphophones and the Grafonola)*, Woodland Hills, Stationery X-Press, 1996.

_____: *Look for the Dog*, Los Angeles, Mulholland Press, 1996.

_____: *The Victor Data Book*, Los Angeles, Mulholland Press, 2003.

Bayly, E., ed.: *The EMI Collection*, Bournemouth (GB), Talking Machine Review, 1977.

_____: *The Talking Machine Review International*, Bournemouth (GB), various issues beginning 1970.

Bayly, Ernie and Michael Kinnear: *The Zon-o-phone Record*, Heidelberg (Victoria, Australia), Michael Kinnear, 2001.

Bill, Edward Lyman, ed.: *The Talking Machine World*, New York, various issues 1905-1929.

Brooks, Tim, ed.: *Little Wonder Records, a History and Discography*, St. Johnsbury (VT), New Amberola Phonograph Company, 2000.

Bryan, Martin, ed.: *The New Amberola Graphic*, St. Johnsbury (VT), various issues 1973 to present.

Charosh, Paul, compiled by: *Berliner Gramophone Records, American Issues 1892-1900*, Westport, Greenwood Press, 1995.

City of London Phonograph and Gramophone Society, various ed.s: *The Hillandale News* (beginning in 2002 *For the Record*), various issues 1960s to present.

Conot, Robert: *A Streak of Luck*, New York, Seaview Books, 1979.

Dethlefson, Ronald, ed.: *Edison Blue Amberol Recordings, 1912-1914*, Brooklyn,

APM Press, 1980.

_____: *Edison Blue Amberol Recordings, 1912-1914* (paper cover), Woodland Hills (CA), Stationery X-Press, 1997.

_____: *Edison Blue Amberol Recordings, 1915-1929*, Brooklyn, APM Press, 1981.

_____: *Edison Blue Amberol Recordings, 1915-1929*, second edition, Los Angeles, Mulholland Press, 1999.

Dethlefson, Ronald and George Copeland: *Pathé Records and Phonographs in America, 1914-1922*, Los Angeles, Mulholland Press, 2000.

Edge, Ruth and Leonard Petts: *The Collector's Guide to 'His Master's Voice' Nipper Souvenirs*, London, EMI Group, 1997.

Fabrizio, Timothy C. and George F. Paul: *Antique Phonograph Accessories and Contraptions*, Atglen (PA), Schiffer Publishing, Ltd., 2003.

_____: *Antique Phonograph Advertising, an Illustrated History*, Atglen (PA), Schiffer Publishing, Ltd., 2002

_____: *Antique Phonograph Gadgets, Gizmos and Gimmicks*, Atglen (PA), Schiffer Publishing, Ltd., 1999.

_____: *Discovering Antique Phonographs*, Atglen (PA), Schiffer Publishing, Ltd., 2000.

_____: *Phonographs With Flair, a Century of Style In Sound Reproduction*, Atglen (PA), Schiffer Publishing, Ltd., 2001.

_____: *The Talking Machine, an Illustrated Compendium 1877-1929*, Atglen (PA), Schiffer Publishing, Ltd., 1997.

Fagan, Ted, compiled by: *The Encyclopedic Discography of Victor Recordings, Pre-Matrix Series,*

Westport, Greenwood Press, 1983.
_____, compiled by: *The Encyclopedic Discography of Victor Recordings*, Westport, Greenwood Press, 1986.

Frow, George L.: *Edison Cylinder Phonograph Companion*, Woodland Hills, Stationery X-Press, 1994.

_____: *The Edison Disc Phonographs and the Diamond Discs*, Sevenoaks, Kent (GB), George L. Frow, 1982.

_____: *The Edison Disc Phonographs and the Diamond Discs*, Los Angeles, Mulholland Press, paperback reprint 2001.

Gaisberg, Frederick W.: *The Music Goes Round*, North Stratford (NH), Ayer Company Publishers, Inc., 1977 (1942 reprint).

Gracyk, Tim, with Frank Hoffmann: *Popular American Recording Pioneers, 1895-1925*, Binghamton (NY), Haworth Press, 2000.

Gracyk, Tim, ed.: *Victrola and 78 Journal*, Roseville (CA), various issues 1994-1998.

Hatcher, Danny, ed.: *Proceedings of the 1890 Convention of Local Phonograph Companies*, Nashville, Country Music Foundation Press, reprint, 1974.

Hazelcorn, Howard: *A Collector's Guide to the Columbia Spring-Wound Cylinder Graphophone, 1894-1910*, Brooklyn, APM Press, 1976.

_____: *Columbia Phonograph Companion, Volume I: Hazelcorn's Guide to the Columbia Cylinder Graphophone*, Los Angeles, Mulholland Press, 1999.

Hunting, Russell, ed.: *The Phonoscope*, New York, Phonoscope Publishing Company, various issues 1896-1900.

Johnson, E. R. Fenimore: *His Master's Voice Was Eldridge R. Johnson,* Milford (DE), State Media, Inc., 1975.

Koenigsberg, Allen, ed.: *The Antique Phonograph Monthly*, Brooklyn, APM Press, various issues 1972-1993.

_____: *Edison Cylinder Records, 1889-1912, 2nd ed.*, Brooklyn, APM Press, 1988.

_____: *The Patent History of the Phonograph, 1877-1912*, Brooklyn, APM Press, 1991.

Laird, Ross: *Brunswick Records, a Discography of Recordings 1916-1931*, in four volumes, Westport, Greenwood Publishing Group, 2001.

Lorenz, Kenneth M.: *Two-Minute... Cylinders of the Columbia Phonograph Company*, Wilmington (DE), Kastlemusick, Inc., 1981.

Marco, Guy A. and Andrews, Frank, eds.: *Encyclopedia of Recorded Sound in the United States*, New York, Garland Publishing, Inc., 1993.

Martland, Peter: *Since Records Began: EMI: the First Hundred Years*, (distributed by Timber Press, Portland, OR), Amadeus Press, 1997.

Marty, Daniel: *Histoire Illustrée du Phonographe (Illustrated History of the Phonograph),* Lausanne/Paris, Edita/Lazarus, 1979, (reprinted in various English language editions with slight modifications of the title).

Michigan Antique Phonograph Society, various ed.s.: *In the Groove,* various issues to present.

Moore, Jerrold Northrop: *A Matter of Records*, New York, Taplinger Publishing Company, 1977.

Moore, Wendell, ed.: *The Edison Phonograph Monthly*, various anthologies 1903-1916, New Albany (IN), Wendell Moore Publications.

Proudfoot, Christopher: *Collecting Phonographs and Gramophones*, New York, Mayflower Books, 1980.

Reiss, Eric L.: *The Compleat Talking Machine*, Chandler (AZ), Sonoran Publishing, (4th revised) 2003.

Rust, Brian and Tim Brooks: *Columbia Master Book Discography*, in four volumes, Westport, Greenwood Publishing Group, 1999.

Rust, Brian, compiled by: *Discography of Historical Records on Cylinders and 78s*, Westport, Greenwood Press, 1979.

Seiter, Bessie and Floyd, and Robin and Joan Rolfs: *Phonograph Dolls That Talk and Sing*, Hortonville (WI), Audio Antique LLC, 2001.

Sherman, Michael W. (with William R. Moran, and Kurt R. Nauck, III): *Collector's Guide to Victor Records*, Dallas, Monarch Record Enterprises, 1992.

Sherman, Michael W. and Kurt R. Nauck, III: *Note the Notes, An Illustrated History of the Columbia 78 rpm Record Label, 1901-1958*, Monarch Record Enterprises, New Orleans, 1998.

Sutton, Allan: *Directory of American Disc Record Brands and Manufacturers, 1891-1943*, Westport, Greenwood Press, 1994.

Sutton, Allan and Kurt Nauck: *American Record Labels and Companies, an Encyclopedia (1891-1943)* [with CD-ROM], Denver, Mainspring Press, 2000.

Wile, Raymond R.and Ronald Dethlefson: *Edison Diamond Disc Recreations, Records and Artists 1910-1929,* Brooklyn, APM Press, 1985.

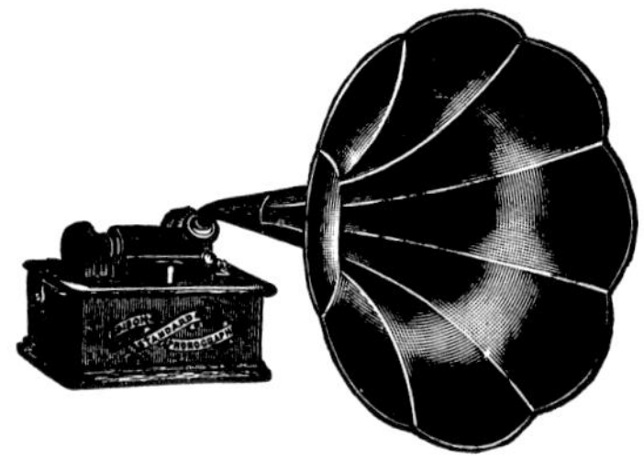

GLOSSARY

AMBEROL: The first four-minute cylinder developed by Edison, 1908-1912. These cylinders are black in color, made of especially hard metallic soap, and were usually sold in green and white containers.

AMBEROLA: The name denoting Edison's line of internal-horn cylinder Phonographs, 1909-1929.

BACK MOUNT: A term used to describe a talking machine which uses a back bracket which supports the horn and thus removes its inertia from the SOUNDBOX.

BEDPLATE: The metal motor mounting plate (often a casting) to which the upper mechanism of a cylinder talking machine is mounted, and from which the motor is suspended.

BLANK: A smooth, grooveless unrecorded cylinder.

BLUE AMBEROL: The blue celluloid four-minute cylinder sold by Edison, 1912-1929. The containers for these cylinders were blue until 1917, orange and blue thereafter.

CARRIAGE: The assembly which "carries" the REPRODUCER of a cylinder talking machine across the recording.

CARRIER ARM: Edison nomenclature for the CARRIAGE. The HALF-NUT and its flat supporting spring are attached directly to the carrier arm on nearly all Edison machines.

COMBINATION ATTACHMENT: Devices offered by Edison to convert pre-1908 Phonographs to play four-minute cylinders in addition to the two-minute variety.

CONCERT: Edison's trade name for the five-inch diameter cylinder and the Edison Phonograph designed to play it.

CRANE: The support used to mount horns of 15-inch length or greater to cylinder talking machines.

CYLINDER: A geometric form on which entertainment recordings were made, 1877-1929. The earliest sheets of tinfoil were followed by self-supporting cylinders of ozocerite-covered cardboard, stearic acid/paraffin, hard metallic soap, and celluloid. Advantages of the cylinder format included a constant surface speed and the ability to make home recordings.

DIAMOND DISC: Edison's line of disc records and machines, 1912-1929. The Diamond Discs were vertically recorded, and the Diamond Disc Phonographs were one of the very few disc talking machines to use a FEEDSCREW.

DIAPHRAGM: The flexible, circular vibrating membrane of a SOUNDBOX or REPRODUCER which converts mechanical energy to acoustic energy or sound waves. Mica, glass, compressed paper, copper and aluminum were commonly used as diaphragms in early talking machines.

ELBOW: The connection between the horn and the SOUNDBOX or TONE ARM of a disc talking machine. Earliest elbows are made of leather, gradually giving way to metal elbows after 1900.

ENDGATE: A swinging arm which supports one end of the MANDREL on some cylinder talking machines. To mount or remove a cylinder it is therefore necessary to open and close the endgate.

FEEDSCREW: A threaded rod which usually drives a half-nut fixed to the carriage of a cylinder talking machine, thus guiding the reproducer across the record grooves. Certain disc machines used FEEDSCREWS to drive the SOUNDBOX across the record or to move the turntable beneath a fixed soundbox. Similarly, some cylinder machines used feedscrews to drive a mandrel longitudinally beneath a fixed reproducer.

FRONT MOUNT: A term used to describe a disc machine where the horn attaches directly to the SOUNDBOX, and the support arm (or mount) runs directly below and parallel to the HORN. In such an arrangement the support arm points in the same general direction as the bell of the horn.

GRAMOPHONE: Originally the name by which Emile Berliner's disc talking machine was known, it became a generic term to denote any disc playing talking machine, but fell out of use in the United States.

GRAND: Columbia's trade name for the five-inch diameter cylinder and the various Columbia machines designed to play it.

GOVERNOR: The mechanical assembly in a talking machine motor which regulates the speed, usually by limiting the outward movement of spinning weights.

HALF-NUT: An internally threaded metal piece usually in the form of a nut which has been cut in half. The threads of the half-nut correspond to the threads of the FEEDSCREW. As the feedscrew revolves, the half-nut will be propelled along it, thus driving the CARRIAGE of a talking machine.

LATERAL RECORDING: The type of disc talking machine recording where the information is encoded in sides of the groove. (see VERTICAL RECORDING)

MANDREL: The tapered drum upon which the cylinder record is placed for playing.

NEEDLE: The point (usually steel) of the SOUNDBOX which rides the grooves of the recording and transmits vibrations to the DIAPHRAGM.

NEEDLE HEAD: A colloquial term for the SOUNDBOX or PICK-UP which accepts vibrations directly from the record.

NEEDLE TIN: A container to store and dispense steel talking machine needles.

NEEDLE SHARPENER (OR CUTTER): A device for repointing disc talking machine needles of fiber, thorn or steel.

PHONOGRAPH: Originally, the name given by Edison to the sound recording/reproducing device he invented in 1877. In American usage, the term has been applied to sound reproducing equipment in general, regardless of age or type of record played.

RECORDER: A device comprising a DIAPHRAGM and cutter stylus for recording cylinder or disc records.

REPEATING ATTACHMENT: A device for cylinder or disc talking machines which returns the NEEDLE or STYLUS to the beginning of the record after the selection has played.

REPRODUCER: The component comprising STYLUS, linkage and DIAPHRAGM, which reproduces the sound from the record grooves. This term is most frequently applied to cylinder machines. (See SOUNDBOX)

SHAVER: On a cylinder talking machine, a device which holds a knife in position to pare the recorded surface from a wax cylinder.

SOUNDBOX: The component comprising NEEDLE, linkage and DIAPHRAGM, which reproduces the sound from the record grooves. This term is most frequently applied to disc machines. (See REPRODUCER)

STYLUS: The point (usually sapphire or diamond) of the REPRODUCER which rides the grooves of the recording and transmits vibrations to the DIAPHRAGM or PICK-UP.

TALKING MACHINE: A general term for sound reproducing devices playing either cylinder or disc records, made prior to the advent of electric reproduction. Specific brands such as Graphophone, Victrola or Amberola all fall within the general category of talking machines. Modern American usage substitutes PHONOGRAPH for the older term "talking machine." For the purposes of historical accuracy, both terms will appear frequently in this book.

TONE ARM: A movable hollow tube which conducts sound to the HORN from the SOUNDBOX.

TRUNNION: Rightly, the sleeve visible on either side of the CARRIAGE which slides over the FEEDSCREW shaft on a Graphophone, but commonly used to denote the entire carriage.

TURNTABLE: The round platter which supports a disc record and revolves while the record is being played. After 1960, that separate component in a SOUND SYSTEM which comprised the TURNTABLE and RECORD CHANGER was often referred to simply as the TURNTABLE.

VERTICAL RECORDING: The type of cylinder or disc recording in which the sound vibrations are encoded in the bottom of the groove. (see LATERAL RECORDING)

WAX: Although the material from which many cylinder records were made is generally referred to as "wax," it was more rightly various formulations of metallic soap. The resemblance of this material to conventional wax promotes the common descriptive term.

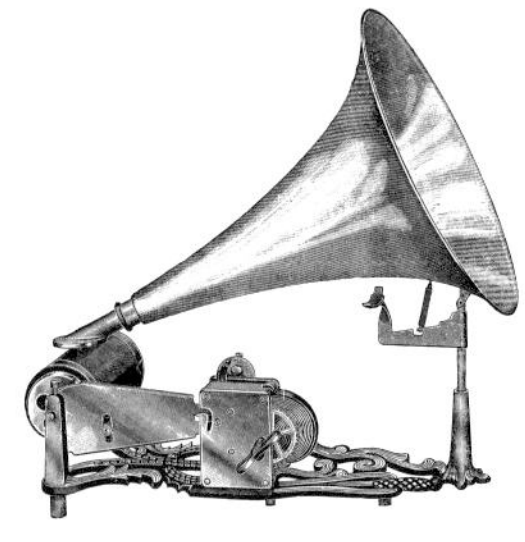

INDEX

Advertising fans: 3-101, 3-102, 4-46, 4-58, 4-71
Advertising mirrors: 3-50
Advertising slide: 3-79
Advertising pencil: 4-12
Advertising thermometers: 3-58, 4-65
Aeolian Vocalion: 3-71
American Graphophone Company: 1-11, 1-19, 1-25
American Record Company: 2-54
American Talking Machine Company: 3-51, 3-58
Amet, Edward H.: 1-28, 1-34
Aretino: 3-5, 3-109
Averill Manufacturing Company: 4-43

Babson, Fred K.: 3-117, 3-118, 5-94
Bacigalupi, Peter: 1-42
Baseball team: 2-113
Batteries, phonograph: 1-51
Berlin, Irving: 3-60
Berliner, Emile: 1-38, 1-52, 1-58, 3-37, 4-63
Berliner Gramophone
coin-slot: 1-52
lever-wind: 1-37
trade mark: 1-52, 2-19, 5-11, 5-14, 5-28
Bettini, Gianni: 1-45, 1-49, 1-50
Blocks, toy: 1-15
Brady, Matthew: 1-1, 1-2
Brunswick: 3-57, 3-82, 3-83, 3-93 through 3-98
Style "40": 3-97
Style "60": 3-97
Style "125": 3-96
Style "150": 3-96
Style "175": 3-95
Style "200": 3-95
Style "225": 3-94
Style "275": 3-94
Style "350": 3-93
Style "1500": 3-93
"Panatrope": 4-59, 4-60
Bryan, William Jennings: 2-96, 2-101
Bubble Books: 3-63 through 3-69
Buffalo Talking Machine Company: 4-65
"Buster Brown": 2-66
"Busy Bee": 2-46 through 2-48, 3-5, 3-109
Buxton and Parker: 5-43

Cabinets, record: 2-86 through 2-89, 2-92
Cabinets, specially decorated: 3-99
"Cabinet That Matches": 2-92
Canadian Edison Phonographs, Limited: 4-32
Capitol Lamp Phonograph: 4-44
Carette, George, & Company: 2-7
Carrying cases: 2-34
Caruso, Enrico: 2-8, 3-48, 3-49, 3-50
Capps motor: 1-28
Celluloid horn: 2-3
Cheney, Forest: 3-90
Cheney Talking Machine Company: 3-90
Chicago Talking Machine Company: 1-16, 1-19, 1-27 through 1-30, 1-32
Aluminum motor: 1-28, 1-31
Choraleon Phonograph Company: 4-11
Class "M" Phonograph: (see "Edison Class "M" Phonograph")
Coin-operated phonographs: p.77, 1-53, 2-56 through 2-75
Columbia Phonograph Company: 1-13, 1-16, 1-19, 2-3, 2-35, 2-36, 2-106, 2-112, 3-37, 3-44, 3-60, 3-63, 4-58, 5-77
Acquisition of Okeh: 4-61
Banner: 4-9, 5-95
Calendars: 3-113, 4-10
Catalogue cover: 2-85, 3-84
Christmas advertising: 3-55, 3-56, 3-119
Counter plaques: 4-15, 4-16
Educational Records: 3-44
Electrotype: 4-23
Figurine: 4-14
Grafonola oil: 4-17
Indestructible Records: 2-91
Price tags: 3-59
Record box: 3-111
The Columbia Record: 2-104
Compagnie Francaise du Gramophone: 2-19
Compagnie Internationale Phonique: 2-33
Consolidated Talking Machine Company: 3-109
Conventions, talking machine: 4-3 through 4-7
Cornish Phonographs: 3-110
Cummings, A.B.: 4-34 through 4-36

"Da-Lite" electric displays: 4-25 through 4-27
Dal Monte, Toti: 4-68
DeFrance, C.Q.: 3-39, 3-41
Deutshen Grammophon Gesellschaft: 2-110
Douglass, Leon F.: 1-16
Duplex Phonograph Company: 2-83, 2-84, 3-39 through 3-42
Dusters, record: 4-42

Eastern Phonograph Company: 5-47
Eastern Talking Machine Company: 3-51
Easton, Edward D.: 1-13, 3-37
Echophone: 1-34
Ediphone: 3-112, 4-64
Edison Amberola "IA": 2-108
Edison Amberola "III": 3-38
Edison Amberola "IV": 3-38
Edison "Apex": 2-38
Edison-Bell: 2-7
Edison banner: 5-41
Edison Blue Amberol records: 2-4, 3-117, 3-118
Edison "C-250": 3-102, 3-103, 4-4
Edison Caravan Camel: 4-6
Edison Class "M" Phonograph: 1-5, 1-7, 1-11, 1-12, 2-50, 5-3, 5-4
Edison Continuous Exchange Proposition: 2-109
Edison Diamond Amberolas: 5-112
Edison Diamond Points: 4-3
Edison Disc Phonographs/Records: 3-72, 3-102, 3-103, 4-3, 4-19, 5-100, 5-112, 5-113
Edison electrotypes: 4-22
Edison "Excelsior": 2-38
Edison "Fireside" Phonograph: 2-107
Edison "Gem" Phonograph: 5-49, 5-57
Edison "H" reproducer: 2-95
Edison "Home" Phonograph: 1-32, 1-35, 2-18, 3-38, 5-42, 5-53, 5-56, 5-57, 5-63, 5-64, 5-91
Edison "Imitation" disc record: 4-21
Edison Kinetoscope: 5-4
Edison, "Let Us Not Forget": 3-104, 3-105
Edison "Majestic": 2-38
Edison Masonic Hymns: 2-114
Edison "Mood Music": 4-20
Edison menu: 4-6
Edison "Military" or "Portable" Phonograph: 5-1
Edison paperweights: 4-3, 4-4
Edison Phonograph Hall: 3-23
Edison Phonograph Monthly: 2-103, 2-104, 2-109, 3-54
Edison Phonograph Works: 1-5
Edison "Phono-Kinetoscope": 1-54
Edison pinback: 4-5
Edison "Special Amberol" records: 2-111
Edison "Special License": 3-4
Edison "Spring Motor" Phonograph: 1-28, 1-35, 5-5
Edison "Standard" Phonograph: 2-95, 3-38, 5-15, 5-16, 5-39, 5-57, 5-94
Edison Talking Doll: 5-1
Edison, Thomas A.: 1-1, 1-3, 1-4, 1-8, 1-11, 1-14, 2-8, 3-33, 3-37, 3-104, 3-105
Edison, Thomas A., Incorporated: 2-103, 3-4, 4-64
Edison "Tinfoil" Phonograph: 1-54, 5-1
Edison "Triumph" Phonograph: 3-4
Edison watermark: 3-73
Edison window decal: 2-76
Edison "Windsor": 2-38
English Columbia Phonograph Company, General: 2-77
Everlasting Indestructible records: 2-91

Ferrar, Geraldine: 3-21
Flemish art: 2-30
Floor stands: 2-40
Funeral home: 4-45

Gaskin, George J.: 1-12
Gerville-Reache, J.: 3-21
Goodale, Franklin C.: 2-116
Goodman Company: 4-15
Gramophone Company, Limited: 3-27 through 3-29, 5-35
Graphophone
Cylinder, "Twentieth Century": 1-57
Cylinder, Type "E": 1-24, 1-25
Cylinder, Type "P": 1-24
"Baby Regent": 3-34
Grafonola "Grand": 3-37, 5-73
"Lyric": 5-98
"Multiplex Grand": 1-57
Phonograph-Graphophone: 1-6, 1-7
Price tags: 3-59
"Princess": 3-34
"Regent": 3-79
"Symphony Grand/Upright": 2-78 through 2-80
"Thornward": 2-5
"Toy Graphophone": 2-6
Type "A": 1-29, 1-39, 1-41, 2-5, 5-6, 5-7
Type "AA": 5-13
Type "AB": 2-2
Type "AD": 2-2
Type "AF": 2-2
Type "AG": 2-2
Type "AH": 5-24
Type "AK": 2-12, 2-24
Type "AN": 1-32, 1-39
Type "AS": 1-53
Type "AT": 1-59, 5-8, 5-10, 5-98
Type "AU": 5-36
Type "BC": 5-59
Type "BE": 5-61
Type "BET": 2-112
Type "BKT": 5-51
Type "BS": 1-53, 5-17
Type "C": 1-23
Type "E": 1-24
Type "F": 1-30
Type "G": 1-21
Type "H": 1-21
Type "HG": 2-2
Type "I": 1-23, 1-30
Type "K": 1-23, 1-30, 5-2
Type "KF": 1-30
Type "N": 2-5
Type "Q": 2-47, 2-49, 2-51, 3-37
Type "S": 1-24
Type "T": 1-21
Type "U": 1-22
Grub's Music Store: 2-55

Harmony Talking Machine: 2-106, 3-109
Harper & Brothers: 3-63
Harthorn's music store: 3-34
Hartmann, Robert: 1-45
Hawthorne & Sheble: 1-39, 1-48 through 1-51, 2-34, (and Prescott: 2-54), 3-8, 5-39, 5-94
Compartment Needle Box: 2-84
Henry's Music Store: 3-115
Herzog Art Furniture Company: 2-86 through 2-89, 5-65
Heppe, C.J. & Son: 3-58
Hiller Talking Clock: 2-116
Hogan Collapsible Horn: 2-3
Horn cranes: 2-42, 2-43
Hunt Show Card Company: p.77, 2-57, 2-58, 2-61, 2-69, 2-71, 2-73, 2-75

Imperial Talking Machine: 2-106
Indestructible Record Company: 2-91
Inspector's Handbook of the Phonograph: 1-5
International Graphophone Company: 1-9

Johnson, Eldridge: 1-58, 3-37
Johnson, George W.: 1-12

Kansas Phonograph Company: 5-4
Keen, Morris, 3-7, 3-13, 3-15
Keen-O-Phone Company: 3-7 through 3-17
Hornless: 3-12
Keenolet Jr.: 3-11
Model "I": 3-14
Model "II": 3-13
Model "IX": 3-16
Model "XXX": 3-7 through 3-9
Model "XXV": 3-10
Model "XX": 3-10, 3-11
Kelley Cigar Company: 1-46
Kilties Band records: 2-10
Kimball: 3-70
Kipp Brothers Company: 2-37
Kraemer, Thomas: 3-8, 3-14

Lambert Indestructible Records: 2-4, 2-91
Lamb's Music House: 3-2,
Lamb's Music Store: 3-24
Lawton show card: 2-65
Letter openers: 3-51, 4-12
Lewis, George Andrew: 2-49 through 2-51
L'Illustration: 2-101
Lioret "Le Merveilleux": 1-43
Little Wonder: 3-60 through 3-63
Lyon & Healy: 1-26

Macdonald, Thomas H.: 1-30
"Madame Hendren" doll records: 4-43
"MAKAPHONE": 4-11
Mandel: 3-106
Massani, Pompeo: 2-8
Metropolitan Phonograph Company: 3-102
McKinley Music Company: 3-17
McKinley, William: 1-56
Menier chocolate: 1-43
Modern Phonograph & Supply Company: 4-11
Montgomery Ward: 1-52, 2-5
Murdoch, John G.: 2-100
"Mystic Oracle" 4-59

National Phonograph Company: 2-103, 2-109
National Toy Company: 3-53 (see also 4-34 through 4-36)
Nipper figurine: 4-48
North American Phonograph Company: 1-5, 1-7, 1-10, 1-11, 1-32

Okeh: 4-61
O'Neill, Arthur J.: 3-5, 3-109
O'Neill-James Company: 2-47, 3-109
Outcault, R.F.: 2-66
"Oval Horn": 2-44

Palm Fechteler & Company: 2-76
Pathe: 2-32, 3-30, 3-82, 3-86 through 3-89, 3-116, 4-2, 5-9, 5-85
Penn Horn Crane: 1-39
Penn Phonograph Company: 3-21
Pennsburg Furniture Company: 3-92
Phonograph and Talking Machine Weekly: 4-25
Phrynis: 2-44
Plate: 2-15
Playing cards: 4-41
Polyphone: 1-44
Pooley Furniture Company: 3-9, 3-10, 3-16
Portable phonographs: 4-39, 4-40, 4-66, 4-67, 4-70, 5-118 through 5-121
Postcards: 4-62
Puck: 2-7, 5-23
Puritan: 3-91
Puzzle: 1-15

Quinn, Dan W.: 1-12

"Ragtime Rastus": 3-53
Reading Phonograph Company: 3-1
Record cabinets: 2-86 through 2-89, 2-92
Record dusters: 4-42
"Rectorphone": 2-45
Regina Music Boxes: 2-37
Repp, Clinton B.: 3-45
Rex Talking Machine Corporation: 3-17
Rishell display cards: 3-78
Romain, Manuel: 2-97
Roosevelt, Theodore: 2-85, 3-26, 3-117, 3-118, 5-53
Royal Talking Machine: 5-71

Schwegler Brothers: 4-41
Searchlight Horn Company: 5-63
Sears, Roebuck & Company: 1-40, 1-41, 2-12, 3-16, 3-62, 3-107
Sheet music: 1-36, 3-36 (see also "Victor sheet music")
Signboards, coin-slot: 1-53, 2-56 through 2-75
"Silvertone": 3-107
"Simp": 2-83
Smoking set: 3-3
Sonora: 3-92, 4-24, 4-45, 4-66, 4-67
Sound Modifier: 2-82
Standard Disc Talking Machine: 2-106, 3-6, 3-109, 5-58, 5-67, 5-83, 5-87
Standard Talking Machine Company: 3-5
Starr: 4-24
St. Louis Talking Machine Company: 2-113
"Symphonion": 2-100
Syracuse wire record rack: 5-57, 5-97

Taft, William Howard: 2-96, 2-101, 3-26
Talking Book Corporation: 4-13
Talking Machine Company of Chicago: 1-44, 2-92
Talking Machine News: 2-83, 2-100
Talking Machine World: 3-10, 3-17, 5-65
Talk-O-Phone: 2-21, 5-27
Tape measurers: 3-57
Tea Tray Company: 2-39 through 2-44
Tewksbury, George: 5-4
Trolley cards: 3-100, 4-55, 4-56, 4-57
Turek, S.K.: 3-85

United Talking Machine: 3-109
United Phonographs Corporation: 3-91
United States Marine Band: 1-13
Universal Talking Machine (and Manufacturing) Company: 1-52, 2-81

"Valuphone": 4-31
Victor Distributing & Export Company: 2-82
Victor Talking Machine Company: 1-58, 2-8, 2-19, 2-53, 2-81, 2-102, 3-40, 3-90, 4-1, 4-69
Automatic Needle Clamp: 2-28
Blotter: 4-37
Banner: 5-70
Brochure: 4-49
Cabinet Finishes Samples: 3-108
Calendar: 4-8
Catalogue, "New Victor Line" of Orthophonics: 4-50 through 4-54
Catalogue cover: 2-9, 2-90
Cigarette dispenser: 4-29
Convention program: 4-7
"D": 3-29
"Da-Lite" electric displays: 4-25 through 4-27 (see also 4-28)
"Exhibition" sound box: 2-27
Gift Certificate: 4-33
"Improved Concert" sound box: 3-20
Kilties Band records: 2-10
"Monarch": 2-16, 2-19, 2-20, 2-41
"Monarch Junior": 2-16, 2-20, 2-52, 5-19, 5-20
"Monarch Special": 2-16, 3-29
Packing slips: 3-80, 3-81
Photogravures: 3-48, 3-49
Ready Made Window Display: 3-19, 4-47
Record Albums: 2-115
Record display stand: 4-38
Record shipping box: 3-52
Repairman's toolbox: 2-117
"Rigid Arm": 2-17
"Royal": 2-16, 2-17, 2-52, 5-12
Royalty Statement: 4-68
Sheet music: 2-53, 3-35
Show card: 3-18
Spring removal tool: 2-26
Teapot and tray: 4-30
Victor Book of the Opera: 3-31, 3-32
Victor "A": 5-117
Victor "I": 2-17, 2-23, 2-52, 2-105
Victor "II": 2-20, 2-52, 2-105, 5-37, 5-48, 5-81
Victor "III": 2-20, 2-52, 3-14, 5-79, 5-109
Victor "V": 2-20, 2-25, 2-52, 5-42, 5-44
Victor "VI": 2-22, 2-25, 2-52, 5-66, 5-82
Victor/Victrola "XXV": 3-43
Victrola: 2-78, 2-108, 3-35, 4-58
"IV": 5-89, 5-90
"X": 3-34, 4-18
"XI": 3-2, 3-85, 5-107
"Louis XV": 2-93, 2-94
"XVI": 3-49, 5-74, 5-117
"XVII": 3-100
"XX": 2-93-2-94
Vernis-Martin: 3-25, 5-74
"50": 4-39, 4-40
"80": 4-18
"110": 5-117
"260": 4-42
"8-30" or "Credenza": 4-51, 4-58
Voice of the Victor: 2-102, 2-104
Wooden sign: 3-114
World War I poster: 3-74
Yiddish/Hebrew records: 3-77
Vitaphone: 3-45 though 3-47

Waterson, Henry: 3-60
Werle, A.W.: 2-117
Wilson, Woodrow: 3-26
Window decal, Edison: 2-76
Wisconsin Chair Company: 3-91
Wittich Store: 3-57
Wooden signs: 3-114, 3-115
World War I: 3-74 through 3-76, 3-84
Wotton, D.E.: 5-112

Yankee Prince: 3-5
"Yellow Kid": 2-66, 2-68

Zonophone: 2-11, 2-81, 5-18

The MacG
RECORD
No.
SELECTION
155 Orange Street.
BARNEY'S MUSIC STORE
154 THAMES ST.
NEWPORT, R.I.
9855.
PHONOGRAPHS RECORDS
AND EVERYTHING
MUSICAL
EDISON
RECORDS
THE A.
PHONOGRAPHS
T. L. MERRILL
CO.
228 MIDDLE ST.
PORTLAND, ME.
COMPLETE LINE OF
PHONOGRAPH
SUPPLIES
RECORD
MINKLER & COLLINS
PHONOGRAPHS
AND
SUPPLIES.
BICYCLES AND SPORTING GOODS
Battle Hymn of
the Republic
23 RAILROAD ST.
GREAT BARRINGTON
MASS.
PHONOGRAPHS, BICYCLES
GUNS, RIFLES,
FISHING TACKLE & SPORTING GOODS.
EDWARD A. NELSON
214 MAIN ST
SPRINGFIELD, MASS.
F. E. LOCKWOOD & CO.
No. 2051
39 WALL ST
PHONOGRAPHS
AND
RECORDS